Applied Signal Processing

Thierry Dutoit · Ferran Marqués

Applied Signal Processing

A MATLAB™-Based Proof of Concept

Thierry Dutoit
Faculté Polytechnique de Mons
Mons
Belgium
thierry.dutoit@ fpms.ac.be

Ferran Marqués
Universitat Politècnica de Catalunya
Barcelona
Spain
ferran@gps.tsc.upc.edu

Additional material to this book can be downloaded from http://extras.springer.com

ISBN 978-0-387-74534-3 e-ISBN 978-0-387-74535-0
DOI 10.1007/978-0-387-74535-0

Library of Congress Control Number: 2008939860

Printed on acid-free paper

springer.com

To Louis and Alice

To Naresh

To the children of the world *

* The authors of this textbook have decided to transfer all royalties to UNICEF

Foreword

Digital signal processing (DSP) has been a major part of the broad area of communications and systems for more than 40 years, having originated in the pioneering efforts of Ben Gold and Charlie Rader at MIT Lincoln Labs in the 1960s along with those of Jim Kaiser and his colleagues at AT&T Bell Laboratories at about the same time. The earliest textbooks in the field of DSP were available in the late 1960s and mid-1970s including the classic *Digital Processing of Signals* by Gold and Rader in 1969, along with the pair of books *Digital Signal Processing* by Oppenheim and Schafer in 1975 (later updated to the book *Discrete-Time Signal Processing* by Oppenheim, Schafer and Buck in 1989 with a second edition in 1999) and *Theory and Application of Digital Signal Processing* by Rabiner and Gold in 1975. Since the time of publication of these early textbooks, hundreds of books have been written and published about virtually every aspect of digital signal processing from simple filter design methods to basic theory to advanced concepts in spectral analysis, etc. It is clear that DSP has become a major discipline within most Electrical and Computer Engineering (ECE) curricula and is universally taught at universities throughout the world.

Over this same 40-year time span, applications of DSP to fields such as speech processing, image and video processing, geological explorations, and biomedical applications have grown up, evolved and have subsequently become the frontiers in signal processing in the twenty-first century. Speech processing applications have often served as the driving force behind advanced signal processing technology, having been the key technology used for evaluating the efficacy of a range of signal processing algorithms. For example, digital speech coders have been deployed in the wireline telecommunications network for more than 50 years in the form of mu-law (or A-law) codecs for pseudo-logarithmic compression of speech amplitudes. Sub-band coding digital filter banks have been used extensively in defining the MP3 (layer 3 audio standard of the MPEG1 video coding standard) and AAC (Advanced Audio Coding standard as part of the MPEG2 video coding standard) audio coding standards that have become the basis for the generation of music coding and compression algorithms and their concomitant audio players, most notably the Apple iPod. More recently, code-excited linear prediction (CELP)-based coders have been the technological basis for the speech codec (encoder–decoder) of more than 2 billion worldwide cell phones in use on a daily basis.

Similarly, image processing technologies have been essential in defining the basic signal processing methods behind the international standards for still images (JPEG, JPEG2000, motion-JPEG) and for video (MPEG1, MPEG2, MPEG4, MPEG7). Although many communication and information technologies are utilized in these standards (e.g., arithmetic or Huffman coding and wavelet representations), the fundamental basis of image and video signal processing is the discrete cosine transform used for processing image and video signals in small blocks along with the algorithms that utilize properties of the resulting transforms for reducing the information rate of the signal without seriously distorting (in a perceptual sense) the ultimate image or video signal. Furthermore, the concept of coding and quantization of a difference signal (thereby taking advantage of short-time signal correlations) is fundamental to the ideas behind motion compensation of video frames or signal correlation between blocks of an image signal.

Along with an understanding of the theory of digital signal processing has come an understanding of the basic signal processing concepts (such as short-time processing, frame-based analyses, statistical methods for signal quantization, aliasing cancellation) and a realization of the importance of a hands-on experience with implementing both the theory and the practice of digital signal processing. Although generic programming languages such as Fortran, C, and C++ were used for quite awhile to implement signal processing algorithms and digital systems, in the mid-1980s a new simulation language, MATLAB, was introduced and quickly became the language of choice for implementing digital systems. Over the past 22 years, the strengths of MATLAB were enhanced and the weaknesses were essentially eliminated, so much so that today almost all researchers implement their ideas in MATLAB code so as to test the efficacy and efficiency of their ideas. Many of the key textbooks in DSP were quick to latch onto MATLAB as a tool that would help students (and active researchers) better understand digital implementations of systems and as a result a number of MATLAB exercise textbooks started to appear in the late 1990s, perhaps the most significant being the book *Computer-Based Exercises for Signal Processing Using MATLAB 5* by McClellan et al. More recently, in 2006, the Third Edition of Mitra's textbook *Digital Signal Processing, A Computer-Based Approach* appeared in print with MATLAB code totally integrated into the book on a section-by-section basis, thereby elevating the importance of concept implementation in viable MATLAB code to the level that it was as important as the theory and concept of digital designs.

With the introduction of this book, Thierry Dutoit and Ferran Marqués have elevated the importance of digital system implementation in MATLAB code to a new level. Here theory is integrated directly into practice through a series of 11 reasonably large and detailed projects, for which a series of cell-based code sections leads the user through all the steps necessary to understand and implement each of the projects. The projects begin with one on speech coding for cell phone and end with one on methods for understanding brain scan images to determine the degree of brain degeneration in certain patients. Along the way we learn how audio CDs can maintain 96 dB signal-to-noise ratios, how MP3 players can compress the data rate of an audio CD by a factor of 11-to-1 without sacrificing perceptual audio quality, how a statistical model of speech can be used to recognize words and sounds, how a sound can be modified in speed or in pitch without degrading the speech quality significantly, how acoustic measurements in the ocean can be used to track whales, how private information can be embedded in music without being detectable or detracting from the music quality, and how image and video signals can be efficiently and accurately coded. Each of the projects is in the form of a section of text that reviews the relevant ideas and signal processing theory in the project implementation, followed by working code that has been carefully designed for presentation in a series of MATLAB cells so as to maintain the highest degree of understanding of both theory and practice associated with each project.

Thierry Dutoit, Ferran Marqués, and all their co-authors have done the signal processing community a great service in both publishing this ground-breaking book and in providing superbly written MATLAB code for each of the 11 projects of the book. The reader will gain much from going through each of the projects on a cell-by-cell basis, examining the resulting plots and printouts, and even playing with the code to make fundamental changes in the algorithms and seeing their effects on the final outputs. It has been a pleasure reading this book and thinking about the implications of having this material so well documented and implemented in MATLAB.

Lawrence Rabiner
Rutgers University, July 2008

Preface

This is not a textbook with accompanying MATLAB (®) code.[1]

On the spirit of this book

Engineering courses tend to be increasingly based on projects for which *ex-cathedra* courses are used as introductions to theoretical foundations, the ultimate goal being to prepare students to use the right tools at the right time in the right way. This "hands-on" approach to engineering education is still made easier for teaching Applied Signal Processing by the generalized use of MATLAB[2] in Engineering Faculties all over the world. Being able to take advantage of the toolboxes provided by MATLAB, as well as by third parties, in the most efficient way, can be set as a major practical goal of courses on applied digital signal processing. In this perspective, one of the most important tasks of professors is to be able to propose motivating projects to their students, with a good balance between reinforcement and exploratory parts, on contemporary problems and using state-of-the-art solutions.

Let us face it, though: although signal processing professors have a deep understanding of signal processing theory and principles, usually each of them specialize into a few (if not one) *applied* area. Image processing experts do not often work on speech; speech experts seldom work on biomedical signals; biomedical specialists rarely solve telecommunications problems. As a result, applied signal processing courses and projects are often biased towards the teachers' most comfortable field (and the authors of the chapters of this volume are no exceptions).

When designing the contents of *"Applied Signal Processing"*, we aimed at letting readers benefit from the teaching background of experts in various applied signal processing fields, presented in a project-

[1] As a matter of fact, it is a series of MATLAB-based proofs of concept, with an accompanying textbook.

[2] MATLAB is a registered trademark of The MathWorks, Inc.

oriented framework. Each chapter exposes how digital signal processing is applied for solving a real engineering. Chapters are typically organized as follows:

- Part A (10–15 pages): Description of the problem in its applicative context and *functional* review of theory related to its solution, with pointers to textbooks for more insight. Equations are only used for a precise description of the problem and its final solution(s). No long theoretical development is given.
- Part B (20–30 pages): Step-by-step MATLAB-based *proof of concept*, with graphs and comments. This typically involves the use of existing or tailored toolboxes and data when available. Solutions are simple enough for readers with general signal processing background to grasp their content. They use state-of-the-art signal processing principles (which usually do 80% of the job and takes 20% of the total design time) but do not implement state-of-the-art optimizations and improvements (the last 20% of the job, which takes 80% of the remaining design time, if not forever). Such details are only outlined with pointers to the literature: the reader is not overwhelmed with implementation details, so as to maintain a tutorial approach.

The ultimate goal of this book is to take the hand of the reader through a journey to essential signal processing applications and give him/her enough assurance to go further by him/herself.

Readership

This book is not a stand-alone introduction to signal processing. Target readers are undergraduate and graduate students with general signal processing background, as well as signal processing professors looking for student projects in various areas of applied signal processing.

Concepts such as recurrence equations, filters, Z transform, discrete Fourier transform and FFT, spectrogram, autocorrelation and its estimators, as well as spectral power density and its estimators, are assumed to be familiar to the reader.

Coverage

Chapter titles have the "*How does systemX perform taskY ?*" form. Each chapter is cowritten by one of the two main authors of this book (Profs Thierry Dutoit and Ferran Marqués), with the help of signal processing experts in *systemX* and/or *taskY*. Chapter themes have been chosen so as to cover a wide range of signal processing applications and tools (although for practical reasons we have had to shorten our initial list of 25 chapters down to 11 chapters, which somehow makes it hard to cover *all* areas of signal processing).

The following list shows which DSP techniques are proposed in each chapter.

Chapter 1: How is speech processed in a cell phone conversation?
Linear Prediction (LP) analysis, F0 analysis, LP synthesis, LPC-10, CELP

Chapter 2: How are bits played back from an audio CD?
Oversampling, noise shaping, sigma-delta conversion

Chapter 3: How is sound processed in an MP3 player?
Sub-band decomposition, transform coding, QMF filters, MDCT, perceptual coding

Chapter 4: How does a dictation machine recognize speech?
Markov chains, Hidden Markov Models (HMMs), *n*-grams

Chapter 5: How does an audio effects processor perform pitch shifting?
Time-scale modification, resampling, phase vocoder, phase-locking

Chapter 6: How can marine biologists track sperm whales in the oceans?
Teager-Kaiser energy operator, TDOA estimation, generalized cross-correlation, adaptive estimation, multilateration

Chapter 7: How could music contain hidden information?
Digital communication design, spread spectrum, audio watermarking, informed watermarking, perceptual shaping filter, Wiener filter

Chapter 8: How are digital images compressed in the web?
YCbCr Color transform, DCT (1D-2D), entropy coding, JPEG

Chapter 9: How are digital TV programs compressed to allow broadcasting?
MJPEG, Inter frame prediction by motion estimation and compensation MPEG

Chapter 10: How does digital cinema compress images?
Wavelets, multi-resolution transform, DWT, JPEG2000, rate-distortion bit allocation

Chapter 11: How can physicians quantify brain degeneration?
Image segmentation, Markov random fields, hidden Markov random fields, statistical image classification

Finally, as once mentioned by Peter Kroon, the book could almost have been entitled "What is happening in my cell phone?", as many of the techniques described here are used in one way or another in today's cell phones (even the chapter on whale spotting introduces the multilateration technique, which can be used for spotting cell phone users).

MATLAB inside

The book was written primarily as a MATLAB-based proof of concept, exposing the main signal processing ideas implemented in widely known applications (many of them commercially available to the general public). MATLAB was used for that purpose, given its wide use in Signal Processing classes and the number of third-party functions available.

We made use of the MATLAB Signal Processing, Image Processing, and Wavelets toolboxes for obvious reasons. Although it would sometimes have been possible to rewrite some of the functions available in these toolboxes, we believe that part of the tutorial aspect of this book is precisely to show how we use the tools we have at hand in the most efficient way. When the functions we needed were not available in standard toolboxes, we have included in-house code in the MATLAB archive accompanying this book, together with the main script implementing each chapter. A short documentation for these functions is also provided in the main text, before using them.

Each chapter is built around a central `.m` script. This script is best used with MATLAB 7, as it is formatted using *editing cells* which provide a convenient way of accessing parts of MATLAB scripts without having to set breakpoints. In order to enable cell-based processing of a script, the following steps must be made:

1. Open the script from inside MATLAB (file/open);
2. Set the current MATLAB directory to that of the chosen script (this can be done using the `cd` command or by using the CurrentDirectory tab in the MATLAB editor);
3. Enable the “cell mode” in the Editor window (Cell/Enable Cell Mode): when this is done, each "%%" comment in the script appears as bold and defines a new cell;
4. Put the cursor at the beginning of the script file (which will appear with a pink background to show the currently active cell);
5. Run cells one after the other by pressing Ctrl+Shift+Enter and read the corresponding code, comments, and results (in the Command Window and on the plots generated by the current cell).

Alternatively, but less conveniently, the script can be used with the `playshow` function of MATLAB, which automatically formats the displays into a slide show, with pauses and comments. As an example, playing the `ASP_cell_phone` script associated with Chapter 1 can be done from the MATLAB command line[3] as:

```
>>playshow ASP_cell_phone echo
```

Acknowledgments

The first seven chapters of this book were mostly written while I was on a 6-months Sabbatical, from Feb to Aug 2007, with financial support from Faculté Polytechnique de Mons, Belgium, to whom I want to express all my gratitude.

I am proud to have Ferran's name with mine on the cover of this book. He quickly supported the idea of this book and provided invaluable help for its image processing content.

Special thanks to my wife, Catherine, for her love and understanding and for having inspired me to take this Sabbatical after seeing me write the contents of this book 3 years ago and struggle for finding the time to write it. I also want to thank my mother and my grandparents for having always been there. I owe them a lot.

T. Dutoit

First of all, I want to thank Thierry for offering me the opportunity of collaborating with him on this book. From the beginning, I felt enthusiastic with the book concept and I have enjoyed developing it with him. Finally, special thanks to my family, especially my wife Nayana, and my friends for their understanding during the time devoted to complete my writing.

Thanks to all of them for their support.

F. Marqués

We want to thank our coauthors for their involvement and sense of humor while discussing, writing, and reviewing our chapters.

This book would probably not be without the SIMILAR European Network of Excellence initiated by our colleague Prof. B. Macq. This network was a wonderful trigger for all sorts of collaborations. Several of the coauthors of this book were members of SIMILAR. SIMILAR also partly

[3] Note that the current directory should first be set to the directory containing the `ASP_cell_phone.m` script.

funded the travel expenses related to the coordination of the book through its *troubadour* grants, a wonderful invention: one researcher visits several labs in a row and sings the song of each previously visited lab, establishing living connections between them.

We are also grateful to The Mathworks for having assisted the writing of this book through their book support program.

Last but not least, we have been honored by the kind encouragements sent to us by our foreword writer, Larry Rabiner, during the last months of this writing.

T. Dutoit and F. Marqués

Contents

List of Contributors

Meritxell Bach Cuadra
Ecole Polytechnique Fédérale de Lausanne, Switzerland

Cléo Baras
GIPSA-lab (Grenoble Image Parole Signal et Automatique), Grenoble, France

Hervé Bourlard
Ecole Polytechnique Fédérale de Lausanne, Switzerland

Laurent Couvreur
Faculté Polytechnique de Mons, Belgium

Antonin Descampe
intoPIX S.A., Belgium

Christophe De Vleeschouwer
Université Catholique de Louvain, Belgium

Thierry Dutoit
Faculté Polytechnique de Mons, Belgium

Laurent Jacques
Université Catholique de Louvain, Belgium
Ecole Polytechnique Fédérale de Lausanne, Switzerland

Varvara Kandia
Foundation for Research and Technology-Hellas, Heraklion, Greece

Peter Kroon
LSI, Allentown, Pennsylvania, USA

Jean Laroche
Creative Labs, Inc., Scotts Valley, California, USA

Ferran Marqués
Universitat Politècnica de Catalunya, Spain

Manuel Menezes
Instituto Superior de Ciências do Trabalho e da Empresa, Portugal

Nicolas Moreau
Ecole Nationale Supérieure des Télécommunications, Paris, France

Javier Ruiz-Hidalgo
Universitat Politècnica de Catalunya, Spain

Richard Schreier
Analog Devices, Inc., Toronto, Canada

Yannis Stylianou
University of Crete, Heraklion, Greece

Jean-Philippe Thiran
Ecole Polytechnique Fédérale de Lausanne, Switzerland

Chapter 1

How is speech processed in a cell phone conversation?

T. Dutoit(°), N. Moreau(*), P. Kroon(+)

(°) Faculté Polytechnique de Mons, Belgium
(*) Ecole Nationale Supérieure des Télécommunications, Paris, France
(+) LSI, Allentown, PA, USA

Every cell phone solves 10 linear equations in 10 unknowns every 20 milliseconds

Although most people see the cell phone as an extension of conventional wired phone service or POTS (plain old telephone service), the truth is that cell phone technology is extremely complex and a marvel of technology. Very few people realize that these small devices perform hundreds of millions of operations per second to be able to maintain a phone conversation. If we take a closer look at the module that converts the electronic version of the speech signal into a sequence of bits, we see that for every 20 ms of input speech, a set of speech model parameters is computed and transmitted to the receiver. The receiver converts these parameters back into speech. In this chapter, we will see how linear predictive (LP) analysis–synthesis lies at the very heart of mobile phone transmission of speech. We first start with an introduction to linear predictive speech modeling and follow with a MATLAB-based proof of concept.

1.1 Background – Linear predictive processing of speech

Speech is produced by an excitation signal generated in our throat, which is modified by resonances produced by different shapes of our vocal, nasal,

T. Dutoit, F. Marqués (eds.), *Applied Signal Processing*,
DOI 10.1007/978-0-387-74535-0_1, © Springer Science+Business Media, LLC 2009

and pharyngeal tracts. This excitation signal can be the glottal pulses produced by the periodic opening and closing of our vocal folds (which creates *voiced* speech such as the vowels in "*voice*"), or just some continuous air flow pushed by our lungs (which creates *unvoiced* speech such as the last sound in "*voice*"), or even a combination of both at the same time (such as the first sound in "*voice*").

The periodic component of the glottal excitation is characterized by its fundamental frequency F_0 (Hz) called *pitch*[1]. The resonant frequencies of the vocal, oral, and pharyngeal tracts are called *formants*. On a spectral plot of a speech frame, pitch appears as narrow peaks for fundamental and harmonics; formants appear as wide peaks of the envelope of the spectrum (Fig. 1.1).

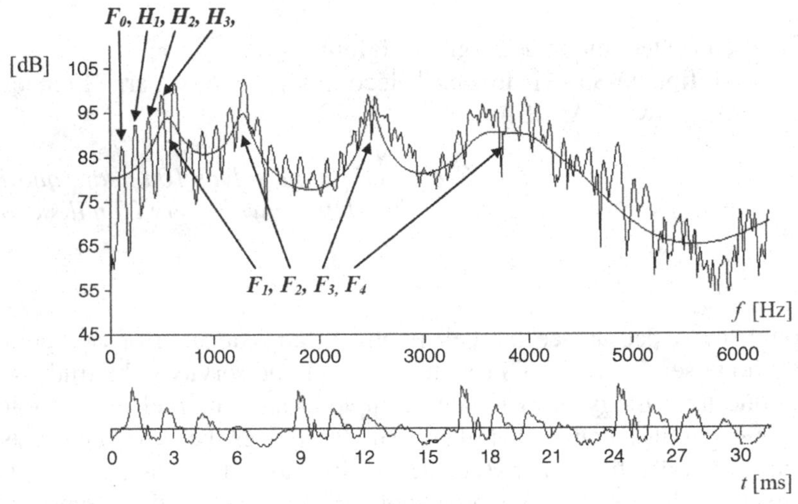

Fig. 1.1 A 30- ms frame of voiced speech (*bottom*) and its spectrum (shown here as the magnitude of its FFT). Harmonics are denoted as H_1, H_2, H_3, etc.; formants are denoted as F_1, F_2, F_3, etc. The spectral envelope is shown here for convenience; it implicitly appears only in the regular FFT

1.1.1 The LP model of speech

As early as 1960, Fant proposed a linear model of speech production (Fant 1960), termed as the *source-filter model*, based on the hypothesis that the

[1] Strictly speaking, pitch is defined as the *perceived* fundamental frequency.

glottis and the vocal tract are fully uncoupled. This model led to the well-known *autoregressive* (AR) or *linear predictive* (LP)[2] model of speech production (Rabiner and Shafer 1978), which describes speech $s(n)$ as the output $\tilde{s}(n)$ of an *all-pole* filter $1/A(z)$ excited by $\tilde{e}(n)$:

$$\tilde{S}(z)=\tilde{E}(z)\frac{1}{\sum_{i=0}^{p} a_i z^{-i}}=\tilde{E}(z)\frac{1}{A_p(z)} \qquad (a_0=1) \tag{1.1}$$

where $\tilde{S}(z)$ and $\tilde{E}(z)$ are the Z transforms of the speech and excitation signals, respectively, and p is the *prediction order*. The excitation of the LP model (Fig. 1.2) is assumed to be either a sequence of regularly spaced pulses (whose period T_0 and amplitude σ can be adjusted) or white Gaussian noise (whose variance σ^2 can be adjusted), thereby implicitly defining the so-called voiced/unvoiced (V/UV) decision. The filter $1/A_p(z)$ is termed as the *synthesis filter* and $A_p(z)$ is called the *inverse filter.*

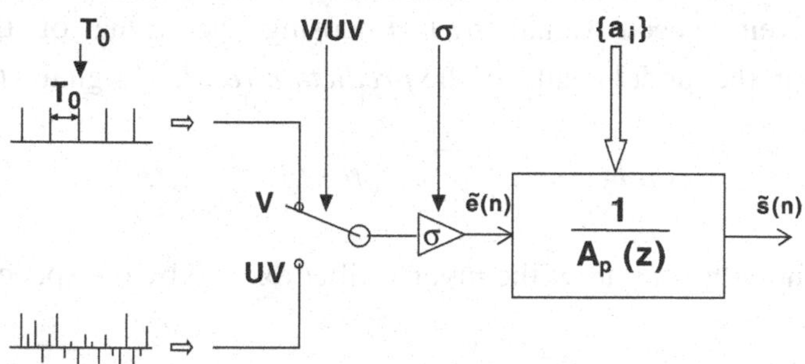

Fig. 1.2 The LP model of speech production

Equation (1.1) implicitly introduces the concept of linear predictability of speech (hence the name of the model), which states that each speech sample can be expressed as a weighted sum of the p previous samples, plus some excitation contribution:

$$\tilde{s}(n)=\tilde{e}(n)-\sum_{i=1}^{p} a_i \tilde{s}(n-i) \tag{1.2}$$

[2] Sometimes it is denoted as the *LPC model (linear predictive coding)* because it has been widely used for speech coding.

1.1.2 The LP estimation algorithm

From a given signal, a practical problem is to find the best set of prediction coefficients – that is, the set that minimizes modeling errors – by trying to minimize audible differences between the original signal and the one that is produced by the model of Fig. 1.2. This implies to estimate the value of the LP parameters: pitch period T_0, gain σ, V/UV switch position, and prediction coefficients $\{a_i\}$.

Pitch and voicing (*V/UV*) determination is a difficult problem. Although speech seems periodic, it is never truly the case. Glottal cycle amplitude varies from period to period (*shimmer*) and its period itself is not constant (*jitter*). Moreover, the speech waveform reveals only filtered glottal pulses rather than glottal pulses themselves. This makes a realistic measure of T_0 even more complex. In addition, speech is rarely completely voiced; its additive noise components make pitch determination even harder. Many techniques have been developed to estimate T_0 (see Hess 1992; de la Cuadra 2007).

The estimation of σ and of the prediction coefficients can be performed simultaneously and fortunately independently of the estimation of T_0.

For a given speech signal *s(n)*, imposing the value of the $\{a_i\}$ coefficients in the model results in the *prediction residual* signal, *e(n)*:

$$e(n) = s(n) + \sum_{i=1}^{p} a_i s(n-i) \tag{1.3}$$

which is simply the output of the inverse filter excited by the speech signal (Fig. 1.3).

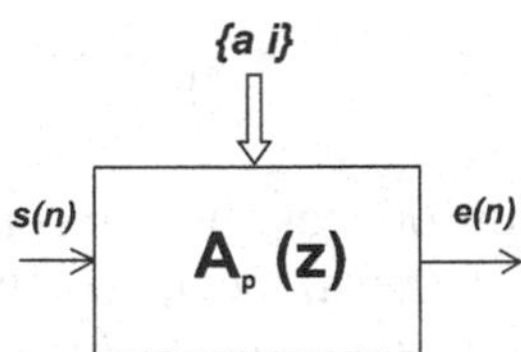

Fig. 1.3 Inverse filtering of speech

The principle of AR estimation is to choose the set $\{a_1, a_2, \ldots a_p\}$, which minimizes the expectation $E(e^2(n))$ of the residual energy:

$$\{a_i\}^{opt} = \underset{a_i}{\arg\min}\left(E(e^2(n))\right) \tag{1.4}$$

As a matter of fact, it can be shown that, if *s(n)* is stationary, the synthetic speech $\tilde{s}(n)$ produced by the LP model (Fig. 1.2) using this specific set of prediction coefficients in Equation (1.2) will exhibit the same spectral envelope as *s(n)*. Since the excitation of the LP model (pulses or white noise) has a flat spectral envelope, this means that the frequency response of the synthesis filter will approximately match the spectral envelope of *s(n)* and that the spectral envelope of the LP residual will be approximately flat. In a word, inverse filtering decorrelates speech.

Developing the LMSE (least mean squared error) criterion (1.4) easily leads to the so-called set of *p* Yule-Walker linear equations:

$$\begin{bmatrix} \phi_{xx}(0) & \phi_{xx}(1) & \dots & \phi_{xx}(p-1) \\ \phi_{xx}(1) & \phi_{xx}(0) & \dots & \phi_{xx}(p-2) \\ \dots & \dots & \dots & \dots \\ \phi_{xx}(p-1) & \phi_{xx}(p-2) & \dots & \phi_{xx}(0) \end{bmatrix} \begin{bmatrix} a_1 \\ a_2 \\ \dots \\ a_p \end{bmatrix} = - \begin{bmatrix} \phi_{xx}(1) \\ \phi_{xx}(2) \\ \dots \\ \phi_{xx}(p) \end{bmatrix} \tag{1.5}$$

in which $\phi_{xx}(k) \quad (k = 0 \dots p)$ are the *p*+1 first autocorrelation coefficients of *s(n)*. After solving this set of equations, the optimal value of σ is then given by the following equation:

$$\sigma^2 = \sum_{i=0}^{p} a_i \phi_{xx}(i) \tag{1.6}$$

It should be noted that since Equations (1.5) are based only on the autocorrelation function of *s(n)*, the model does not try to imitate the exact speech waveform, but rather its spectral envelope (based on the idea that our ear is more sensitive to the amplitude spectrum than to the phase spectrum).

1.1.3 LP processing in practice

Since speech is nonstationary, the LP model is applied on speech *frames* (typically 30 ms long, with an overlap of 20 ms; Fig. 1.4) in which the signal is assumed to be stationary given the inertia of the articulatory muscles[3].

Speech samples are usually weighted using a *weighting window* (typically a 30-ms-long Hamming window). This prevents the first samples of each frame, which cannot be correctly predicted, from having too much weight in Equation (1.4) by producing higher values of $e^2(n)$.

The $\phi_{xx}(k)$ $(k=0...p)$ autocorrelation coefficients are then estimated on a limited number of samples (typically 240 samples, for 30 ms of speech with a sampling frequency of 8 kHz). The prediction order p (which is also the number of poles in the all-pole synthesis filter) is chosen such that the resulting synthesis filter has enough degrees of freedom to copy the spectral envelope of the input speech. Since there is approximately one formant per kilohertz of bandwidth of speech, at least $2B$ poles are required (where B is the signal bandwidth in kHz, i.e., half the sampling frequency). Two more poles are usually added for modeling the glottal cycle waveform (and also empirically, because the resulting LPC speech sounds better). For telephone-based applications, working with a sampling frequency of 8 kHz, this leads to $p=10$.

Although Equation (1.5) can be solved with any classical matrix inversion algorithm, the so-called *Levinson–Durbin* algorithm is preferred for its speed, as it takes into account the special structure of the matrix (all elements on diagonals parallel to the principal diagonal are equal; this characterizes a *Toeplitz* matrix). See Rabiner and Schafer (1978) or Quatieri (2002) for details.

The *prediction coefficients* $\{a_i\}$ are finally computed for every frame (i.e., typically every 10–20 ms).

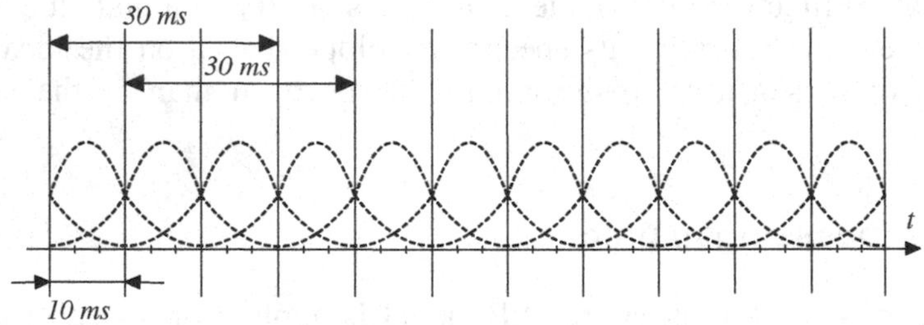

Fig. 1.4 Frame-based processing of speech (shown here with a frame length of 30 ms and a shift of 10 ms)

[3] In practice, this is only an approximation, which tends to be very loose for plosives, for instance.

The complete block diagram of an LPC speech analysis–synthesis system is given in Fig. 1.5.

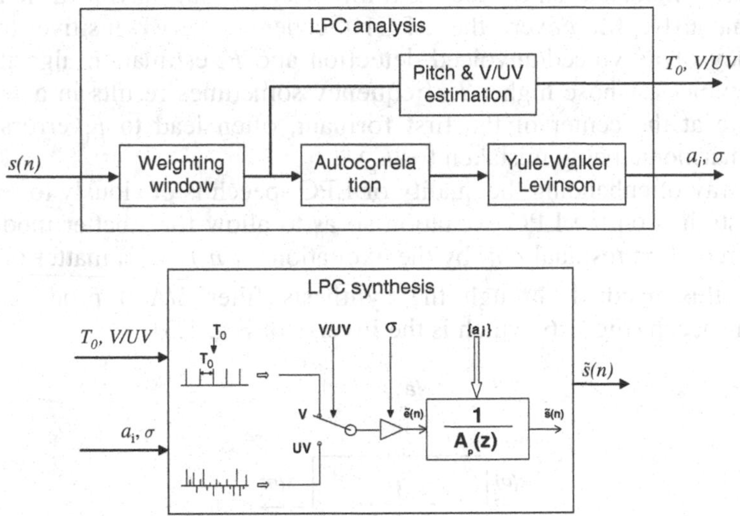

Fig. 1.5 A linear predictive speech analysis–synthesis system

1.1.4 Linear predictive coders

The LPC analysis–synthesis system, which has been described above, is not exactly the one embedded in cell phones.

It is, however, implemented in the so-called NATO LPC10 standard (NATO, 1984), which was used for satellite transmission of speech communications until 1996. This norm makes it possible to encode speech with a bit rate as low as 2,400 bits/s (frames are 22.5 ms long, and each frame is coded with 54 bits: 7 bits for pitch and *V/UV* decision, 5 bits for the gain, and 42 bits for the prediction coefficients[4]). In practice, prediction coefficients are actually not used as such; the related *reflection coefficients* or *log area ratios* are preferred, since they have better quantization properties. Quantization of prediction coefficients can result in unstable filters.

The number of bits in LPC10 was chosen such that it does not bring audible artifacts to the LPC speech. The example LPC speech produced in Section 1.2 is therefore a realistic example of typical LPC10 speech.

[4] Advanced LP coders, such as CELP, have enhanced prediction coefficients coding down to 30 bits.

Clearly this speech coder suffers from the limitations of the poor (and binary!) excitation model. Voiced fricatives, for instance, cannot be adequately modeled since they exhibit voiced *and* unvoiced features simultaneously. Moreover, the LPC10 coder is very sensitive to the efficiency of its voiced/unvoiced detection and F_0 estimation algorithms. Female voices, whose higher F_0 frequency sometimes results in a second harmonic at the center of the first formant, often lead to F_0 errors (the second harmonic being mistaken for F_0).

One way of enhancing the quality of LPC speech is obviously to reduce the constraints on the LPC excitation so as to allow for a better modeling of the prediction residual $e(n)$ by the excitation $\tilde{e}(n)$. As a matter of fact, passing this residual through the synthesis filter $1/A(z)$ produces the original speech (Fig. 1.6, which is the inverse of Fig. 1.3).

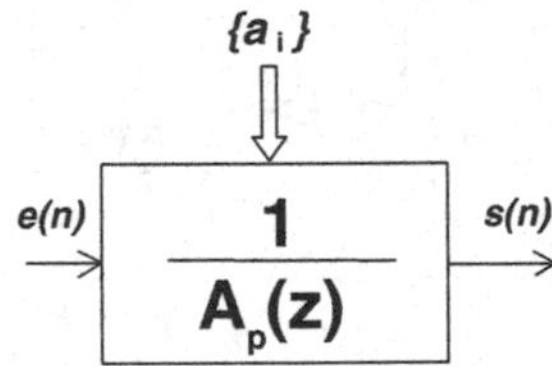

Fig. 1.6 Passing the prediction residual through the synthesis filter produces the original speech signal

The *multipulse excited* (MPE; Atal and Remde 1982) was an important step in this direction, as it was the first approach to implement an analysis-by-synthesis process (i.e., a closed loop) for the estimation of the excitation features. The MPE excitation is characterized by the positions and amplitudes of a limited number of pulses per frame (typically 10 pulses per 10 ms frame; Fig. 1.7). Pitch estimation and voiced/unvoiced decision are no longer required. Pulse positions and amplitudes are chosen iteratively (Fig. 1.8) so as to minimize the energy of the modeling error (the difference between the original speech and the synthetic speech). The error is filtered by a *perceptual filter* before its energy is computed:

$$P(z) = \frac{A(z)}{A(z/\gamma)} \tag{1.7}$$

The role of this filter, whose frequency response can be set to any intermediate between all pass response (γ=1) and the response of the inverse filter (γ=0), is to reduce the contributions of the formants to the estimation of the error. The value of γ is typically set to 0.8.

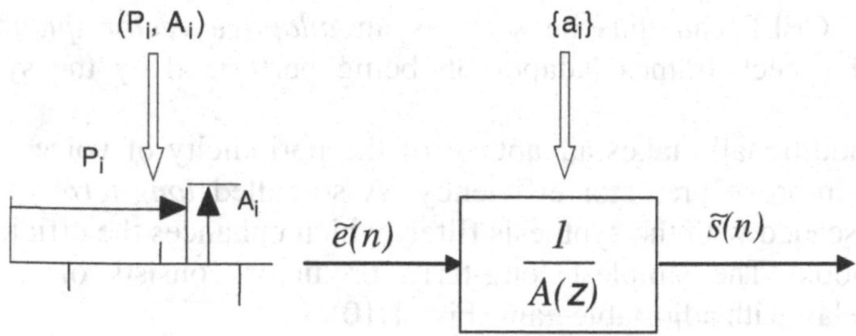

Fig. 1.7 The MPE decoder

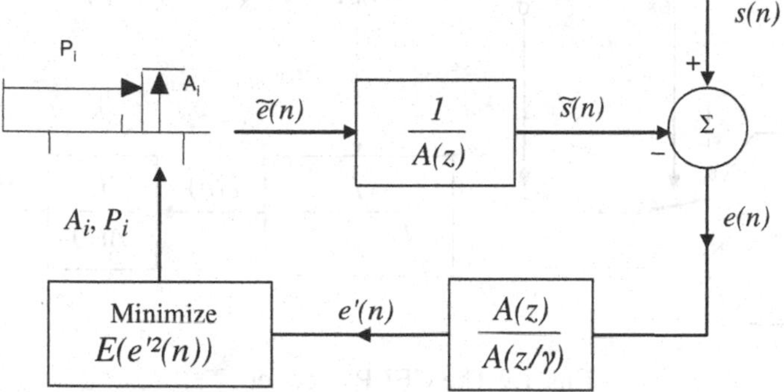

Fig. 1.8 Estimation of the MPE excitation by an analysis-by-synthesis loop in the MPE encoder

The *code-excited linear prediction* (CELP) coder (Schroeder and Atal, 1985) further extended the idea of analysis-by-synthesis speech coding by using the concept of *vector quantization* (VQ) for the excitation sequence. In this approach, the encoder selects one excitation sequence from a predefined *stochastic codebook* of possible sequences (Fig. 1.9) and sends only the index of the selected sequence to the decoder, which has a similar codebook. Although the lowest quantization rate for scalar quantization is 1 bit per sample, VQ allows fractional bit rates. For example, quantizing two samples simultaneously using a 1-bit codebook will result in 0.5 bits per sample. More typical values are a 10-bit codebook with codebook vectors of dimension 40, resulting in 0.25 bits per sample. Given the very high variability of speech frames, however (due to changes in glottal excitation *and* vocal tract), vector-quantized speech frames would be possible only with a very large codebook. The great idea of CELP is precisely to perform VQ on LP residual sequences: as we have seen in Section 1.1.2, the LP residual has a flat spectral envelope, which makes it easier to produce a small but somehow exhaustive codebook of LP residual sequences. CELP can thus be seen as an *adaptive vector quantization* scheme of speech frames (adaptation being performed by the synthesis filter).

CELP additionally takes advantage of the periodicity of voiced sounds to further improve predictor efficiency. A so-called *long-term predictor* filter is cascaded with the synthesis filter, which enhances the efficiency of the codebook. The simplest long-term predictor consists of a simple variable delay with adjustable gain (Fig. 1.10).

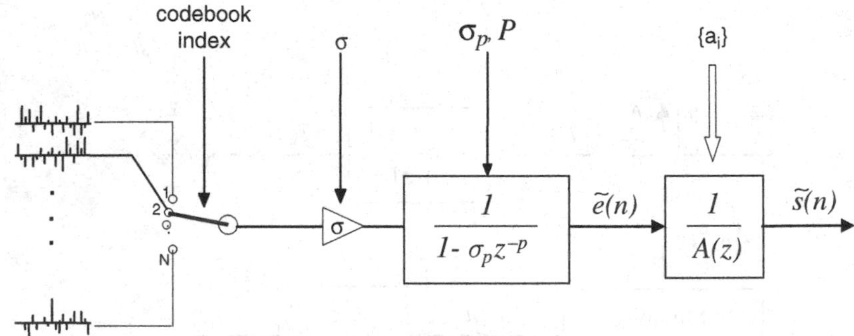

Fig. 1.9 The CELP decoder

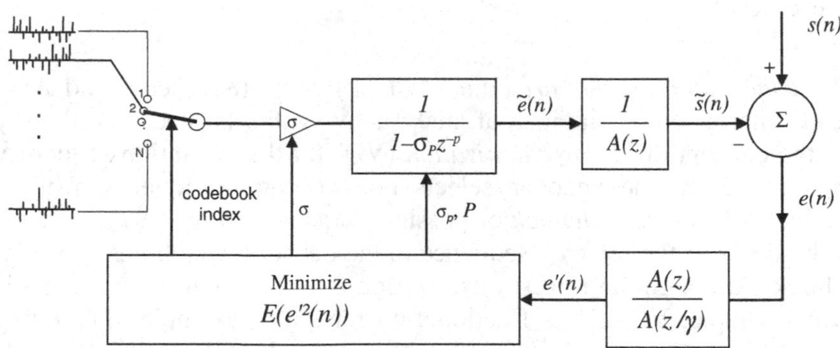

Fig. 1.10 Estimation of the CELP excitation by an analysis-by-synthesis loop in the CELP encoder

Various coders have been developed after MPE and CELP using the same analysis-by-synthesis principle with the goal of enhancing CELP quality while further reducing bit rate, among which are the mixed-excitation linear prediction (MELP; McCree and Barnwell, 1995) and the harmonic and vector excitation coding (HVXC; Matsumoto et al. 1997). In 1996, LPC-10 was replaced by MELP to be the United States Federal Standard for coding at 2.4 kbps.

From 1992 to 1996, GSM (global system for mobile communication) phones embedded a particular form of MPE, the *RPE-LPC* (regular pulse excited; Kroon et al. 1986) coder, with additional constraints on the positions of the pulses: the RPE pulses were evenly spaced (but their amplitude, as well as the position of the first pulse, is left open). Speech is divided into 20 ms frames, each of which is encoded as 260 bits, giving a total bit rate of 13 kbps. In 1996, this so-called *full-rate* (FR) codec was replaced by the *enhanced full-rate* (EFR) codec, implementing a variant of CELP termed as algebraic-CELP (ACELP, Salami et al. 1998). The ACELP codebook structure allows efficient searching of the optimal codebook index thereby eliminating one of the main drawbacks of CELP which is its complexity. The EFR coder operates at 11.2 kbps and produces better speech quality than the FR coder at 13 kb/s. A variant of the ACELP coder has been standardized by ITU-T as G.729 for operation at a bit rate of 8 kbps. Newer generations of coders that are used in cell phones are all based on the CELP principle and can operate at bit rates as low as 4.75 – 11.2 kbps.

1.2 MATLAB proof of concept : ASP_cell_phone.m

We will first examine the contents of a speech file (Section 1.2.1) and perform LP analysis and synthesis on a voiced (Section 1.2.2) and an unvoiced frame (Section 1.2.3). We will then generalize this approach to the complete speech file by first synthesizing all frames as voiced and imposing a constant pitch (Section 1.2.4), then by synthesizing all frames as unvoiced (Section 1.2.5), and finally by using the original pitch[5] and voicing information as in LPC10 (Section 1.2.6). We will conclude this section by changing LPC10 into CELP (Section 1.2.7).

[5] By "original pitch," we mean the pitch that can be measured on the original signal.

1.2.1 Examining a speech file

Let us load file "speech.wav," listen to it, and plot its samples (Fig. 1.11). This file contains the sentence "Paint the circuits" sampled at 8 kHz, with 16 bits.[6]

```
speech=wavread('speech.wav');
plot(speech)
xlabel('Time (samples)'); ylabel('Amplitude');
sound(speech,8000);
```

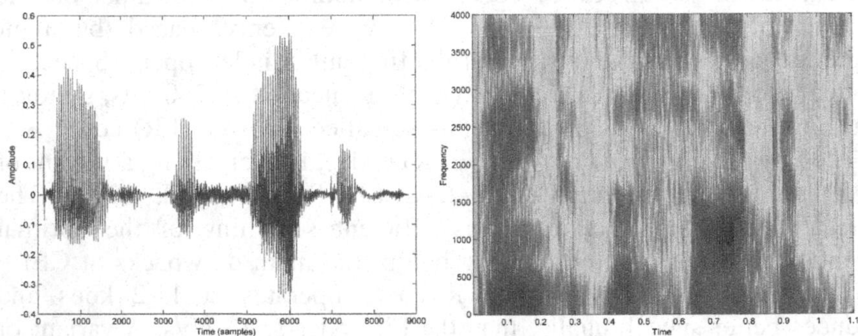

Fig. 1.11 Input speech: the speech.wav file (*left*: waveform; *right*: spectrogram)

The file is about 1.1 s long (9,000 samples). One can easily spot the position of the four vowels appearing in this plot, since vowels usually have higher amplitude than other sounds. The vowel "e" in "the", for instance, is approximately centered on sample 3,500.

As such, however, the speech waveform is not "readable," even by an expert phonetician. Its information content is hidden. In order to reveal it to the eyes, let us plot a spectrogram of the signal (Fig. 1.11). We then choose a *wideband spectrogram*[7] by imposing the length of each frame to be approximately 5 ms long (40 samples) and a hamming weighting window.

```
specgram(speech,512,8000,hamming(40))
```

In this plot, pitch periods appear as vertical lines. As a matter of fact, since the length of analysis frames is very small, some frames fall on the

---

6 This sentence was taken from the Open Speech Repository on the web.

7 A wideband spectrogram uses a small amount of samples (typically less than the local pitch period) so as to better reveal formants.

peaks (resp., on the valleys) of pitch periods and thus appear as a darker (resp., lighter) vertical lines.

In contrast, formants (resonant frequencies of the vocal tract) appear as dark (and rather wide) horizontal traces. Although their frequency is not easy to measure with precision, experts looking at such a spectrogram can actually often read it (i.e., guess the corresponding words). This clearly shows that formants are a good indicator of the underlying speech sounds.

1.2.2 Linear prediction synthesis of 30 ms of voiced speech

Let us extract a 30-ms frame from a voiced part (i.e., 240 samples) of the speech file and plot its samples (Fig. 1.12).

```
input_frame=speech(3500:3739);
plot(input_frame);
```

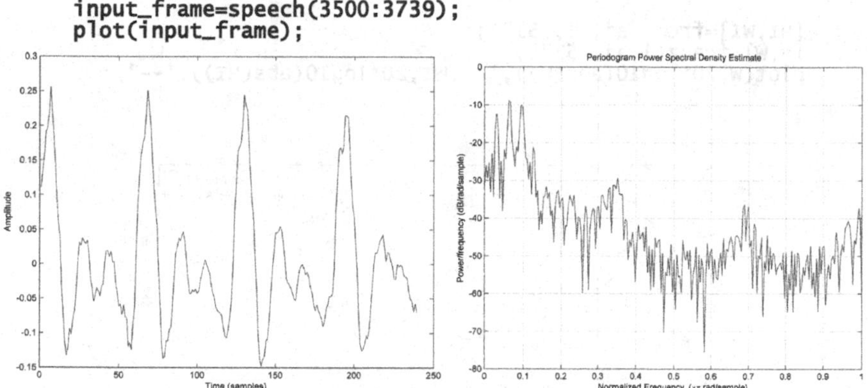

Fig. 1.12 A 30-ms-long voiced speech frame taken from a vowel (*left*: waveform; *right*: periodogram)

As expected this sound is approximately periodic (period=65 samples, i.e., 80 ms; fundamental frequency = 125 Hz). Note, though, that this is only apparent; in practice, no sequence of samples can be found more than once in the frame.

Now let us see the spectral content of this speech frame (Fig. 1.12) by plotting its *periodogram* on 512 points (using a normalized frequency axis; remember π corresponds to $F_s/2$, i.e., to 4,000 Hz here).

```
periodogram(input_frame,[],512);
```

The fundamental frequency appears again at around 125 Hz. One can also roughly estimate the position of formants (peaks in the spectral envelope) at ± 300, 1,400, and 2,700 Hz.

Let us now fit an LP model of order 10 to our voiced frame.[8] We obtain the prediction coefficients (`ai`) and the variance of the residual signal (`sigma_square`).

```
[ai, sigma_square]=lpc(input_frame,10);
sigma=sqrt(sigma_square);
```

The estimation parameter inside LPC is called the Levinson–Durbin algorithm. It chooses the coefficients of an FIR filter *A(z)* so that when passing the input frame into *A(z)*, the output, termed as the prediction residual, has minimum energy. It can be shown that this leads to a filter which has anti-resonances wherever the input frame has a formant. For this reason, the *A(z)* filter is termed as the "inverse" filter. Let us plot its frequency response (on 512 points) and superimpose it to that of the "synthesis" filter *1/A(z)* (Fig. 1.13).

```
[HI,WI]=freqz(ai, 1, 512);
[H,W]=freqz(1,ai, 512);
plot(W,20*log10(abs(H)),'-',WI,20*log10(abs(HI)),'--');
```

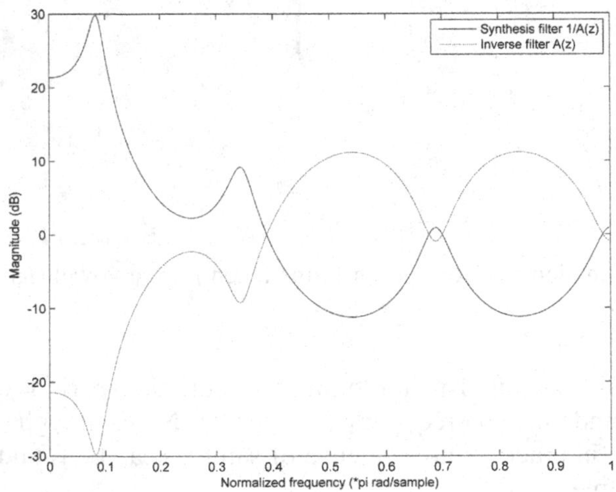

Fig. 1.13 Frequency responses of the inverse and synthesis filters

[8] We do not apply windowing prior to LP analysis now, as it has no tutorial benefit. We will add it in subsequent sections.

In other words, the frequency response of the filter *1/A(z)* matches the spectral amplitude envelope of the frame. Let us superimpose this frequency response to the periodogram of the vowel (Fig. 1.14).[9]

```
periodogram(input_frame,[],512,2)
hold on;
plot(W/pi,20*log10(sigma*abs(H)));
hold off;
```

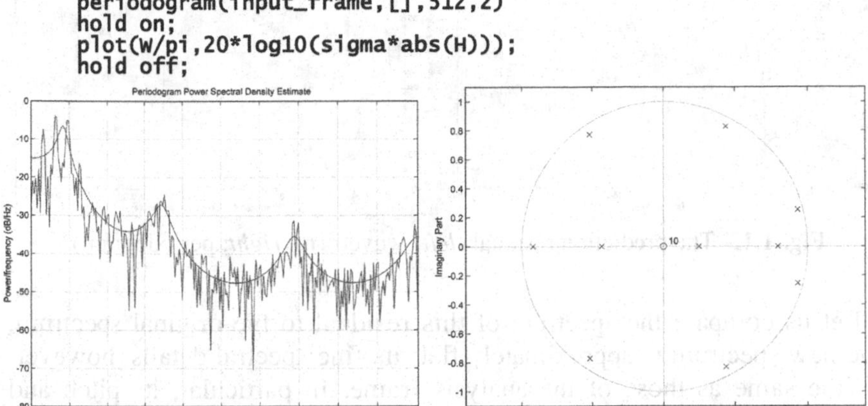

Fig. 1.14 *Left*: Frequency response of the synthesis filter superimposed with the periodogram of the frame; *right*: poles and zeros of the filter

In other words, the LPC fit has automatically adjusted the poles of the synthesis filter close to the unit circle at angular positions chosen to imitate formant resonances (Fig. 1.14).

```
zplane(1,ai);
```

If we apply the inverse of this filter to the input frame, we obtain the prediction residual (Fig. 1.15).

```
LP_residual=filter(ai,1,input_frame);
plot(LP_residual)
periodogram(LP_residual,[],512);
```

[9] The `periodogram` function of MATLAB actually shows the so-called *one-sided periodogram*, which has twice the value of the two-sided periodogram in [0, $F_s/2$]. In order to force MATLAB to show the real value of the two-sided periodogram in [0, $F_s/2$], we claim $F_s = 2$.

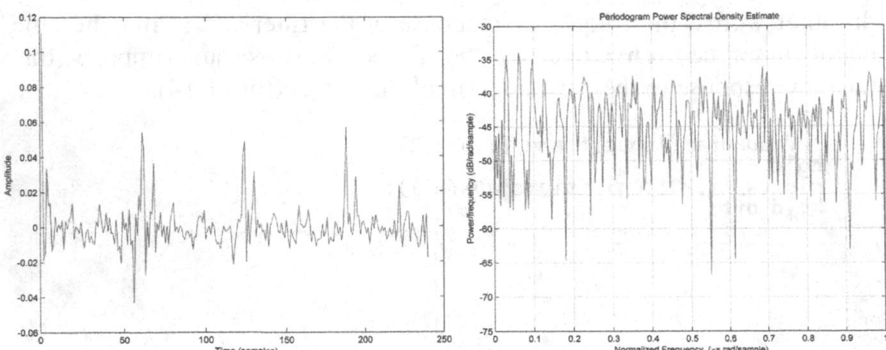

Fig. 1.15 The prediction residual (*left*: waveform; *right*: periodogram)

Let us compare the spectrum of this residual to the original spectrum. The new spectrum is approximately flat; its fine spectral details, however, are the same as those of the analysis frame. In particular, its pitch and harmonics are preserved.

For obvious reasons, applying the synthesis filter to this prediction residual results in the analysis frame itself (since the synthesis filter is the inverse of the inverse filter).

```
output_frame=filter(1, ai,LP_residual);
plot(output_frame);
```

The LPC model actually models the prediction residual of voiced speech as an impulse train with adjustable pitch period and amplitude. For the speech frame considered, for instance, the LPC ideal excitation is a sequence of pulses separated by 64 zeros (so as to impose a period of 65 samples; Fig. 1.16). Note we multiply the excitation by some gain so that its variance matches that of the residual signal.

```
excitation = [1;zeros(64,1);1;zeros(64,1);1;zeros(64,1);...
              1;zeros(44,1)];
gain=sigma/sqrt(1/65);
plot(gain*excitation);
periodogram(gain*excitation,[],512);
```

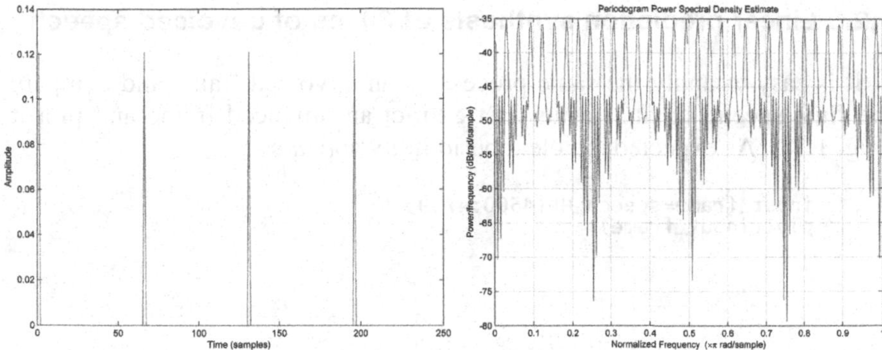

Fig. 1.16 The LPC excitation (*left*: waveform; *right*: periodogram)

Clearly, as far as the waveform is concerned, the LPC excitation is far from similar to the prediction residual. Its spectrum (Fig. 1.16), however, has the same broad features as that of the residual: flat envelope and harmonic content corresponding to F_0. The main difference is that the excitation spectrum is "over-harmonic" compared to the residual spectrum.

Let us now use the synthesis filter to produce an artificial "e."

```
synt_frame=filter(gain,ai,excitation);
plot(synt_frame);
periodogram(synt_frame,[],512);
```

Although the resulting waveform is obviously different from the original one (this is due to the fact that the LP model does not account for the phase spectrum of the original signal), its spectral envelope is identical. Its fine harmonic details, though, also widely differ: the synthetic frame is actually "over-harmonic" compared to the analysis frame (Fig. 1.17).

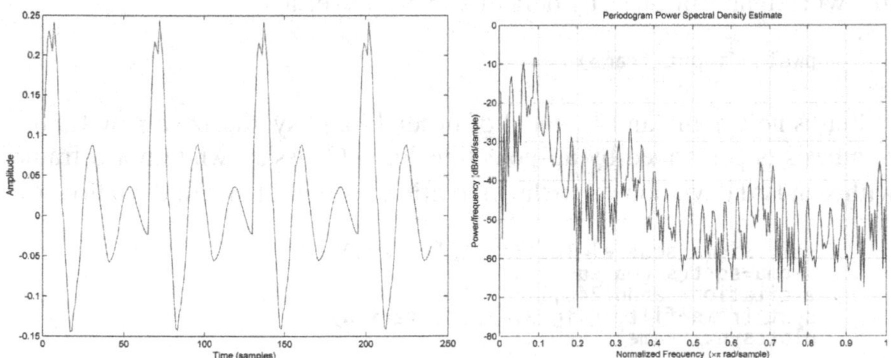

Fig. 1.17 Voiced LPC speech (*left*: waveform; *right*: periodogram)

1.2.3 Linear prediction synthesis of 30 ms of unvoiced speech

It is easy to apply the same process to an unvoiced frame and compare the final spectra again. Let us first extract an unvoiced frame and plot it (Fig. 1.18). As expected, no clear periodicity appears.

```
input_frame=speech_HF(4500:4739);
plot(input_frame);
```

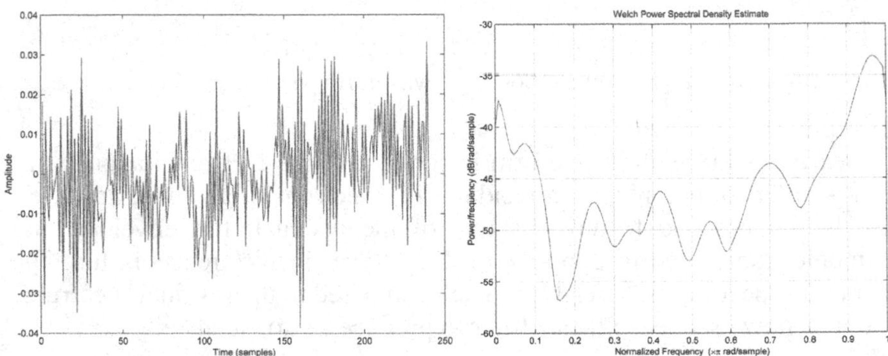

Fig. 1.18 A 30-ms-long frame of unvoiced speech (*left*: waveform; *right*: power spectral density)

Now let us see the spectral content of this speech frame. Note that, since we are dealing with noisy signals, we use the *averaged periodogram* to estimate power spectral densities, although with less-frequency resolution than using a simple periodogram. The MATLAB `pwlech` function does this with eight subframes by default and 50% overlap.

```
pwelch(input_frame);
```

Let us now apply an LP model of order 10 and synthesize a new frame. Synthesis is performed by all-pole filtering a Gaussian white noise frame with standard deviation set to the prediction residual standard deviation, σ.

```
[ai, sigma_square]=lpc(input_frame,10);
sigma=sqrt(sigma_square);
excitation=randn(240,1);
synt_frame=filter(sigma,ai,excitation);
plot(synt_frame);

pwelch(synt_frame);
```

The synthetic waveform (Fig. 1.19) has no sample in common with the original waveform. The spectral envelope of this frame, however, is still similar to the original one, enough at least for both the original and synthetic signals to be perceived as the same colored noise.[10]

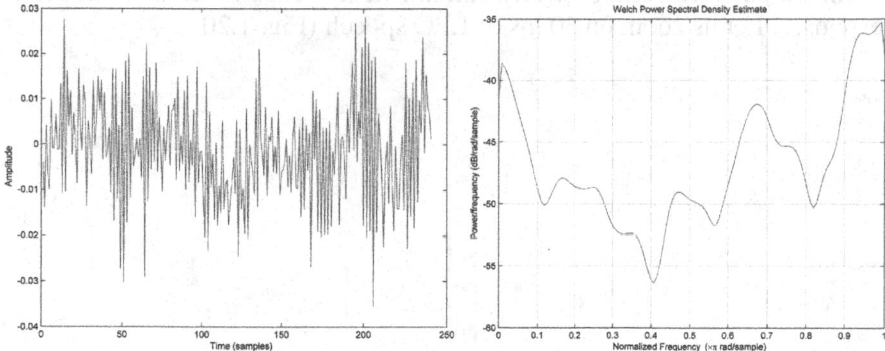

Fig. 1.19 Unvoiced LPC speech (*left*: waveform; *right*: psd)

1.2.4 Linear prediction synthesis of a speech file, with fixed F_0

We will now loop the previous operations for the complete speech file using 30 ms analysis frames overlapping by 20 ms. Frames are now weighted with a Hamming window. At synthesis time, we simply synthesize 10 ms of speech and concatenate the resulting synthetic frames to obtain the output speech file. Let us choose 200 Hz as synthesis F0, for convenience: this way each 10-ms excitation frame contains exactly two pulses.

```
for i=1:(length(speech)-160)/80; % number of frames

    % Extracting the analysis frame
    input_frame=speech_HF((i-1)*80+1:(i-1)*80+240);

    % Hamming window weighting and LPC analysis
    [ai, sigma_square]=lpc(input_frame.*hamming(240),10);
    sigma=sqrt(sigma_square);

    % Generating 10 ms of excitation
    % = 2 pitch periods at 200 Hz
    excitation=[1;zeros(39,1);1;zeros(39,1)];
    gain=sigma/sqrt(1/40);

    % Applying the synthesis filter
```

[10] Although both power spectral densities have identical spectral slopes, one should not expect them to exhibit a close match in terms of their details, since only LPC modeling reproduces the smooth spectral envelope of the original signal.

```
        synt_frame=filter(gain, ai,excitation);

        % Concatenating synthesis frames
        synt_speech_HF=[synt_speech_HF;synt_frame];

    end
```

The output waveform basically contains a sequence of LP filter impulse responses. Let us zoom on 30 ms of LPC speech (Fig. 1.20).

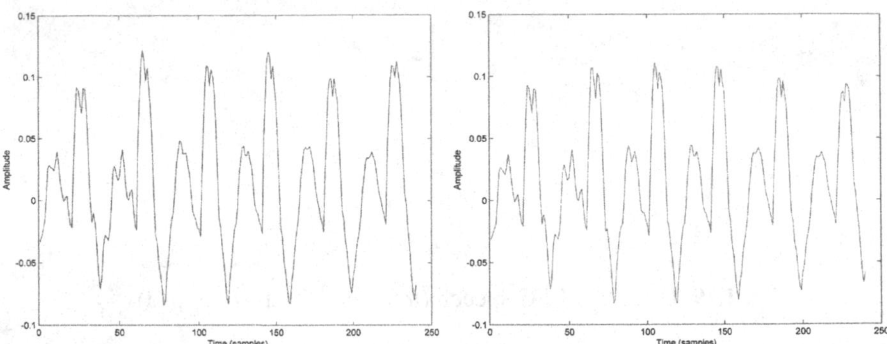

Fig. 1.20 Zoom on 30 ms of LPC speech (*left*: with internal variable reset; *right*: with internal variable memory)

It appears that in many cases the impulse responses have been cropped to the 10-ms synthetic frame size. As a matter of fact, since each synthesis frame was composed of two identical impulses, one should expect our LPC speech to exhibit pairs of identical pitch periods. This is not the case, because for producing each new synthetic frame, the internal variables of the synthesis filter are implicitly reset to zero. We can avoid this problem by maintaining the internal variables of the filter from the end of each frame to the beginning of the next one.

We initialize a vector `z` with 10 zeros and change the synthesis code into

```
% Applying the synthesis filter
% Taking care of the internal variables of the filter
gain=sigma/sqrt(1/40);
[synt_frame,z]=filter(gain, ai, excitation, z);
```

This time the end of each impulse response is properly added to the beginning of the next one, which results in more smoothly evolving periods (Fig. 1.20).

If we want to synthesize speech with constant pitch period length different from a submultiple of 80 samples (say, `N0`=65 samples), we additionally need to take care of a possible pitch period `offset` in the excitation signal. After initializing this `offset` to zero, we simply change the excitation code into

```
% Generating 10 ms of excitation
% taking a possible offset into account

% if pitch period length > excitation frame length
if offset>=80
    excitation=zeros(80,1);
    offset=offset-80;
else
    % complete the previously unfinished pitch period
    excitation=zeros(offset,1);
    % for all pitch periods in the remaining of the frame
    for j=1:floor((80-offset)/N0)
        % add one excitation period
        excitation=[excitation;1;zeros(N0-1,1)];
    end;
    % number of samples left in the excitation frame
    flush=80-length(excitation);
    if flush~=0
        % fill the frame with a partial pitch period
        excitation=[excitation;1;zeros(flush-1,1)];
        % remember to fill the remaining of the period in
        % next frame
        offset=N0-flush;
    else offset=0;
    end
end
gain=sigma/sqrt(1/N0);
```

1.2.5 Unvoiced linear prediction synthesis of a speech file

Synthesizing the complete speech file as LPC unvoiced speech is easy. Periodic pulses are simply replaced by white noise, as in Section 1.2.3.

```
% Generating 10 ms of excitation
excitation=randn(80,1); % White Gaussian noise
gain=sigma;
```

As expected, the resulting speech sounds like whisper.

1.2.6 Linear prediction synthesis of a speech file, with original F_0

We will now synthesize the same speech using the original F_0. We will thus have to deal with the additional problems of pitch estimation (on a frame-by-frame basis), including voiced/unvoiced decision. This approach is similar to that of the LPC10 coder (except that we do not quantize coefficients here). We change the excitation generation code into

```
% local synthesis pitch period (in samples)
N0=pitch(input_frame);
```

```
% Generating 10 ms of excitation
if N0~=0 % voiced frame
    % Generate 10 ms of voiced excitation
    % taking a possible offset into account

    (same code as in Section 1.2.4)

else
    % Generate 10 ms of unvoiced voiced excitation

    (same code as in Section 1.2.5)

    offset=0; % reset for subsequent voiced frames
end;
```

MATLAB function involved:

- `T0=pitch(speech_frame)` returns the pitch period T_0 (in samples) of a speech frame (T_0 is set to zero when the frame is detected as unvoiced). T_0 is obtained from the maximum of the (estimated) autocorrelation of the LPC residual. Voiced/unvoiced decision is based on the ratio of this maximum to the variance of the residual. This simple algorithm is not optimal but will do the job for this proof of concept.

The resulting synthetic speech (Fig. 1.21) is intelligible. It shows the same formants as the original speech. It is therefore acoustically similar to the original except for the additional buzziness that has been added by the LP model.

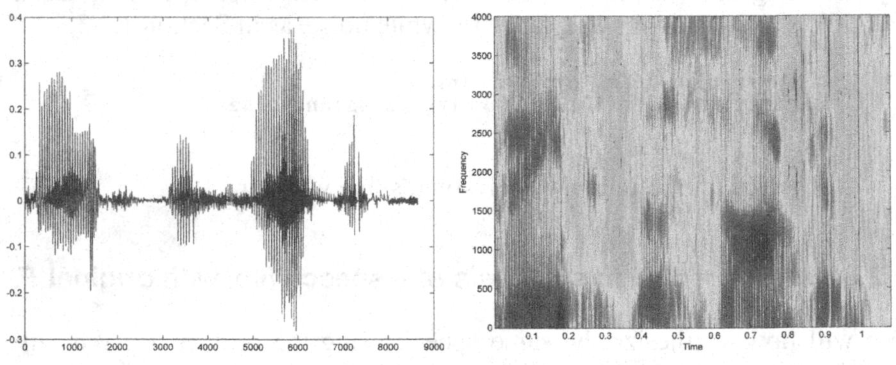

Fig. 1.21 LPC10 speech

1.2.7 CELP analysis–synthesis of a speech file

Our last step will be to replace the LPC10 excitation by a more realistic code-excited linear prediction (CELP) excitation obtained by selecting the best linear combination of excitation components from a codebook. Component selection is performed in a closed loop so as to minimize the difference between the synthetic and original signals.

We start with 30-ms LP analysis frames, shifted every 5 ms, and a codebook size of 512 vectors from which 10 components are chosen for every 5-ms synthesis frame.[11]

MATLAB function involved:

- `[gains, indices] = find_Nbest_components(signal, ...`
  `codebook_vectors, codebook_norms , N)`

This function finds the `N` best components of signal from the vectors in `codebook_vectors`, so that the residual error
`error = signal - codebook_vectors(indices)*gains`
is minimized. Components are found one-by-one using a greedy algorithm. When components in `codebook_vectors` are not orthogonal, the search is therefore suboptimal.

```
frame_length=240; % length of the LPC analysis frame
frame_shift=40; % length of the synthesis frames
codebook_size = 512; % number of vectors in the codebook
N_components= 10;  % number of codebook components per frame

speech=wavread('speech.wav');

% Initializing internal variables
z_inv=zeros(10,1);  % inverse filter
z_synt=zeros(10,1); % synthesis filter
synt_speech_CELP=[];

% Generating the stochastic excitation codebook
codebook = randn(frame_shift,codebook_size);

for i=1:(length(speech)-frame_length+frame_shift)/frame_shift;

    input_frame=speech((i-1)*frame_shift+1:...
                       (i-1)*frame_shift+frame_length);

    % LPC analysis of order 10
    ai = lpc(input_frame.*hamming(frame_length), 10);

    % Extracting frame_shift samples from the LPC analysis frame
    speech_frame = input_frame((frame_length-frame_shift)/2+1:...
```

---

[11] These values actually correspond to a rather high bit rate, but we will show in the next paragraphs how to lower the bit rate while maintaining the quality of synthetic speech.

```
                    (frame_length-frame_shift)/2+frame_shift);

    % Filtering the codebook (all column vectors)
    codebook_filt = filter(1, ai, codebook);

    % Finding speech_frame components in the filtered codebook
    % taking into account the transient stored in the internal
    % variables of the synthesis filter
    ringing = filter(1, ai, zeros(frame_shift,1), z_synt);
    signal = speech_frame - ringing;
    [gains, indices] = find_Nbest_components(signal, ...
        codebook_filt, N_components);

    % Generating the corresponding excitation as a weighted sum
    % of codebook vectors
    excitation = codebook(:,indices)*gains;

    % Synthesizing CELP speech, and keeping track of the
    % synthesis filter internal variables
    [synt_frame, z_synt] = filter(1, ai, excitation, z_synt);
    synt_speech_CELP=[synt_speech_CELP;synt_frame];

end
```

Note that this analysis–synthesis simulation is implemented as mentioned in Section 1.2.4 as an adaptive vector quantization system. This is done by passing the whole codebook through the synthesis filter, for each new frame, and searching for the best linear decomposition of the speech frame in terms of filtered codebook sequences.

Also note our use of `ringing`, which stores the natural response of the synthesis filter due to its nonzero internal variables. This response should not be taken into account in the adaptive VQ.

The resulting synthetic speech sounds more natural than in LPC10. Plosives are much better rendered, and voiced sounds are no longer buzzy, but speech sounds a bit noisy. Note that pitch and V/UV estimation are no longer required.

One can see that the closed-loop optimization leads to excitation frames, which can somehow differ from the LP residual, while the resulting synthetic speech is similar to its original counterpart (Fig. 1.22).

In the above script, though, each new frame was processed independently of past frames. Since voiced speech is strongly self-correlated, it makes sense to incorporate a long-term prediction filter in cascade with the LPC (short-term) prediction filter. In the example below, we can reduce the number of stochastic components from 10 to 5 while still increasing speech quality, thanks to long-term prediction.

```
N_components= 5;  % number of codebook components per frame
```

Since CELP excitation frames are only 5 ms long, we store them in a 256 samples circular buffer (i.e., a bit more than 30 ms of speech) for finding the best long-term prediction delay in the range [0–256] samples.

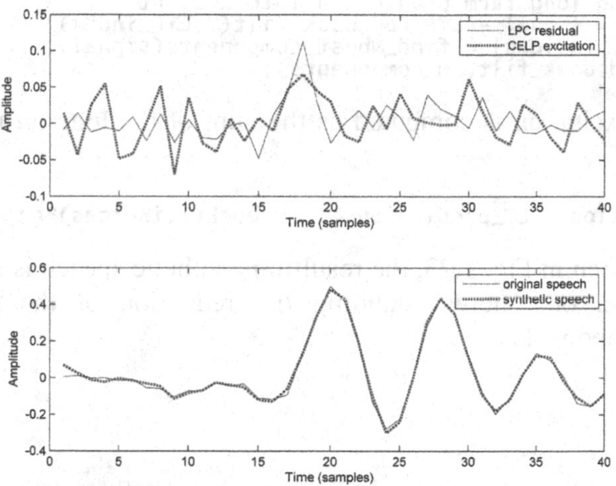

Fig. 1.22 CELP analysis–synthesis of frame #140. *Top*: CELP excitation compared to linear prediction residual. *Bottom*: CELP synthetic speech compared to original speech

```
LTP_max_delay=256; % maximum long-term prediction delay
excitation_buffer=zeros(LTP_max_delay+frame_shift,1);
```

Finding the delay itself (inside the frame-based loops) is achieved in a way very similar to finding the N best stochastic components in our previous example: we create a long-term prediction codebook, pass it through the synthesis filter, and search for *the* best excitation component in this filtered codebook.

```
% Building the long-term prediction codebook and filtering it
for j = 1:LTP_max_delay
    LTP_codebook(:,j) = excitation_buffer(j:j+frame_shift-1);
end
LTP_codebook_filt = filter(1, ai, LTP_codebook);

% Finding the best predictor in the LTP codebook
ringing = filter(1, ai, zeros(frame_shift,1), z_synt);
signal = speech_frame - ringing;
[LTP_gain, LTP_index] = find_Nbest_components(signal, ...
    LTP_codebook_filt, 1);

% Generating the corresponding prediction
LT_prediction= LTP_codebook(:,LTP_index)*LTP_gain;
```

Stochastic components are then searched *in the remaining signal* (i.e., the original signal minus the long-term predicted *signal*).

```
% Finding speech_frame components in the filtered codebook
% taking long term prediction into account
signal = signal - LTP_codebook_filt(:,LTP_index)*LTP_gain;
[gains, indices] = find_Nbest_components(signal, ...
    codebook_filt, N_components);
```

The final excitation is computed as the sum of the long-term predicted *excitation*.

```
excitation = LT_prediction + codebook(:,indices)*gains;
```

As can be seen in Fig. 1.23, the resulting synthetic speech is still similar to the original one, notwithstanding the reduction of the number of stochastic components.

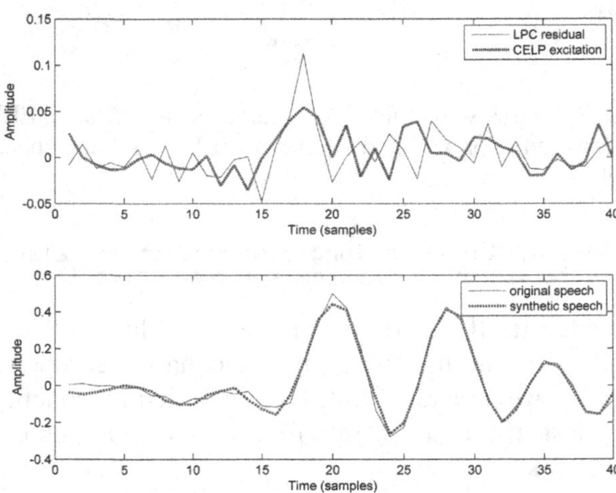

Fig. 1.23 CELP analysis–synthesis of frame #140 with long-term prediction and only five stochastic components. *Top*: CELP excitation compared to linear prediction residual. *Bottom:* CELP synthetic speech compared to original speech

Although the search for the best components in the previous scripts aims at minimizing the energy of the difference between original and synthetic speech samples, it makes sense to use the fact that the ear will be more tolerant to this difference in parts of the spectrum that are louder and vice versa. This can be achieved by applying a perceptual filter to the error, which enhances spectral components of the error in frequency bands with less energy and vice versa (Fig. 1.24).

In the following example, we still decrease the number of components from 5 to 2, with the same overall synthetic speech quality.

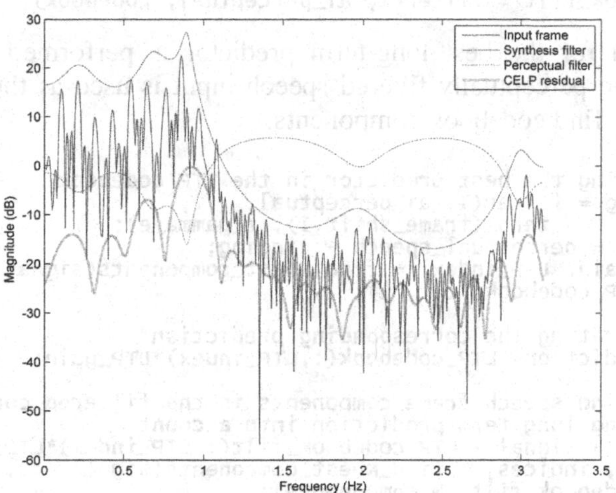

Fig. 1.24 CELP analysis–synthesis of frame #140: the frequency response of the perceptual filter approaches the inverse of that of the synthesis filter. As a result, the spectrum of the CELP residual somehow follows that of the speech frame

```
N_components= 2;  % number of codebook components per frame
```

We will apply perceptual filter $A(z)/A(z/\gamma)$ to the input frame, and filter $1/A(z/\gamma)$ to the stochastic and long-term prediction codebook vectors.[12] We will therefore need to handle their internal variables.

```
gamma = 0.8;    % perceptual factor
z_inv=zeros(10,1);  % inverse filter
z_synt=zeros(10,1); % synthesis filter
z_gamma_s=zeros(10,1); % perceptual filter for speech
z_gamma_e=zeros(10,1); % perceptual filter for excitation
```

Finding the coefficients of $A(z/\gamma)$ is easy.

```
ai_perceptual = ai.*(gamma.^(0:(length(ai)-1)) );
```

One can then filter the input frame and each codebook.

[12] In the previous examples, the input frame was not perceptually filtered, and codebooks were passed through the synthesis filter 1/A(z).

```
% Passing the central 5ms of the input frame through
% A(z)/A(z/gamma)
[LP_residual, z_inv] = filter(ai, 1, speech_frame, z_inv);
[perceptual_speech, z_gamma_s] = filter(1, ...
                    ai_perceptual, LP_residual, z_gamma_s);

% Filtering both codebooks
LTP_codebook_filt = filter(1, ai_perceptual, LTP_codebook);
codebook_filt = filter(1, ai_perceptual, codebook);
```

The search for the best long-term predictor is performed as before, except that the perceptually filtered speech input is used as the reference from which to find codebook components.

```
% Finding the best predictor in the LTP codebook
ringing = filter(1, ai_perceptual, ...,
            zeros(frame_shift,1), z_gamma_e);
signal = perceptual_speech - ringing;
[LTP_gain, LTP_index] = find_Nbest_components(signal, ...
    LTP_codebook_filt, 1);

% Generating the corresponding prediction
LT_prediction= LTP_codebook(:,LTP_index)*LTP_gain;

% Finding speech_frame components in the filtered codebook
% taking long term prediction into account
signal = signal - LTP_codebook_filt(:,LTP_index)*LTP_gain;
[gains, indices] = find_Nbest_components(signal, ...
    codebook_filt, N_components);
```

Last but not least, one should not forget to update the internal variables of the perceptual filter applied to the excitation.

```
[ans, z_gamma_e] = filter(1, ai_perceptual, excitation, ...
    z_gamma_e);
```

While using less stochastic components than in the previous example, synthetic speech quality is maintained, as revealed by listening. The synthetic speech waveform also looks much more similar to original speech than its LPC10 counterpart (Fig. 1.25).

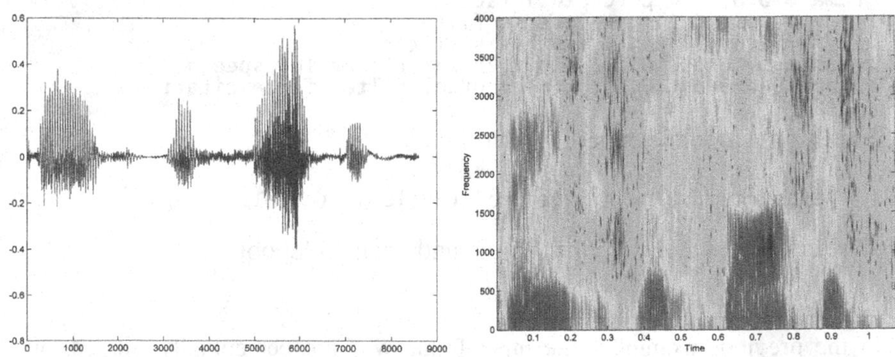

Fig. 1.25 CELP speech

One can roughly estimate the corresponding bit rate. Assuming 30 bits are enough for the prediction coefficients and each gain factor is quantized on 5 bits, we have to send for each frame: 30 bits [ai] + 7 bits [LTP index] + 5 bits [LTP gain] + 2 [stochastic components] *(9 bits [index] + 5 bits [gain]) = 70 bits every 5 ms, i.e., 14 kbps.

Note that G729 reaches a bit rate as low as 8 kbps by sending prediction coefficients only once every four frame.

1.3 Going further

Various tools and interactive tutorials on LP modeling of speech are available on the web (see Fellbaum 2007, for instance).

MATLAB code by A. Spanias for the LPC10e coder can be found on the web (Spanias and Painter 2002).

Another interesting MATLAB-based project on LPC coding, applied to wideband speech this time, can be found on the dspexperts.com website (Khan and Kashif 2003).

D. Ellis provides interesting MATLAB-based audio processing examples on his web pages (Ellis 2006), among which are a sinewave speech analysis/synthesis demo (including LPC) and a spectral warping of LPC demo.

For a broader view of speech coding standards, one might refer to Woodard (2007) or to the excellent book by Goldberg and Riek (2000).

1.4 Conclusion

We now understand how every cell phone solves a linear system of 10 equations in 10 unknowns every 20 ms which is the basis of the estimation of the LP model through Yule-Walker equations. The parameters that are actually sent from one cell phone to another are vocal tract coefficients related to the frequency response of the vocal tract and source coefficients related to the residual signal.

The fact that the vocal tract coefficients are very much related to the geometric configuration of the vocal tract for each frame of 10 ms of speech calls for an important conclusion: cell phones, in a way, transmit a picture of our vocal tract rather than the speech it produces.

In fact, the reach of LP speech modeling goes far beyond the development of cell phones. As shown by Gray (2006), its history is intermixed with that of Arpanet, the ancestor of Internet.

References

Atal BS, Remde JR (1982) A New Model LPC Excitation for Producing Natural Sounding Speech at Low Bit Rates. In: Proc. ICASSP'82, pp 614–617

de la Cuadra P (2007) Pitch Detection Methods Review [online] Available: http://www-ccrma.stanford.edu/~pdelac/154/m154paper.htm [20/2/1007]

Ellis D (2006) Matlab Audio Processing Examples [online] Available: http://www.ee.columbia.edu/%7Edpwe/resources/matlab/ [20/2/2007]

Fant G (1960) Acoustic Theory of Speech Production. The Hague: Mouton

Fellbaum K (2007) Human Speech Production Based on a Linear Predictive Vocoder [online] Available: http://www.kt.tu-cottbus.de/speech-analysis/ [20/2/2007]

Goldberg RG, Riek L (2000) Speech Coders. CRC Press: Boca Raton, FL

Gray RM (2006) Packet speech on the Arpanet: A history of early LPC speech and its accidental impact on the Internet Protocol [online] Available: http://www.ieee.org/organizations/society/sp/ Packet_Speech.pdf [20/2/2007]

Hess W (1992) Pitch and Voicing Determination. In: Advances in Speech Signal Processing, S. Furui, M. Sondhi, eds., Dekker, New York, pp 3–48

Khan A, Kashif F (2003) Speech Coding with Linear Predictive Coding (LPC) [online] Available: http://www.dspexperts.com/dsp/projects/lpc [20/2/2007]

Kroon P, Deprettere E, Sluyter R (1986) Regular-pulse excitation – A novel approach to effective and efficient multipulse coding of speech. IEEE Transactions on Acoustics, Speech, and Signal Processing 34(5): 1054–1063

Matsumoto J, Nishiguchi M, Iijima K (1997) Harmonic Vector Excitation Coding at 2.0 kbps. In: Proc. IEEE Workshop on Speech Coding, pp 39–40

McCree AV, Barnwell TP (1995) A mixed excitation LPC vocoder model for low bit rate speech coding. IEEE Transactions on Speech and Audio Processing, 3(4):242–250

NATO (1984) Parameters and coding characteristics that must be common to assure interoperability of 2400 bps linear predictive encoded speech. NATO Standard STANAG-4198-Ed1

Quatieri T (2002) Discrete-Time Speech Signal Processing: Principles and Practice. Prentice-Hall, Inc.: Upper Saddle River, NJ

Rabiner LR, Schafer RW (1978) Digital Processing of Speech Signals. Prentice-Hall, Inc.: Englewood Cliffs, NJ

Salami R, Laflamme C, Adoul J-P, Kataoka A, Hayashi S, Moriya T, Lamblin C, Massaloux D, Proust S, Kroon P, Shoham, Y (1998) Design and description of CS-ACELP: A toll quality 8 kb/s speech coder, IEEE Transactions on Speech and Audio Processing 6(2): 116–130

Schroeder MR, Atal B (1985) Code-Excited Linear Prediction(CELP): High Quality Speech at Very Low Bit Rates. In: Proc. IEEE ICASSP-85, pp 937–940

Spanias A, Painter T (2002) Matlab simulation of LPC10e vocoder [online] Available: http://www.cysip.net/lpc10e_FORM.htm [19/2/2007]

Woodard J (2007) Speech coding [online] Available: http://www-mobile.ecs.soton.ac.uk/speech_codecs/ [20/2/2007]

Chapter 2

How are bits played back from an audio CD?

T. Dutoit (°), R. Schreier (*)

(°) Faculté Polytechnique de Mons, Belgium
(*) Analog Devices, Inc., Toronto, Canada

An audio digital-to-analog converter adds noise to the signal, by requantizing 16-bit samples to one-bit. It does it…on purpose.

Loading a CD player with one's favorite CD has become an ordinary action. It is taken for granted that the stream of 16-bit digital information it contains can easily be made available to our ears, i.e., in the analog world in which we live. The essential tool for this is the digital-to-analog converter (DAC).

In this chapter we will see that, contrary to what might be expected, many audio DACs (including those used in CD and MP3 players, for instance, or in cell phones) first requantize the 16-bit stream into a *1-bit* stream[1] with very high sampling frequency, using a signal processing concept known as delta–sigma[2] ($\Delta\Sigma$) modulation, and then convert the resulting bipolar signal back to an audio waveform.

The same technique is used in ADCs for digitizing analog waveforms. It is also the heart of the direct stream digital (DSD) encoding system implemented in super audio CDs (SACDs).

---

[1] In practice, 1-bit $\Delta\Sigma$ DACs have been superseded by multiple-bit DACs, but the principle remains the same.

[2] Sometimes also referred to as *sigma–delta.*

T. Dutoit, F. Marqués (eds.), *Applied Signal Processing*,
DOI 10.1007/978-0-387-74535-0_2, © Springer Science+Business Media, LLC 2009

2.1 Background – Delta–sigma modulation

An N-bit DAC converts a stream of discrete-time linear PCM[3] samples of N bits at sample rate F_s to a continuous-time voltage. This can be achieved in many ways. Conventional DACs (Section 2.1.2) directly produce an analog waveform from the input PCM samples. Oversampling DACs (Section 2.1.3) start by increasing the sampling frequency using digital filters and then make use of a conventional DAC with reduced design constraints. Adding noise shaping makes it possible to gain resolution (Section 2.1.4). Delta–sigma DACs (Section 2.1.5) oversample the input PCM samples and then requantize them to a 1-bit data stream, whose low-frequency content is the expected audio signal.

Before examining these DACs, we start with a review of uniform quantization (Section 2.1.1), as it will be used throughout the chapter.

2.1.1 Uniform quantization: Bits vs. SNR

Quantization lies at the heart of digital signal processing. An N-bit uniform quantizer maps each sample $x(n)$ of a signal to one out of 2^N equally spaced values $X(n)$ in the interval $(-A,+A)$, separated by the quantization step $q=2A/2^N$. This operation (Fig. 2.1) introduces an error $e(n)$:

$$e(n)=X(n)-x(n) \qquad\qquad -q/2\leq e(n)\leq q/2 \tag{2.1}$$

If the number of bits is high enough and the input signal is complex, the quantization error is equivalent to uniform white noise in the range[$(-q/2, +q/2)$]. It is easy to show that its variance is then given by the following equation:

$$\sigma^2{}_{ee}=\frac{q^2}{12} \tag{2.2}$$

The main result of uniform quantization theory, which can be found in most signal processing textbooks, is the standard “1 bit = 6 dB” law, which gives the expression of the signal-to-quantization-noise ratio:

$$SNR(dB)=10*\log_{10}\left(\frac{\sigma_{xx}{}^2}{\sigma_{ee}{}^2}\right) \tag{2.3}$$

as a function of N and A, in the absence of saturation:

[3] PCM stands for pulse code modulated.

$$SNR(dB) = 6.02N + 4.77 + 20\log_{10}(\Gamma) \qquad (\Gamma = A/\sigma_{xx}) \tag{2.4}$$

where Γ is the load factor defined as the saturation value of the quantizer normalized by the standard deviation of the input signal.

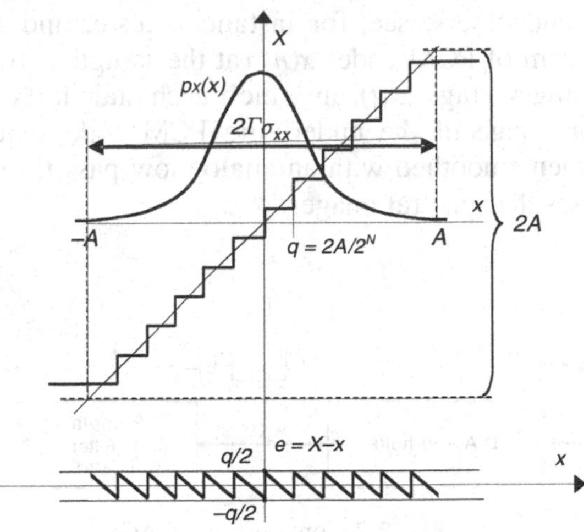

Fig. 2.1 Uniform quantization[4]

When the amplitude of the input signal approaches the quantization step, the quantization error becomes correlated with the signal. If the signal itself is not random, the quantization error can then be heard as nonlinear audio distortion (rather than as additive noise).

This can be avoided by *dithering*, which consists of adding real noise to the signal before quantizing it. It has been shown that a triangular white noise (i.e., white noise with a triangular probability density function) in the range $[-q,+q]$ is the best dither: it decorrelates the quantization noise with the signal (it makes the mean and variance of the quantization noise independent of the input signal; see Wannamaker 1997) while adding the least possible noise to the signal. Such a noise is easy to obtain by summing two independent, uniform white noise signals in the range $[-q/2,+q/2]$, since the probability density function of the sum of two independent random variables is the convolution of their respective pdfs.

[4] We actually show a *mid-rise* quantizer here, which quantizes a real number x into (floor(x/q)+0.5)*q. Mid-thread quantizers, which compute floor(x/q+0.5)*q, are also sometimes used (see Section 3.2.5, for instance).

As Wannamaker puts it: "Appropriate dithering prior to (re)quantization is as fitting as appropriate anti-aliasing prior to sampling – both serve to eliminate classes of signal-dependent errors."

2.1.2 Conventional DACs

A conventional DAC uses analog circuitry (R/2R ladders, *thermometer* configuration, and others; see, for instance, Kester and Bryant 2003) to transform a stream of PCM codes $x(n)$ (at the sampling frequency F_s) into a staircase analog voltage $x^*(t)$, in which each stair lasts $T_s=1/F_s$ seconds and is a lincar image of the underlying PCM code sequence. Staircase signal $x^*(t)$ is then smoothed with an analog low-pass filter $S(f)$ (Fig. 2.2), which suppresses the spectral images.

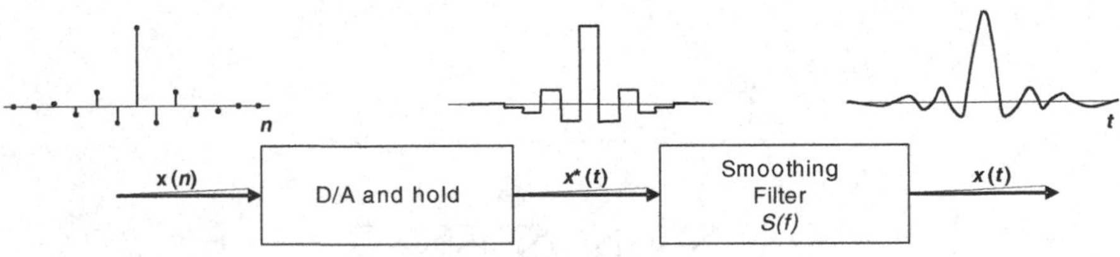

Fig. 2.2 Conventional DAC

The first operation can be seen as convolving a sequence of Dirac pulses $x^+(t)$ obtained from PCM code with a rectangular wave of length T_s, hence filtering the Dirac pulses with the corresponding low-pass filter (Fig. 2.3). The second operation completes the smoothing job.

Conventional DACs have several drawbacks. First, they require high-precision analog components and are very vulnerable to noise and interference. In a 16-bit DAC with 3 V reference voltage, for instance, one half of the *least significant bit* corresponds to 2^{-17} 3 V=23 μV. What is more, they impose hard constraints on the design of the analog smoothing filter, whose transition band must fit within $[F_m, F_s-F_m]$ (where F_m is the maximum frequency of the signal; Fig. 2.3), so as to efficiently cancel spectral images.

2.1.3 Oversampling DACs

An oversampling DAC first performs digital K times oversampling of the input signal $x(n)$ by first inserting $K-1$ zeros between each sample of $x(n)$ and then applying digital low-pass filtering with passband equal to

$[0,F_s/K]$ (Fig. 2.4). The resulting signal, $x(n/K)$, is then possibly requantized on N' bits (with $N'<N$), by simply keeping the N' most significant bits from each interpolated sample, and the requantized signal $x'(n/K)$ is sent to a conventional DAC clocked at $K*F_s$ Hz with N' bits of resolution.

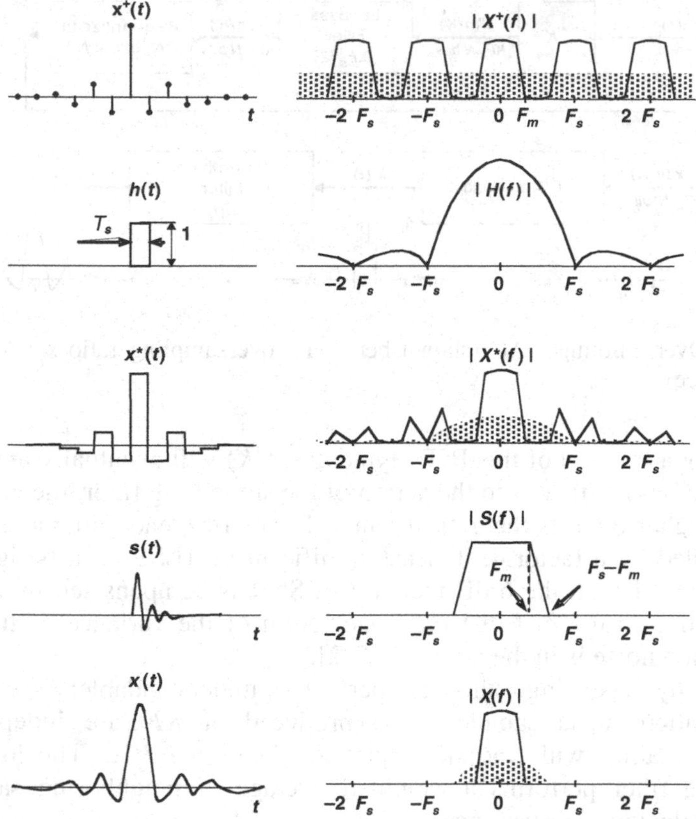

Fig. 2.3 Digital-to-analog conversion seen as double filtering: $x^+(t)$ is convolved with $h(t)$ to produce $x^*(t)$, which is smoothed by $S(f)$. Quantization noise is shown as superimposed texture

In principle, requantizing to less than the initial N bits lowers the SNR by 6 dB for each bit lost. However, although the variance of quantization noise $e'(n)$ generated by N'-bit requantization is higher than that of the initial N-bit quantization noise $e(n)$, it is now spread over a (K times) larger frequency range. Its power spectral density (PSD, in V²/Hz) is thus given by the following equation:

$$S_{e'e'}(f) = \frac{\sigma_{e'e'}^{\ 2}}{K\,F_s} \tag{2.5}$$

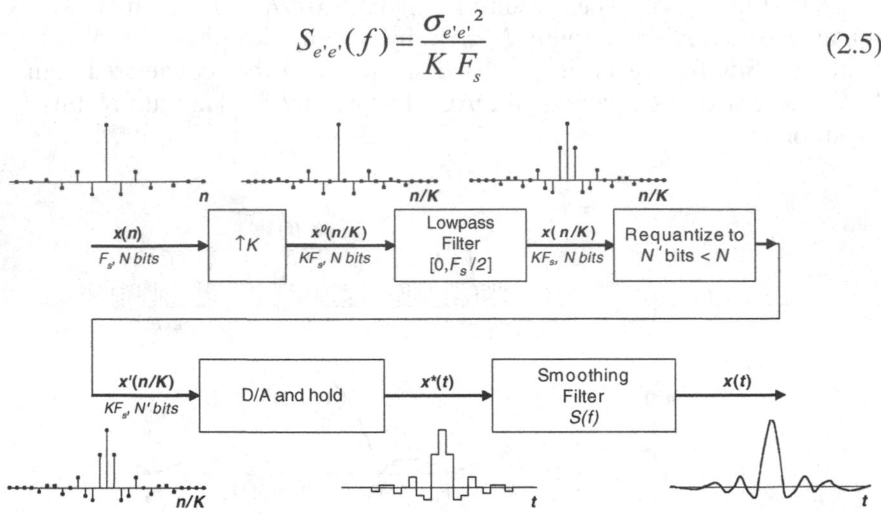

Fig. 2.4 Oversampling DAC (shown here with oversampling ratio set to 2, for convenience)

Since only a fraction of this PSD (typically, $1/K$) will eventually appear in the analog output, thanks to the action of the smoothing filter, the effective SNR is higher than its theoretical value. In practice, each time a signal is oversampled by a factor 4, its least significant bit (LSB) can be ignored. As a matter of fact, the 6 dB increase in SNR is compensated by a 6 dB decrease due to the fact that only one-fourth of the variance of the new quantization noise is in the range $[0, F_s/2]$.[5]

Physically speaking, this is perfectly understandable; successive requantization noise samples $e'(n)$ produced at KF_s are independent random variables with variance equal to $q'^2/12 > q^2/12$. The low-pass smoothing filter performs a weighted average of neighboring samples, thereby reducing their variance.

This effect is very important in practice, as it allows a bad resolution DAC (N' bits$<N$) to produce high-resolution signals.

As a result of oversampling, the smoothing filter is also allowed a larger transition bandwidth $[0, KF_s - F_m]$ (Fig. 2.5). What is more, the same low-pass filter can now be used for a large range or values for F_s (as required for the sound card of a computer, for instance).

[5] Strictly speaking, in $[-F_s/2, F_s/2]$. In this discussion, we consider only positive frequencies, for convenience.

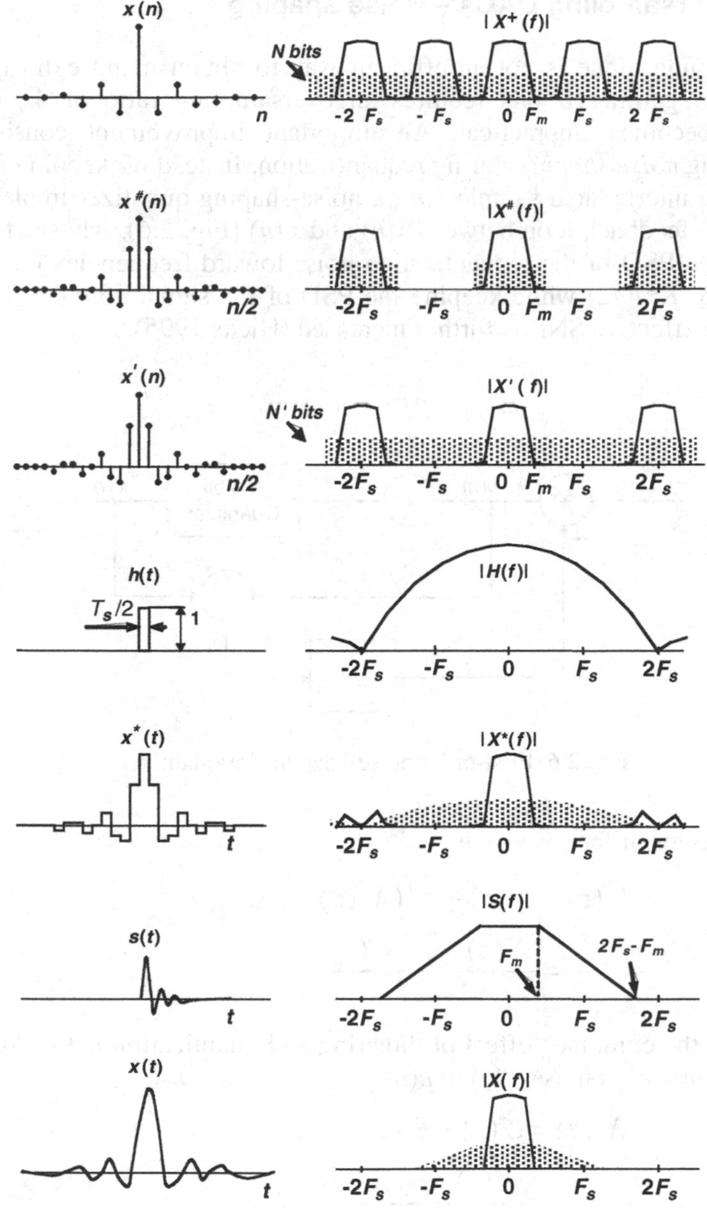

Fig. 2.5 Oversampling (shown here with an oversampling factor of 2) prior to digital-to-analog conversion. Quantization and requantization noise are shown as superimposed texture

2.1.4 Oversampling DACs – Noise shaping

Oversampling alone is not an efficient way to obtain many extra bits of resolution: gaining B bits requires an oversampling ratio of 4^B, which quickly becomes impractical. An important improvement consists of performing *noise shaping* during requantization. Instead of keeping N' bits from each interpolated sample $x(n)$, a noise-shaping quantizer implements a negative feedback loop between $x(n)$ and $x'(n)$ (Fig. 2.6), whose effect is to push the PSD of the requantization noise toward frequencies far above $F_s/2$ (up to $K^*F_s/2$) while keeping the PSD of the signal untouched. As a result, the effective SNR is further increased (Hicks 1995).

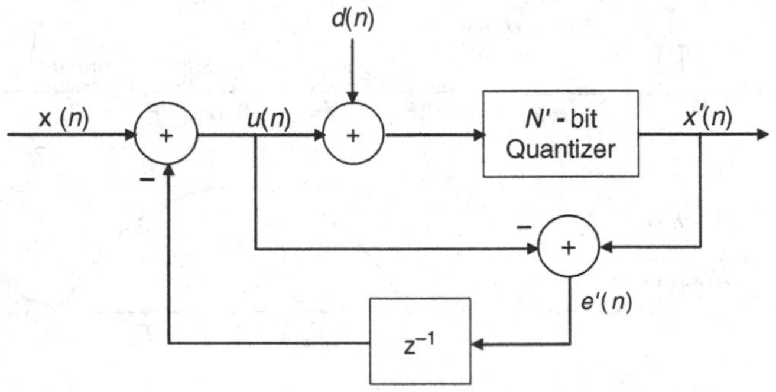

Fig. 2.6 First-order noise-shaping (re)quantizer

As a matter of fact, we have

$$\begin{aligned} U(z) &= X(z) - z^{-1}(X'(z) - U(z)) \\ &= \frac{X(z) - z^{-1}X'(z)}{1 - z^{-1}} \end{aligned} \tag{2.6}$$

and since the combined effect of dithering and quantization is to add some white quantization noise $e'(n)$ to $u(n)$

$$\begin{aligned} X'(z) &= U(z) + E'(z) \\ &= \frac{X(z) - z^{-1}X'(z)}{1 - z^{-1}} + E'(z) \\ &= X(z) + (1 - z^{-1})E'(z) \end{aligned} \tag{2.7}$$

which shows that the output of the noise-shaping requantizer is the input signal plus the first derivative of the white quantization noise $e'(n)$

produced by requantization. This effectively results in colored quantization noise $c(n)=e'(n)-e'(n-1)$, with most of its PSD in the band $[F_s/2, KF_s/2]$ (Fig. 2.7, to be compared to Fig. 2.5), where it will be filtered out by the smoothing filter. As the noise-shaping function $(1-z^{-1})$ is of first order, this configuration is termed as a first-order noise-shaping cell.

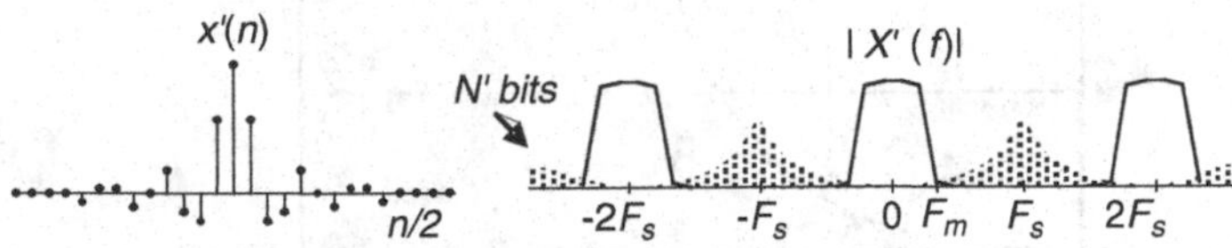

Fig. 2.7 The effect of noise shaping combined with oversampling (by a factor 2) on signal $x'(n)$ at the output of the quantizer

Noise shaping does increase the power of quantization noise, as the variance of colored noise $c(n)$ is given by the following equation:

$$
\begin{aligned}
\sigma_{cc}^2 &= \frac{\sigma_{e'e'}^2}{K F_s} \int_{-KF_s/2}^{-KF_s/2} |1-e^{-j2\pi f}|^2 \, df \\
&= \frac{\sigma_{e'e'}^2}{K F_s} \int_{-KF_s/2}^{-KF_s/2} (1+|e^{-j\theta}|^2) \, df \\
&= 2\sigma_{e'e'}^2
\end{aligned}
\tag{2.8}
$$

But again, since this variance is mostly pushed in the $[F_s/2, KF_s/2]$ band, the effective SNR can be lowered (Fig. 2.8).

This technique makes it possible to gain 1 bit every time the signal is oversampled by a factor 2. It was used in early CD players when only 14-bit hardware D/A converters were available at low cost. By combining oversampling and noise shaping (in the digital domain), a 14-bit D/A converter was made comparable to a 16-bit D/A converter.

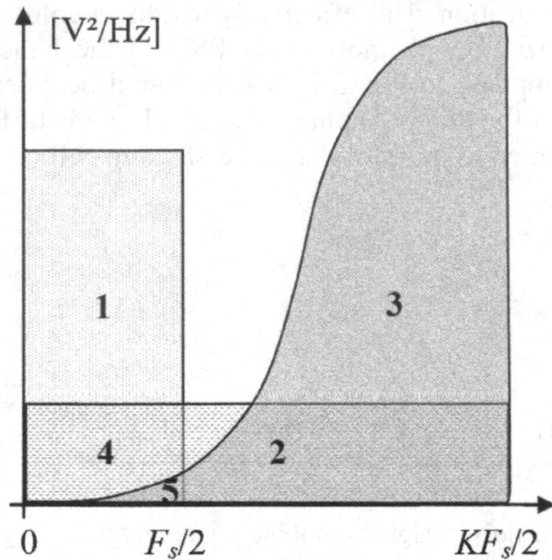

Fig. 2.8 A comparison of quantization noise power density functions for the same number of bits. (1) With a conventional DAC; effective noise variance = area 1; (2), With an oversampling DAC, effective noise variance = area 4; (3) With an oversampling DAC using noise shaping, effective noise variance = area 5.

2.1.5 Delta–sigma DACs

The delta–sigma architecture is the ultimate extension of the oversampling DAC and is used in most voiceband and audio signal processing applications requiring a D/A conversion. It makes use of a very high oversampling ratio, which makes it possible to requantize the digital signal to 1 bit only. This 1-bit signal is then converted to a purely bipolar analog signal by the DAC, whose output switches between equal positive and negative reference voltages (Fig. 2.9).

The bipolar signal is sometimes referred to as "pulse-density modulated" (PDM), as the density of its binary transitions is a function of the amplitude of the original signal.

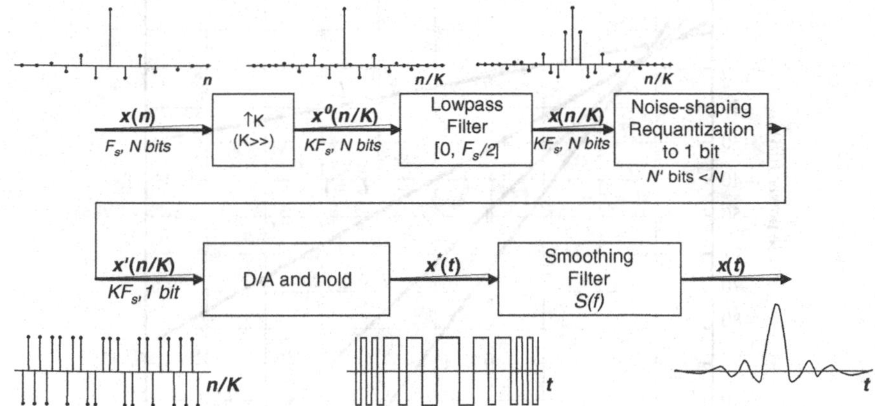

Fig. 2.9 Delta–Sigma DAC (shown here with oversampling ratio set to 2, for convenience; in practice, much higher ratios are used)

In CD and MP3 players, this implies a gain of 15 bits of resolution. Improved noise shaping is therefore required, such as second-order noise shaping cells (whose noise-shaping function is $(1-z^{-1})^2$) or cascades of first-order noise-shaping cells (termed as MASH: multi-stage noise shaping; Matsuya et al. 1987). Deriving a general noise-shaping quantizer with noise-shaping function $H(z)$ from that of Fig. 2.6 is easy: one simply needs to replace the delay by $1-H(z)$ (Fig. 2.10).

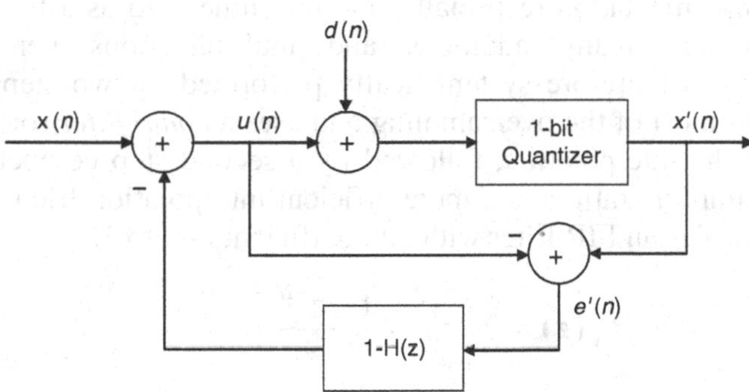

Fig. 2.10 General digital delta–sigma modulator

The influence of the oversampling ratio and of the order of the noise-shaping filter on the noise power in the signal bandwidth is given in Fig. 2.11.

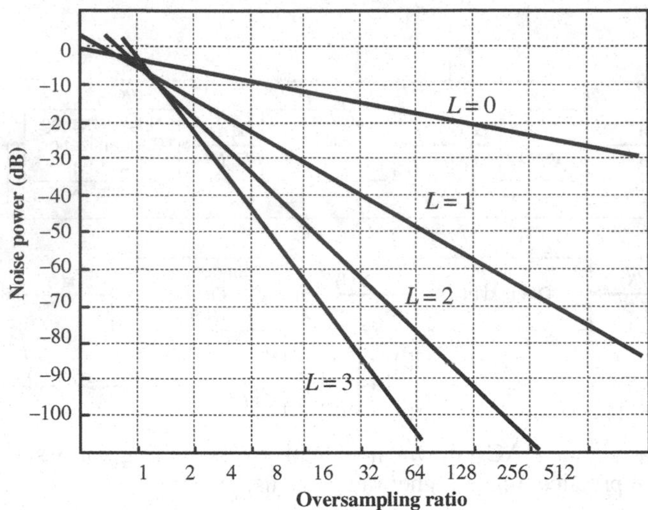

Fig. 2.11 Power of the noise in the signal bandwidth as a function of the oversampling ratio for various noise-shaping orders *L*. 0 dB of noise level corresponds to that of PCM sampled at the Nyquist rate. All quantizers use the same level spacing (after Candy (1997))

The very limited bandwidth allowed for the interpolation filter ($[0, F_s/2]$ at a sampling frequency of KF_s) theoretically implies a very high order for its digital implementation (especially if it is synthesized as a linear-phase FIR filter), i.e., many additions and multiplications per sample. Oversampling is therefore systematically performed in two steps: a first step that does most of the oversampling and uses a *comb filter* (or cascades thereof), for the interpolation, followed by a second step competing with lower oversampling ratio and a more efficient interpolation filter. A comb filter of order *N* is an FIR filter with all coefficients set to 1:

$$H_N(z) = \sum_{n=0}^{N-1} z^{-n} = \frac{1-z^{-N}}{1-z^{-1}} \tag{2.9}$$

This filter has linear phase, while being straightforwardly implemented as

$$y(n) = x(n) - x(n-N) + y(n-1) \tag{2.10}$$

which implies no multiplication. Comb filters, however, cannot provide very efficient frequency responses (for an example, see Fig. 2.16-left). Their transition band is not sharp. This is counterbalanced by the second interpolation stage, which produces a sharp cut-off and additionally compensates for the frequency droop of the comb filter around $F_s/2$.

In summary, delta–sigma DACs are a wonderful example of mixed A–D signal processing constraints. Most of their load is in the digital domain, while the analog output smoothing filter usually reduces to a simple RC lowpass. These converters can thus be fabricated on a wide range of IC processes, which implies low cost and robustness to time and temperature drifts. This is achieved at the expense of speed: since the hardware has to operate at the high oversampling rate, sigma–delta converters are usually limited to sampling rates below 1 MHz.

From 1990 to 2000, the consumer audio industry has produced a large number of 1-bit sigma–delta-based converters, mainly because this technique leads to cheaper manufacturing. It was argued, however, that 1-bit sigma–delta converters were not suitable for high-quality audio applications (Lipschitz and Vanderkooy 2001), as they allow only partial dithering to be performed.[6] Most audio DACs made since 2000 implement *multi-bit* sigma–delta converters, which take the best of both the 1-bit and the 20+-bit worlds (go to the Analog Devices web page, for instance, and search for "audio DAC").

2.2 MATLAB proof of concept: ASP_audio_cd.m

In this Section, we will first revise the basics of uniform quantization (Section 2.2.1), including the important part played by dithering (Section 2.2.2). We will then compare the internals of a conventional DAC (Section 2.2.3) to those of more advanced DACs using oversampling (Section 2.2.4), noise-shaping (Section 2.2.5), and delta–sigma modulation (Section 2.2.6).

2.2.1 Uniform quantization

In order to perform D/A conversion, we first need to create a PCM signal. We will first check the main results of uniform quantization theory, which we will use later.

[6] This led to an interesting controversy in several papers published in conventions of the Audio Engineering Society.

Let us first generate 8000 samples of uniform white noise with zero mean, which we will assume to have been sampled at F_s=8,000 Hz. For convenience, we set its variance to 1 (0 dB) by imposing its range to [–*peak*, +*peak*], with $peak = \sqrt{12}/2 \approx 1.73$. This is confirmed on the power spectral density of the signal (Fig. 2.12).

```
signal_std=1;
peak=sqrt(12*signal_std)/2;
signal=(rand(1,8000)-0.5)*2*peak;

pwelch(signal,[],[],[],2); 7
```

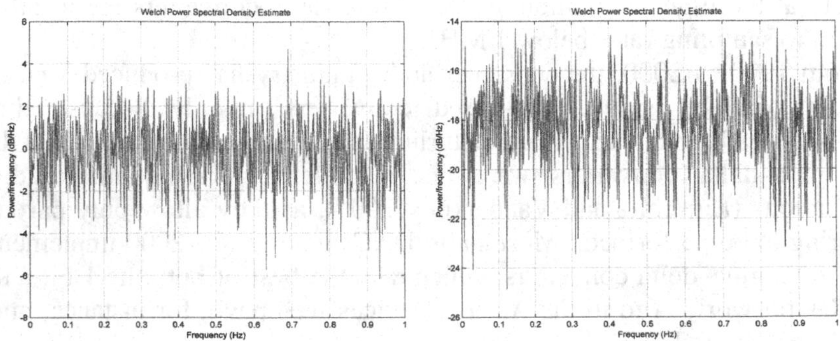

Fig. 2.12 Power spectral density of the input signal (*left*) and of the 6-bit quantized error (*right*)

Quantizing this signal uniformly on *N* bits is easy. Let us choose *N*=3 for convenience. We use mid-rise quantization (hence the +q/2), which is best suited for an even number of quantization steps.

```
N=3;
quantizer_saturation=peak;
q=(2*quantizer_saturation)/2^N;
signal_quantized=floor(signal/q)*q+q/2; % mid-rise
error=signal-signal_quantized;
```

The quantization noise is also white, as revealed by power spectral density estimation. The variance of the noise can thus be read (in dB) on the PSD plot (Fig. 2.12). From the load factor (i.e., the saturation value of the quantizer normalized by the standard deviation of the signal), we can

[7] MATLAB's `pwelch` function shows the PSD in the range [0, F_s/2]. The PSD values it shows are such that their integral over [0, F_s/2] equals the variance of the signal. Claiming F_s=2 as an input argument therefore shows the variance as the average PSD over [0, 1].

compute the theoretical SNR due to quantization. It matches to the SNR computed from the samples and corresponds to the variance of the noise (with opposite sign), since the variance of the signal is 0 dB.

MATLAB function involved:

• `snr = snr(signal, signal_plus_noise, max_shift, showplot)` returns the signal-to-noise ratio computed from the input signals. `Max_shift` gives the maximum time shift (in samples) between `signal` and `signal_plus_noise`. The actual time shift (obtained from the maximum of the cross-correlation between both signals) is taken into account to estimate the noise. If `showplot` is specified, then the `signal`, `signal_plus_noise`, and `error` are plotted, and the SNR is printed on the plot.

```
pwelch(error,[],[],[],2);
load_factor=quantizer_saturation/signal_std;
snr_theory=6.02*N + 4.77 - 20*log10(load_factor)
snr_measured=snr(signal,signal_quantized,0)
```

```
snr_theory =  18.0588

snr_measured =  18.1119
```

2.2.2 Dithering

Let us apply this to a sine wave with fundamental frequency f_0=200 Hz, sampled at F_s=8,000 Hz, and set its variance to 1 (0 dB) by imposing its peak to $\sqrt{2}$.

```
signal_std=1;
peak=sqrt(signal_std*2);
signal=peak*sin(2*pi*200*(0:1/8000:2));
var_signal_dB=10*log10(var(signal))
```

```
var_signal_dB = -2.8930e-015
```

When we quantize it uniformly to 3 bits in the range [–2,+2], we note that the quantization error does not look like real noise (Fig. 2.13).

```
N=3;
quantizer_saturation=2;
q=2*quantizer_saturation/2^N;
signal_quantized=floor(signal/q)*q+q/2; % quantization
error=signal-signal_quantized;

plot(signal(1:50),'-.'); hold on;
plot(error(1:50));hold off
```

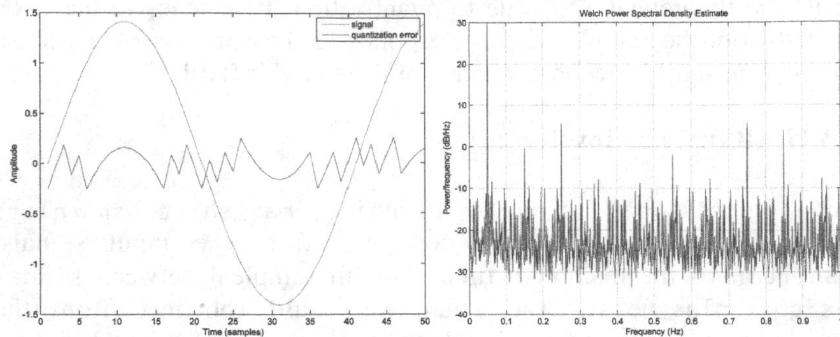

Fig. 2.13 Quantization error (*left*) and power spectral density of the quantized signal (*right*) without dithering. Frequencies are normalized with respect to $F_s/2$

Harmonics of the original frequency appear in the spectrum of the quantized signal (Fig. 2.13) and sound very unpleasant. The SNR is not directly readable on the PSD plot because of these harmonics.

```
snr_quantized=snr(signal,signal_quantized,0)
pwelch(signal_quantized,[],[],[],2);
```

```
snr_quantized = 16.9212
```

In such a case, it may be interesting to whiten the quantization error by adding real noise to it. This operation is termed as dithering.

Let us add a dithering noise with triangular pdf in the range $[-q, +q]$, i.e., two times the quantization step. Such a noise is easily obtained by adding two uniform white noises in the range $[-q/2, q/2]$.

```
dither=(rand(size(signal))-0.5)*q+(rand(size(signal))-0.5)*q;
signal_dithered=signal+dither;
```

The resulting quantization error looks indeed more noise-like (Fig. 2.14)[8].

```
signal_dithered_quantized=floor(signal_dithered/q)*q+q/2;
error_dithered_quantized=signal-signal_dithered_quantized;

plot(signal(1:50),'-.'); hold on;
plot(error_dithered_quantized(1:50));hold off
```

[8] Note that we do not account for quantizer saturation, which may occur when dither is added to a signal that approaches the saturation levels of the quantizer. This is not the case in our example.

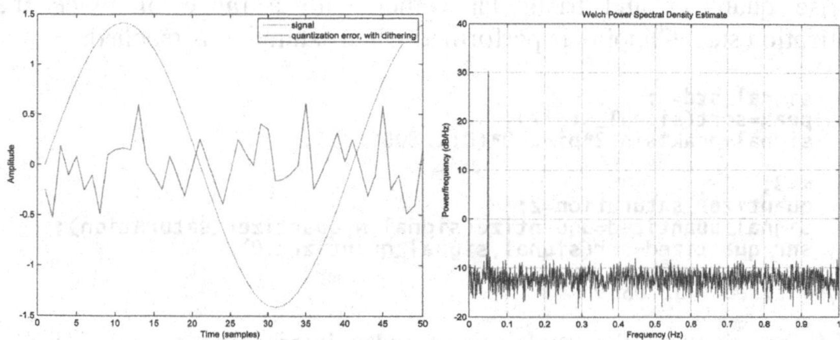

Fig. 2.14 Quantization error (*left*) and power spectral density of the quantized signal (*right*) with dithering

The power spectral density of the quantized signal now appears as white noise added to the initial sine wave.

```
pwelch(signal_dithered_quantized,[],[],[],2);
```

Dithering clearly degrades the SNR (by about 4.8 dB) but results in perceptually more acceptable quantization error.

```
load_factor=quantizer_saturation/signal_std;
snr_theory=6.02*N + 4.77 - 20*log10(load_factor)
snr_dithered_quantized=snr(signal,signal_dithered_quantized,0)
```

snr_theory = 16.8094

snr_dithered_quantized = 12.0562

2.2.3 Conventional DAC

A conventional DAC first creates a staircase signal from the sequence of samples by converting each sample into an analog voltage and holding this voltage for $1/F_s$ seconds. This is called zero-order (analog) interpolation. This operation is critical: the higher the number of bits in the PCM code to convert, the higher the precision required for creating the staircase signal!

We will demonstrate this on a sine wave with N=6 bits.

MATLAB function involved:

- `signal_quantized = uquantize(signal,N,saturation)` quantizes `signal` uniformly to `N` bits in the range [–saturation, +saturation], using a mid-rise quantizer and triangular dither with a range of twice the quantization step. Clipping is performed when saturation is reached.

```
signal_std=1;
peak=sqrt(signal_std*2);
signal=peak*sin(2*pi*200*(0:1/8000:2));

N=3;
quantizer_saturation=2;
signal_quantized=uquantize(signal,N,quantizer_saturation);
snr_quantized=snr(signal,signal_quantized,0)
```

snr_quantized = 12.0809

We can *simulate* the analog zero-order interpolation in the digital domain by working with a much higher sampling frequency than F_s (say, $F_s'=10\ F_s$, i.e., 80 kHz). Zero-order interpolation is then equivalent to inserting nine zeros in between each sample and convolving the resulting signal with a sequence of 10 unity samples (Fig. 2.15).[9]

```
signal_pulses=zeros(1,10*length(signal_quantized));
signal_pulses(1:10:10*length(signal))=...
                                signal_quantized;
hold_impresp=ones(1,10);
signal_staircase=conv(signal_pulses,hold_impresp);

plot(1:10:391, signal_quantized(1:40),'o'); hold on;
plot(signal_staircase(1:400)); hold off;
```

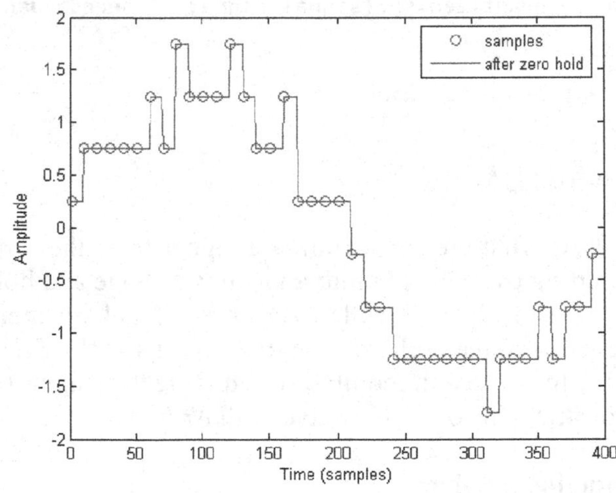

Fig. 2.15 Analog staircase signal after zero-order interpolation

[9] This is actually a very rudimentary form of numerical interpolation.

Note that zero-order interpolation acts as a low-pass filter performing a first attenuation of the spectral images of the signal at integer multiples of F_s. This directly affects the spectrum of the resulting "analog" signal. One can clearly see spectral images weighted by the effect of the interpolation (Fig. 2.16).

```
freqz(hold_impresp,1, 256, 80000);
pwelch(signal_staircase,[],[],[],20);
```

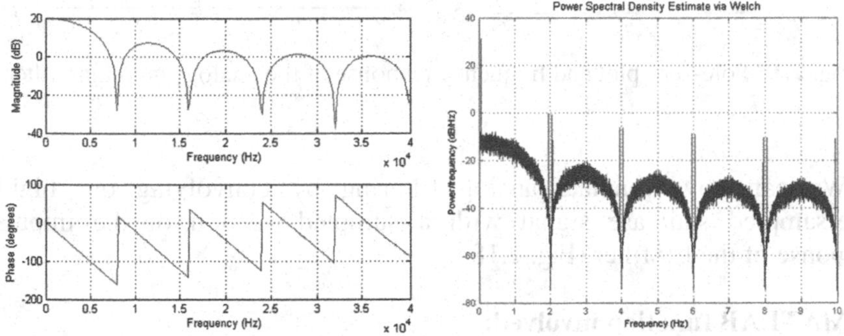

Fig. 2.16 Frequency response of the zero-order analog interpolation (*left*) and power spectral density of the resulting staircase signal (*right*)

The DAC then feeds the resulting staircase signal to an analog low-pass smoothing filter, which removes the spectral images due to sampling. The passband of this filter is limited by the maximum frequency of the signal, F_m, and its stopband must not be greater than F_s–F_m. Let us assume that the PCM signal we have is a telephone signal, with F_m=3,400 Hz and F_s=8,000 Hz, and perform the approximation of the filter with 0.1 dB of ripple in the passband and –60 dB in the stopband. We use Chebyshev approximation so as to keep the order of the filter low.

```
[order,Wn]=cheb1ord(2*pi*3400,2*pi*4600,0.1,60,'s');
[Num_LP,Den_LP]=cheby1(order,0.1,Wn,'s');
zplane(Num_LP,Den_LP);
```

The frequency response of this 12th-order filter meets our requirements (Fig. 2.17).[10]

```
freqs(Num_LP,Den_LP);
```

[10] Note that its phase is nonlinear, but for audio applications this is not a problem.

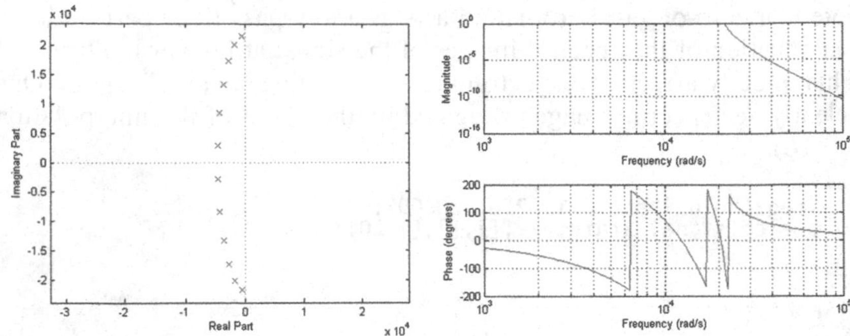

Fig. 2.17 Pole-zero plot and frequency response of the analog smoothing filter

We can now simulate analog filtering by convolving our highly oversampled staircase signal with a sampled version of the impulse response of the LP filter (Fig. 2.18).

MATLAB function involved:

- `y = filters(N,D,x,Fs)` simulates analog filtering of the data in vector `x` with the filter N(s)/D(s) described by vectors `N` and `D` to create the filtered data `y`. `Fs` is the sampling frequency of the input (and output). Filtering is performed by convolving the input with an estimate of the impulse response of the filter obtained by partial fraction expansion.

```
analog_output=filters(Num_LP,Den_LP,signal_staircase,80000);
plot(analog_output(1000:2000));
```

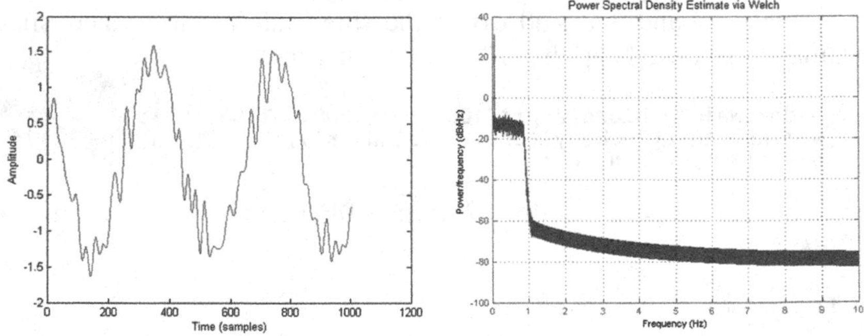

Fig. 2.18 Analog signal produced by the traditional DAC (*left*) and corresponding power spectral density (*right*)

The noise level has been reduced by the analog LP filter, as revealed by spectral analysis (Fig. 2.18). The final SNR is a bit higher than what we had before D/A conversion, but this is due to the fact that the zero-order interpolation attenuates the noise at the upper edge of its passband.

Note, though, that the samples sent to the MATLAB's `soundsc` function are made available to your ears by ... another (real) DAC. One should therefore consider the final quality of this audio sample with care.

```
pwelch(analog_output,[],[],[],20);
signal_sampled_at_10Fs=peak*sin(2*pi*200*(0:1/80000:2));
snr_analog =snr(signal_sampled_at_10Fs,analog_output,400)
                                          % 400=max possible delay
soundsc(analog_output,80000);
```

```
snr_analog =   13.5954
```

2.2.4 Oversampling DAC

By oversampling the digital signal prior to sending it to the analog part of the DAC (which is responsible for creating a staircase analog signal before final analog low-pass filtering), we can broaden the transition band of the analog smoothing filter, thereby strongly decreasing its order (and hardware complexity).

What is more, if we requantize the signal on N–1 bits (1 bit less, i.e., 6 more decibels of total noise power) after multiplying F_s by 4, only one quarter of the resulting power spectral density will contribute to quantization noise in the [0, F_s/2] band; the audible part of the requantization noise power will thus be 10 $\log_{10}$(1/4)dB (i.e., 6 dB) lower than its total power. Hence, this N–1 requantization step will be felt as a new N-bit quantization.

In other words, dropping one bit from four times oversampled digital signals (or k bits for after oversampling by a ratio of k*4) does not do much harm to the signal, while it decreases the required precision of the hardware DAC

Let us check this on our sinusoidal test signal quantized to 6 bits.

```
signal_std=1;
peak=sqrt(signal_std*2);
signal=peak*sin(2*pi*200*(0:1/8000:2));

N=6;
quantizer_saturation=2;
signal_quantized=uquantize(signal,N,quantizer_saturation);

pwelch(signal_quantized,[],[],[],2);
snr_quantized=snr(signal,signal_quantized,0)
```

```
snr_quantized =   30.1654
```

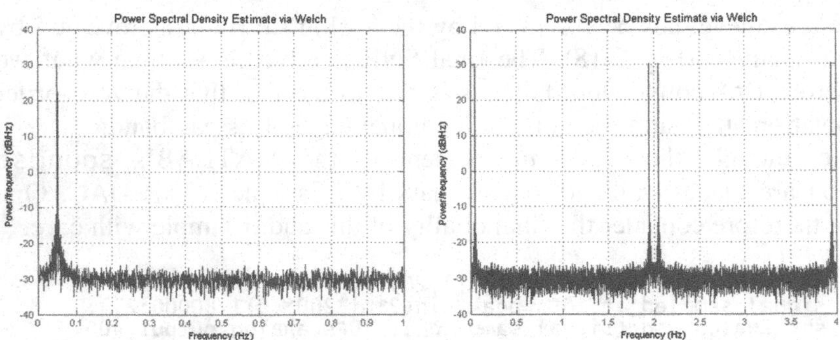

Fig. 2.19 Power spectral density of a sine wave quantized on 6 bits before (*left*) and after four times upsampling (*right*)

We now upsample this quantized signal by a factor of 4 by adding 3 zeros in between its samples. The resulting signal has a sampling frequency F_s'=32,000 Hz (Fig. 2.19). The amplification of samples by 4 is required for the final sine wave to have the same peak-to-peak level as the original.

```
signal_pulses=zeros(1,4*length(signal_quantized));
signal_pulses(1:4:4*length(signal_quantized))=4*signal_quantiz
ed;

pwelch(signal_pulses,[],[],[],8);
```

Interpolation is performed by filtering the impulses with a quarter-band digital LP filter (Fig. 2.20). Note that we use here a very sharp filter, which requires a length of 300 coefficients. In oversampling DACs, LP filtering is performed much more crudely for meeting low computational load constraints.

```
lp_fir=firpm(300,[0 .24 .26 1],[1 1 0 0]);
                % Linear phase quarter-band FIR filter
signal_interpolated=filter(lp_fir,1,signal_pulses);

plot(signal_interpolated(1000:1400));
```

The variances of the signal and of the quantization noise (hence the SNR) have not changed.

```
pwelch(signal_interpolated,[],[],[],8);
signal_sampled_at_4Fs=peak*sin(2*pi*200*(0:1/32000:2));
snr_interpolated=snr(signal_sampled_at_4Fs,...
                     signal_interpolated,300)
```

```
snr_interpolated = 30.1699
```

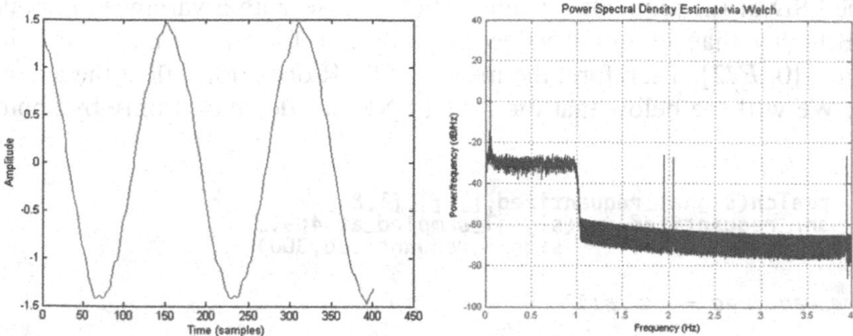

Fig. 2.20 Sine wave after upsampling and interpolation: time domain (*left*) and power spectral density (*right*).

Here comes the first positive effect of oversampling on SNR; (re)quantizing the interpolated signal with N–1 bits (i.e., omitting the least significant bit of the underlying PCM codes) does not much affect the SNR.[11]

Let us first perform 4-bit requantization of the interpolated signal (Fig. 2.21).

```
signal_requantized=uquantize(signal_interpolated,4,...
                             quantizer_saturation);
plot(signal_sampled_at_4Fs(2000:2400)); hold on;
plot(signal_requantized(2151:2550)); hold off; 12
```

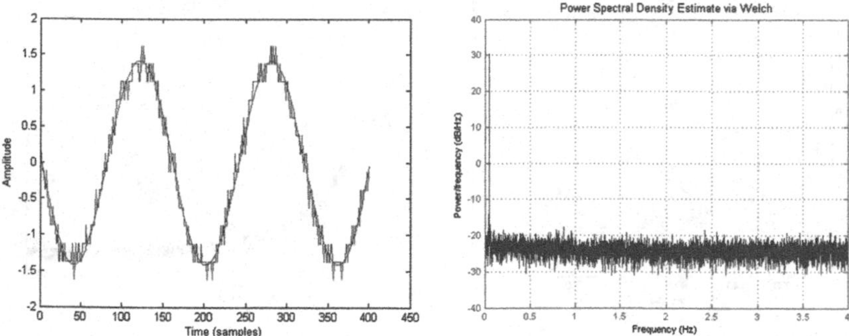

Fig. 2.21 Sine wave after requantization on 4 bits at $4F_s$: time domain (*left*) and power spectral density (*right*)

---

[11] At worst it could degrade the SNR by 3 dB.

[12] Note that we have to shift the oversampled signal when comparing it to the requantized signal; this is due to the delay of the interpolation filter.

The PSD of the requantized signal shows noise with a variance of about 12 dB higher than its initial value, but only one-fourth of that noise is in the band [0, $F_s/2$]. Therefore, the measured SNR does not reflect the actual SNR. We will see below that the actual SNR has decreased only by about 6 dB.

```
pwelch(signal_requantized,[],[],[],8);
snr_requantized=snr(signal_sampled_at_4Fs,...
                    signal_requantized,300)
```

```
snr_requantized =   17.8152
```

Now we can again *simulate* the analog part of D/A conversion (as in Section 2.2.3) using an "analog sampling frequency" of 3*32=96 kHz (Fig. 2.22). Oversampling has a second positive effect on this part: it allows for a much wider transition band for the low-pass smoothing filter: [3,400 Hz, 28,600 Hz]. This results in a simpler filter, of order 4.[13]

```
[order,Wn]=cheb1ord(2*pi*3400,2*pi*28600,0.1,60,'s');
[Num_LP,Den_LP]=cheby1(order,0.1,Wn,'s'); % relaxed filter
signal_pulses=zeros(1,3*length(signal_requantized));
signal_pulses(1:3:3*length(signal_requantized))=...
                                    signal_requantized;
hold_impresp=ones(1,3);
signal_staircase=conv(signal_pulses,hold_impresp);
analog_output=filters(Num_LP,Den_LP,signal_staircase,96000);

plot(analog_output(2000:3300));
pwelch(analog_output,[],[],[],24);
```

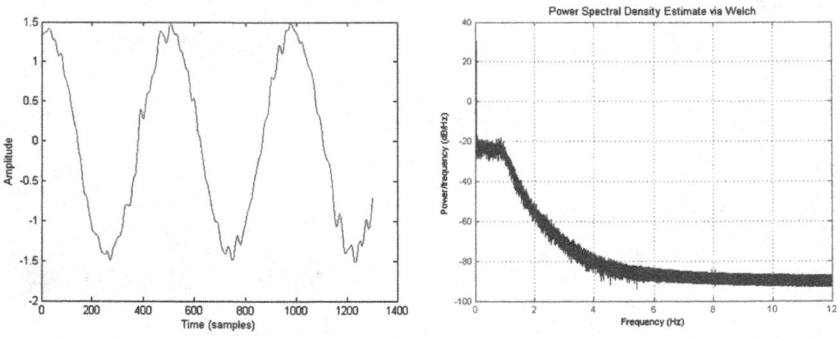

Fig. 2.22 Analog signal produced by the oversampling DAC: time domain (*left*) and power spectral density (*right*)

[13] We use Chebyshev approximation here. For some applications, a Bessel filter could be preferred, as it does not imply important phase nonlinearity. For our sine-based proof of concept, it makes no difference.

As a result of the analog filtering, the apparent SNR is now about 7 dB lower than that of our initial 6-bit quantization, i.e., only 1 dB less than 5-bit quantization at 8 kHz. The lost decibel comes from dithering, and from the fact that the LP filter specifications are loose.

```
signal_sampled_at_12Fs=peak*sin(2*pi*200*(0:1/96000:2));
snr_analog=snr(signal_sampled_at_12Fs,analog_output,4000)
```

snr_analog = 22.9805

In summary, a 6-bit DAC operating at F_s leads to the same SNR as a 5-bit DAC operating at the 4× oversampling rate $4*F_s$.

2.2.5 Oversampling and noise-shaping DAC

We will now show that noise shaping makes it possible to produce 4-bit quantization at $4*F_s$ with the same resolution as 6-bit quantization at F_s.

We start by quantizing to 6 bits and oversampling by 4.

```
signal_std=1;
peak=sqrt(signal_std*2);
signal=peak*sin(2*pi*200*(0:1/8000:2));
%quantization
signal_quantized=uquantize(signal,6,2);
% Oversampling by 4
signal_pulses=zeros(1,4*length(signal_quantized));
signal_pulses(1:4:4*length(signal_quantized))=...
                                        signal_quantized;

lp_fir=firpm(300,[0 .24 .26 1],[1 1 0 0]);
signal_interpolated=filter(lp_fir,1,4*signal_pulses);
```

Let us now perform 4-bit requantization of the interpolated signal with noise shaping as shown in Fig. 2.6.

```
N=4;
quantizer_saturation=2;
q=2*quantizer_saturation/2^N;

delay_memory=0;
signal_requantized=zeros(size(signal_interpolated));
for i=1:length(signal_interpolated)
    u=signal_interpolated(i)+delay_memory;
    % quantization, including dithering
    signal_requantized(i)=uquantize(u,N,quantizer_saturation);
    delay_memory=u-signal_requantized(i);
end;

signal_sampled_at_4Fs=peak*sin(2*pi*200*(0:1/32000:2));
plot(signal_sampled_at_4Fs(2000:2400)); hold on;
plot(signal_requantized(2151:2550)); hold off;
```

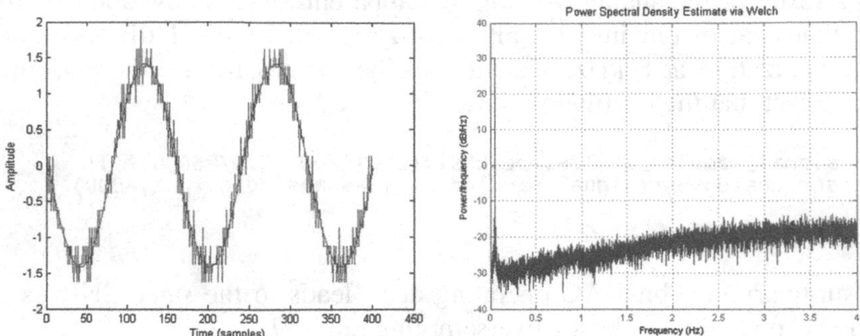

Fig. 2.23 Sine wave after requantization to 4 bits at $4F_s$ using noise shaping: time domain (*left*) and power spectral density (*right*)

The PSD of the requantized signal shows colored quantization noise with most of its energy above $F_s/2$ as a result of noise shaping (Fig. 2.23). Therefore, the measured SNR does not reflect the actual SNR.

```
pwelch(signal_requantized,[],[],[],8);
snr_requantized=snr(signal_sampled_at_4Fs,...
                    signal_requantized,300)
```

snr_requantized = 14.8699

Again, we can *simulate* the analog part of D/A conversion (Fig. 2.24).

```
signal_pulses=zeros(1,3*length(signal_requantized));
signal_pulses(1:3:3*length(signal_requantized))=...
                                    signal_requantized;
hold_impresp=ones(1,3);
signal_staircase=conv(signal_pulses,hold_impresp);

[order,Wn]=cheb1ord(2*pi*3400,2*pi*28600,0.1,60,'s');
[Num_LP,Den_LP]=cheby1(order,0.1,Wn,'s'); % relaxed filter

analog_output=filters(Num_LP,Den_LP,signal_staircase,96000);

plot(analog_output(2000:3300));
```

As a result of analog filtering, the apparent SNR is now only 3 dB lower than that of our initial 6-bit quantization, i.e., only 3 dB less than 4-bit quantization at 8 kHz. Again, the lost decibels come from dithering and from the fact that the LP filter specifications are loose.

```
pwelch(analog_output,[],[],[],24);
signal_sampled_at_12Fs=peak*sin(2*pi*200*(0:1/96000:2));
snr_analog=snr(signal_sampled_at_12Fs,analog_output,4000)
soundsc(analog_output,96000);
```

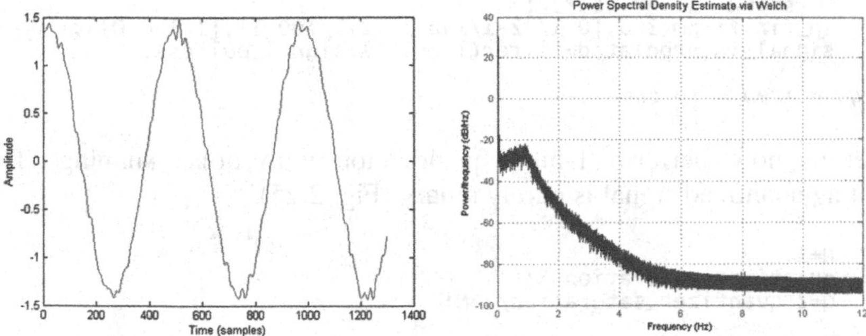

Fig. 2.24 Analog signal produced by the oversampling DAC with noise shaping: time domain (*left*) and power spectral density (*right*)

```
snr_analog =        27.0433
```

2.2.6 Delta–sigma DAC

The principle of delta–sigma modulation is to apply noise shaping and oversampling in such a way that the signal ends up being quantized on 1 bit. This considerably alleviates the task of the hardware D/A converter (which reduces to a switch) and puts most of the load in the digital domain.

In this proof of concept, we will perform 6-to-1 bit requantization using an oversampling ratio of 32 (2^5) and a first-order delta–sigma modulator. Real delta–sigma DACs use higher-order modulators or cascades of first-order modulators and thus do not have to implement oversampling of ratio 2^15!

We start by quantizing to 6 bits and oversampling by 32.[14]

```
signal_std=1;
peak=sqrt(signal_std*2);
signal=peak*sin(2*pi*200*(0:1/8000:1/8));

%quantization
N=6;
quantizer_saturation=3;
signal_quantized=uquantize(signal,N,quantizer_saturation);
snr_quantized=snr(signal,signal_quantized,0)

% Oversampling by 32
signal_pulses=zeros(1,32*length(signal_quantized));
```

[14] We do it here with a single interpolation filter, for convenience. In practice, it would be more computationally efficient to implement oversampling as a cascade of two intermediate oversampling blocks, as mentioned in Section 2.1.5.

```
signal_pulses(1:32:32*length(signal_quantized))=...
    signal_quantized*32;
lp_fir=firpm(300,[0 1/32-1/100 1/32+1/100 1],[1 1 0 0]);
signal_interpolated=filter(lp_fir,1,signal_pulses);
```

```
snr_quantized = 26.6606
```

Let us now perform 1-bit requantization with noise shaping. The resulting quantized signal is purely binary (Fig. 2.25).

```
N=1;
quantizer_saturation=3;
q=2*quantizer_saturation/2^N;

dither=(rand(size(signal_interpolated))-0.5)*q...
      +(rand(size(signal_interpolated))-0.5)*q;

signal_requantized=zeros(size(signal_interpolated));
delay_memory=0;
for i=1:length(signal_interpolated)
    u=signal_interpolated(i)-delay_memory;
    % quantization, including dithering
    signal_requantized(i)=uquantize(u,N,quantizer_saturation);
    % saving the internal variable
    delay_memory=signal_requantized(i)-u;
end;

signal_sampled_at_32Fs=peak*sin(2*pi*200*(0:1/256000:1/8));
plot(signal_sampled_at_32Fs(2000:2600)); hold on;
plot(signal_requantized(2151:2750)); hold off;
xlabel('Time (samples)'); ylabel('Amplitude');
pwelch(signal_requantized,[],[],[],64);
```

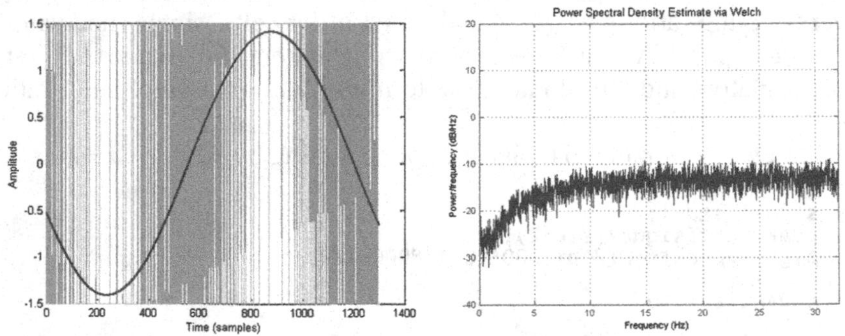

Fig. 2.25 Sine wave after requantization on 1 bit at $32F_s$ using noise shaping: time domain (*left*, superimposed with the original sine wave) and power spectral density (*right*)

Again, we can *simulate* the analog part of D/A conversion (Fig. 2.26). We use an "analog sampling frequency" equal to the one we have reached after 32 times interpolation.

```
[order,Wn]=cheb1ord(2*pi*3400,2*pi*28600,0.1,60,'s');
[Num_LP,Den_LP]=cheby1(order,0.1,Wn,'s'); % relaxed filter
zplane(Num_LP,Den_LP);

analog_output=filters(Num_LP,Den_LP,signal_requantized,...
                      256000);

plot(analog_output(2000:3300));
```

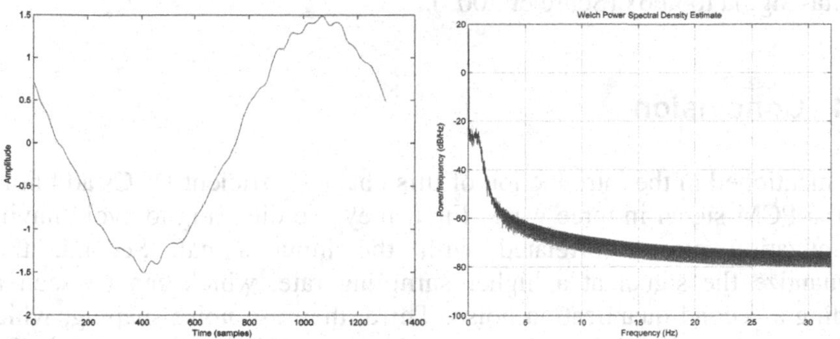

Fig. 2.26 Analog signal produced by the delta–sigma DAC: time domain (*left*) and power spectral density (*right*)

As a result of analog filtering, the apparent SNR is now very close to that of our initial 6-bit quantization! This is confirmed by listening.

```
pwelch(analog_output,[],[],[],64);
snr_analog=snr(signal_sampled_at_32Fs,analog_output,4000)
soundsc(analog_output,256000);
```

```
snr_analog =   25.4677
```

This technique is used in most CD players today[15], for requantizing 16-bits samples to 1 bit, hence with much higher oversampling ratios. Interpolation is therefore performed in several steps for keeping the interpolation filters as simple as possible. Delta–sigma is also the heart of the DSD (direct stream digital) coding used in super audio CDs (SACDs), in which a 1-bit stream is created by the ADC, stored on the CD, and directly converted to sound by the DAC.

[15] Note that all the filters used in this proof of concept use floating-point arithmetics. Using fixed-point arithmetics is more complex.

2.3 Going further

A very simple interactive tutorial on delta–sigma analog-to-digital conversion is available from Analog Devices (2007).

The unescapable reference in delta–sigma modulators is Schreier and Temes (2005). This is a companion book to R. Schreier's MATLAB Delta–Sigma toolbox (Schreier 2003).

2.4 Conclusion

As mentioned in the introduction of this chapter, efficient DACs add noise to the PCM signal in three ways. First, they use dithering to avoid having quantization errors correlated with the input signal. Second, they requantize the signal at a higher sampling rate, which can be seen as adding a second quantization noise. Third, they use noise shaping, which increases the overall quantization noise power but pushes most of it outside the useful band. As a result, and rather unexpectedly, 1-bit DACs are synonymous with high quality.

Besides, 1-bit sigma–delta conversion is the basis of the direct stream digital, a technology used to store audio signals in a digital format, which is used in super audio CD (SACD), the high-resolution CD audio format trademarked by Philips and Sony (Reefman and Janssen 2004). In this technique (Fig. 2.27), the audio waveform for each channel is fed to an analog delta–sigma modulator, which directly samples the signal with a sampling frequency 64 times higher than required and quantizes it to 1 bit. This 82 Mbps-bit stream (to be compared to the 705-kbps bit stream of the classical CD) is recorded on the SACD (it is not low-pass filtered). The SACD reader therefore reduces to a 1-bit DAC producing a bipolar signal, followed by the smoothing filter of an oversampling DAC (i.e., with relaxed constraints).

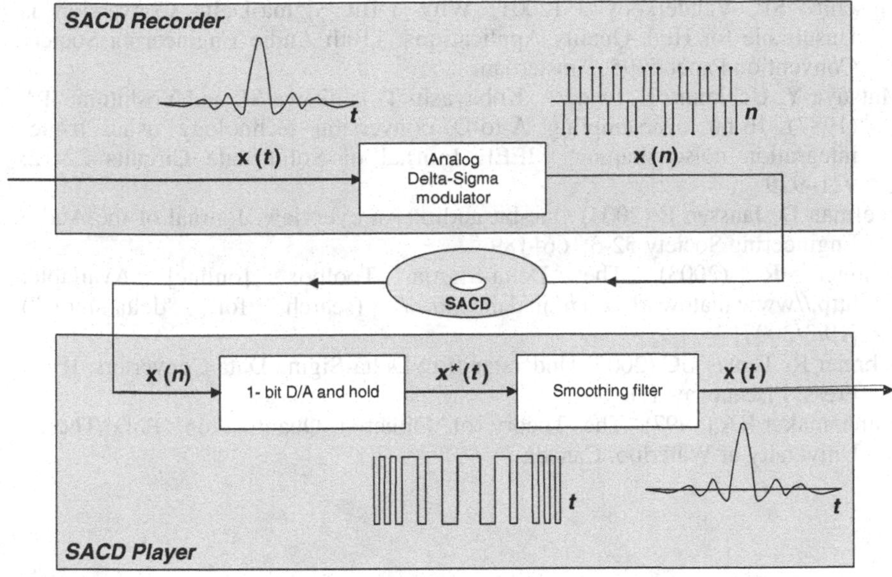

Fig. 2.27 The super audio CD concept

Last but not least, the pulse density modulated (PDM) available after the 1-bit DAC in a DSD player ($x^*(t)$ in Fig. 2.27) can directly be amplified by transistors switching from full ON to full OFF, thereby implementing a fully digital *Class D* power amplifier (Gaalaas 2006).

References

Analog Devices (2007) Interactive Design Tools: Sigma-Delta Analog-to-Digital Converters: Sigma–Delta ADC Tutorial [online] Available: http://www.analog.com/ [19/3/2007]

Candy JC (1997) An overview of basic concepts. In: Norsworthy, Schreier, and Temes (eds) Delta-Sigma Data Converters. IEEE Press: Piscataway, NJ, pp 1–43

Gaalaas E (2006) Class D Audio Amplifiers: What, Why and How. Analog Dialogue 40-06 pp 1–7 [online] Available: http://www.analog.com/library/analogDialogue/archives/40-06/class_d.pdf [15/3/2007]

Hicks C (1995) The application of Dither and Noise-Shaping to Nyquist-Rate Digital Audio: An Introduction [online] Available: www.digitalsignallabs.com/noiseb.ps [15/3/2007]

Kester W, Bryant J (2003) DACs for DSP applications. In: Kester W (ed) Mixed Signal and DSP Design Techniques. Newnes: Amsterdam, pp 99–115.

Lipschitz SP, Vanderkooy J (2001) Why 1-Bit Sigma-Delta Conversion is Unsuitable for High-Quality Applications. 110th Audio Engineering Society Convention Paper 5395, Amsterdam.

Matsuya Y, Uchimura K, Iwata A, Kobayashi T, Ishikawa M, and Yoshitome TA (1987) 16-bit oversampling A-to-D conversion technology using triple-integration noise shaping. IEEE Journal of Solid-State Circuits 22-12: 921–929

Reefman D, Janssen E (2004) One-bit audio: An overview. Journal of the Audio Engineering Society 52-3:166–189

Schreier R (2003) The Delta-Sigma Toolbox [online] Available: http://www.mathworks.com/matlabcentral/ (search for "delta-sigma") [19/3/2007]

Schrcicr R, Temes GC (2005) Understanding Delta-Sigma Data Converters. IEEE Press: Piscataway, NJ.

Wannamaker RA (1997) The Theory of Dithered Quantization. PhD Thesis, University of Waterloo, Canada

Chapter 3

How is sound processed in an MP3 player?

T. Dutoit (°), N. Moreau (*)

(°) Faculté Polytechnique de Mons, Belgium
(*) Ecole Nationale Supérieure des Télécommunications, Paris

Mr. Audio was a bit loose in his PCM tuxedo. Tayloring the suit helped it lose bits and become a lighter MP3 dinner jacket.

In his 1929 painting "La trahison des images," Belgian painter René Magritte highlighted the power of illusions by painting a pipe and commenting it with "Ceci n'est pas une pipe" ("This is not a pipe") as Magritte himself explained: "Try to stuff the painting with tobacco... ."

Perceptual illusions have been used by engineers for designing many consumer products. As an example, one of the most extraordinary imperfections of our visual process, known today as the phi effect (and mistaken for a long time with the persistence of vision), has induced the choice for 24 images per second in the movie pictures industry. In this chapter, we will see how the limitations of the human *auditory* process have been taken into account for the design of *lossy but transparent* perceptual audio coders. In particular, we will focus on the principles underlying the MPEG-1 Layer-I audio coding norm.

3.1 Background – Sub-band and transform coding

Sub-band signal processing is very useful in audio (as well as in video) coding applications. The sub-band encoder (Fig. 3.1) decomposes an input

T. Dutoit, F. Marqués (eds.), *Applied Signal Processing*,
DOI 10.1007/978-0-387-74535-0_3, © Springer Science+Business Media, LLC 2009

signal *x(n)* into *M* different frequency ranges called *sub-bands* through the *analysis filter bank.* Sub-band signals $x_i(n)$ are then downsampled to $y_i(m)=x_i(Mm)$ and quantized. The decoder combines the quantized sub-band signals $\tilde{y}_i(m)$ into an output signal $\tilde{x}(n)$ by upsampling them to $\tilde{x}_i(n)$ (with $\tilde{x}_i(mM)=\tilde{y}_i(m)$ and $\tilde{x}_i(mM+l)=0$ for $l=0,...,M-1$) and passing them through the *synthesis filter bank* (Crochiere and Rabiner 1983, Vaidyanathan 1993). The number of sub-bands and the decimation–interpolation ratio are not independent; in a *critically sampled* filter bank, for which the number of sub-band signal samples is equal to the number of input samples, the number of sub-bands and the decimation–interpolation ratio are identical.

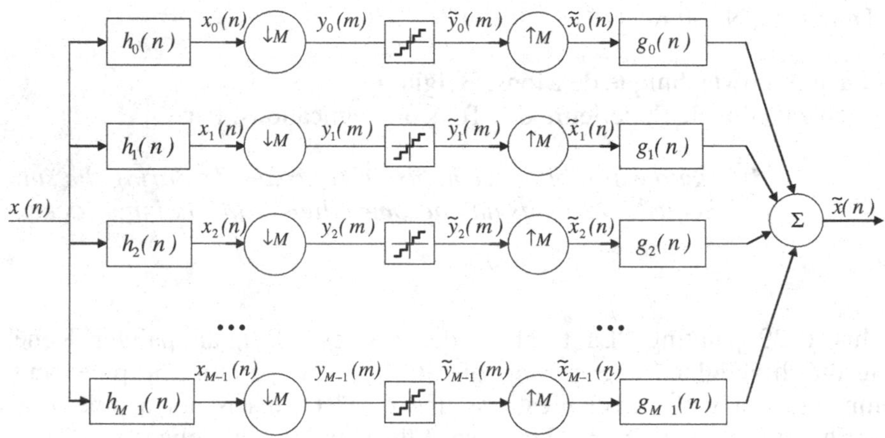

Fig. 3.1 A sub-band coder

The rationale for sub-band signal processing is the need for time- and frequency-dependent SNR control to allow transparent encoding of audio and video signals with very low bit rates. As a matter of fact, when applying uniform quantization to the input signal using a given number of bits per sample, the resulting quantization noise is uniformly distributed in the frequency range of the signal (see Chapter 2). Since the power spectral density (PSD) of the input signal changes with time and frequency, uniform quantization results in time- and frequency-dependent signal-to-noise ratio (*local* SNR). The number of bits per sample (hence the *overall SNR*) is therefore usually set to a very high value (the 16 bits–96 dB of the compact disk) in order for the worst-case local SNRs to be statistically acceptable to the human ear. In contrast, in sub-band signal processing, the number of bits allocated to each sub-band signal can be adapted to the specific SNR required in the corresponding frequency band. This, as we shall see, allows for drastic bit rate reduction.

In this section, we will first examine the conditions required on the analysis–synthesis filters to satisfy the perfect reconstruction condition (Section 3.1.1). We will show that the filter bank approach can also be interpreted in terms of block transform-based processing (Section 3.1.2). The masking properties of the human ear will then be briefly reviewed, and their use in the MPEG1 – Layer I audio coding norm will be explained (Section 3.1.3). We will conclude by mentioning current coding norms and bit rates (Section 3.1.4).

3.1.1 Perfect reconstruction filters

An important feature of the analysis and synthesis filters in a filter bank is that they should allow *perfect reconstruction* (PR) (or, in practice, *nearly* perfect reconstruction) when no quantization is performed, i.e.,

$$\tilde{x}(n) = x(n) \tag{3.1}$$

In order to examine this problem in a simple case, let us focus on a two-channel filter bank (Fig. 3.2) composed of a low-pass (upper) and a high-pass (lower) branch.

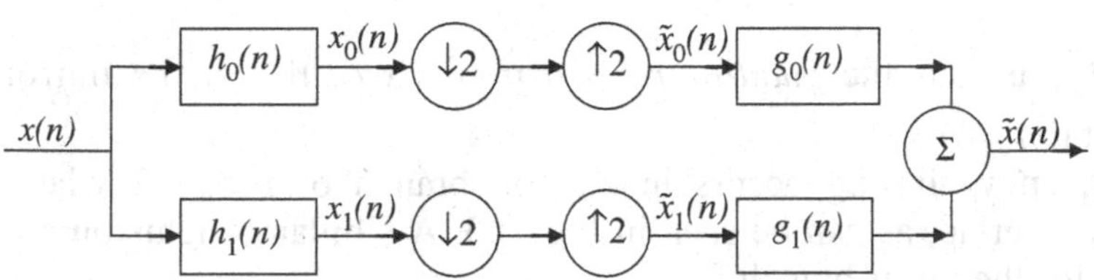

Fig. 3.2 A two-channel filter bank

The PR condition implies to cancel (or at least minimize) three types of distortion, respectively due to the phase and amplitude responses of the filters and to the aliasing that is inevitably introduced by the imperfections of the decimation–interpolation steps. One of the great ideas of sub-band coding, precisely, is that the PR condition does not imply that aliasing should be minimized in each branch separately (which would require close-to-ideal filters) but that some aliasing distortion can be accepted for sub-band signals, *provided all sub-band distortions in adjacent sub-bands are cancelled* by the final summation. As a consequence, analysis and synthesis filters are not chosen independently.

The combined decimation–interpolation steps, which transform each sub-band signal $x_i(n)$ into $\tilde{x}_i(n)$, replace every other sample in $x_i(n)$ $(i=0,1)$ by zero. It is thus equivalent to adding to each sub-band signal $x_i(n)$ a copy of it in which every other sample is sign-reversed:

$$\tilde{x}_i(n) = \frac{1}{2}\left[x_i(n) + x_i^{-}(n)\right] \quad with \quad x_i^{-}(n) = x_i(n)\,(-1)^n$$

$$\text{or } \tilde{X}_i(z) = \frac{1}{2}\left[X_i(z) + X_i^{-}(z)\right] \quad with \quad X_i^{-}(z) = X_i(\;z) \tag{3.2}$$

The added signal $x_i^{-}(n)$ can be seen as a modulation of $x_i(n)$ by exp($jn\pi$), i.e., by a Nyquist rate carrier. Its discrete-time Fourier transform (DTFT) is therefore that of $x_i(n)$, shifted by the Nyquist frequency

$$X_i^{-}(f) = X_i\left(f - \frac{1}{2}\right) \tag{3.3}$$

which implies that

$$|X_i^{-}(f)| = |X_i(\frac{1}{2} - f)| \tag{3.4}$$

$|X_i^{-}(f)|$ is thus the *quadrature mirror* of $|X_i(f)|$, i.e., its mirror with respect to $f = ¼$.

Frequency aliasing occurs in the top branch of Fig. 3.2 when these DTFTs overlap, as can be seen in Fig. 3.3. A similar conclusion could be drawn for the lower branch.

The output of the filter bank can be written as follows:

$$\begin{aligned}\tilde{X}(z) &= \left[\tilde{X}_0(z)G_0(z) + \tilde{X}_1(z)G_1(z)\right] \\ &= \frac{1}{2}[X_0(z)G_0(z) + X_1(z)G_1(z)] \\ &+ \frac{1}{2}[X_0(-z)G_0(z) + X_1(-z)G_1(z)]\end{aligned} \tag{3.5}$$

in which the second term is due to the quadrature mirror images of signals $x_0(n)$ and $x_1(n)$ and is therefore responsible for the possible aliasing.

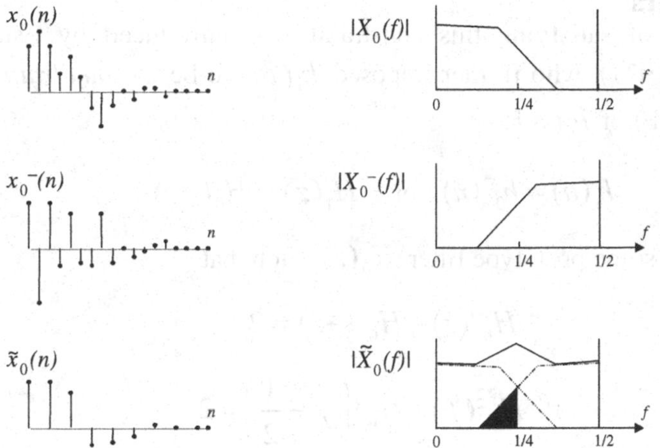

Fig. 3.3 Decimation–interpolation seen as the addition of $x_0^-(n)$ to the low sub-band signal. Due to the overlap of their DTFTs in the black triangle, aliasing occurs, which prevents $|\tilde{X}_0(f)|$ from being equal to $|X_0(f)|$ for f in [0,1/4]

Equation (3.5) can be further expanded as follows:

$$\begin{aligned}\tilde{X}(z) &= \frac{1}{2}X(z)\left[H_0(z)G_0(z)+H_1(z)G_1(z)\right] \\ &+\frac{1}{2}X(-z)\left[H_0(-z)G_0(z)+H_1(-z)G_1(z)\right]\end{aligned} \tag{3.6}$$

which can be identified with (3.1) if

$$\begin{cases} H_0(z)G_0(z)+H_1(z)G_1(z)=2 \\ H_0(-z)G_0(z)+H_1(-z)G_1(z)=0 \end{cases} \tag{3.7}$$

The second condition is easy to satisfy with

$$\begin{aligned} G_0(z) &= H_1(-z) \quad or \quad g_0(n)=h_1^-(n) \\ G_1(z) &= -H_0(-z) \quad or \quad g_1(n)=-h_0^-(n) \end{aligned} \tag{3.8}$$

The first condition in (3.7) then becomes

$$H_0(z)H_1(-z)-H_0(-z)H_1(z)=2 \tag{3.9}$$

QMF filters

One way of satisfying this constraint was introduced by Esteban and Galland (1977), who further imposed $h_1(n)$ to be the *quadrature mirror filter* (QMF) of $h_0(n)$:

$$h_1(n) = h_0^{-}(n) \quad \text{or} \quad H_1(z) = H_0(-z) \tag{3.10}$$

and found some prototype filter $H_0(z)$ such that

$$\begin{aligned} &H_0^2(z) - H_0^2(-z) = 2 \\ \text{or} \quad &H_0^2(f) - H_0^2\left(f - \frac{1}{2}\right) = 2 \end{aligned} \tag{3.11}$$

The final filter bank is given in Fig. 3.4. The beauty of this idea lies in its simplicity.

In practice, $H_0(z)$ is a linear phase (i.e., $h_0(n)$ is symmetrical) FIR filter of length N (N being even[1]), and in order to implement it as a causal filter its impulse response is delayed by $(N-1)/2$. This results in an overall delay of N–1 samples in both branches of the filter bank. Taking into account that

$$H_0(f) = |H_0(f)|\, e^{j2\pi f(N-1)/2} \tag{3.12}$$

one finds easily from (3.11)

$$|H_0(f)|^2 + |H_0(f - \frac{1}{2})|^2 = 2 \tag{3.13}$$

It can be shown that designing $H_0(z)$ such that it satisfies (3.13) while being frequency selective is impossible[2]. Johnston (1980) proposed an iterative approach, which results in frequency-selective filter banks free of phase and aliasing distortion, but with some (controllable) amplitude distortion. Many authors have proposed other PR filters with various properties, such as conjugate quadrature filters (CQF). This scientific discussion also led to the parallel development of the *wavelet theory* (see Chapter 10).

---

[1] It can be shown that FIR QMF filters can be built for N odd but require a modification of (3.10).

[2] Satisfying the QMF constraint while not being frequency selective is not very useful in sub-band coding, as it does not allow efficient frequency-dependent SNR control.

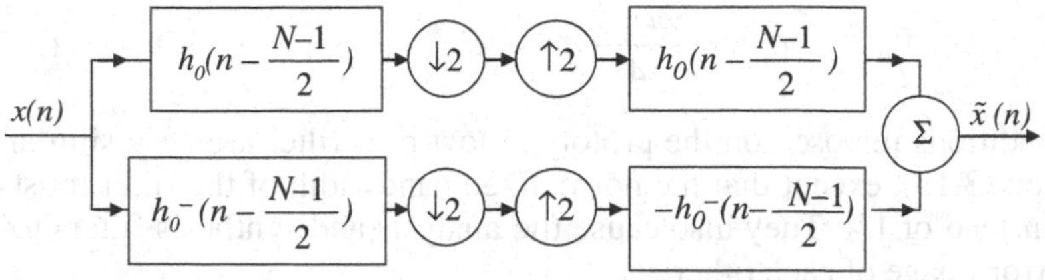

Fig. 3.4 A two-channel QMF filter bank implementing linear phase FIR filters of length *N*

Pseudo-QMF filters

Going from 2 to *M* sub-bands can be done by further splitting each sub-band in Fig. 3.4 into two bands and so on (the so-called *tree-structured* or *hierarchical filter bank* approach), but it is more straightforward to implement the filter bank as *M* parallel passband filters. In the pseudo-QMF (PQMF) approach (Nussbaumer and Vetterli 1984, Vaidyanathan 1993), the analysis filters $H_i(z)$ are obtained by modulating the impulse response *h(n)* of a low-pass prototype filter with *M* carrier frequencies f_i uniformly distributed in the range [0,1/2] (Fig. 3.5):

$$\begin{aligned} h_i(n) &= h(n)\cos(2\pi f_i n + \phi_i) \\ g_i(n) &= h(n)\cos(2\pi f_i n + \theta_i) \end{aligned} \quad with\ f_i = \frac{(2i+1)}{2M}\frac{1}{2} \tag{3.14}$$

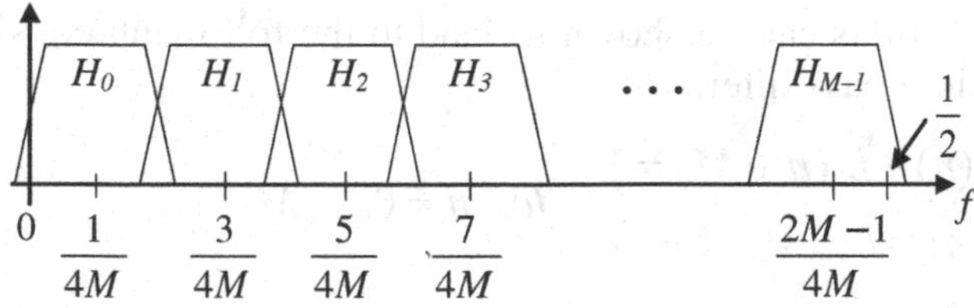

Fig. 3.5 A uniform *M*-channel filter bank (*f* denotes normalized frequency)

The values of ϕ_i and θ_i are chosen so that all significant aliased terms are canceled,[3] which can be shown to happen if

[3] The *pseudo* qualifier in PQMF comes from the fact that only adjacent bands are taken into account in the PR condition. Hence PQMF filters do not strictly satisfy the PR condition.

$$\phi_i = -\frac{2i+1}{4}\pi = -\theta_i \tag{3.15}$$

The conditions imposed on the prototype low-pass filter are very similar to Equation (3.13), except that the normalized bandwidth of the filter must be 1/4*M* instead of 1/4. They also cause the analysis and synthesis filters to be the mirror image of each other:

$$h_i(n) = g_i(N-1-n) \tag{3.16}$$

which forces $H_i(z)G_i(z)$ in all sub-band channels to have linear phase.

PQMF filters are used in the MPEG-1 Layer-I audio coding norm.

MDCT filters

PQMF filters were later generalized to PR cosine-modulated filter banks, with special emphasis on filters whose length N is twice the number of channels, known as the modified discrete cosine transform (MDCT) filter banks (Princen and Bradley 1986). The impulse responses of the analysis filters are given by

$$\begin{aligned} h_i(n) &= h(n)\sqrt{\frac{2}{M}}\cos(2\pi f_i n + \phi_i) \\ g_i(n) &= h_i(2M-1-n) \end{aligned} \quad \textit{with} \quad \begin{aligned} f_i &= \frac{(2i+1)}{2M}\frac{1}{2} \\ \phi_i &= \frac{(2i+1)(M+1)\pi}{4M} \end{aligned} \tag{3.17}$$

The PR conditions can be shown to lead to the following constraints on the prototype low-pass filter:

$$\begin{aligned} h^2(n) + h^2(n+M) &= 1 \\ h(n) &= h(2M-1-n) \end{aligned} \quad \textit{for } n = 0,\ldots,M-1 \tag{3.18}$$

A special case, which is often mentioned, is the modulated lapped transform filter bank (Malvar 1990), in which

$$h(n) = \sin\left[\left(n+\frac{1}{2}\right)\frac{\pi}{2M}\right] \quad \textit{for } n = 0,\ldots,2M-1 \tag{3.19}$$

MDCT filters are used in the MPEG-1 Layer-3 (MP3) as well as in the MPEG-2 advanced audio coding (AAC) and MPEG-4 norms.

3.1.2 Filter banks and lapped transforms

If the length of the filters used in M-channel sub-band filters were exactly M samples, one could advantageously implement the operations performed simultaneously by the M analysis filters as a linear transform of successive non-overlapping M sample frames of the input signal.

Conversely, any linear transform can be seen as implementing a special filter bank (Fig. 3.6). If the linear transform is characterized by its $M \times M$ matrix $\mathbf{H}$ and the input frame is stored in column vector $\mathbf{x}$, then $\mathbf{y}=\mathbf{H}\mathbf{x}$ computes the output of M analysis filters whose impulse responses are the time-reversed rows of $\mathbf{H}$. As a matter of fact, Fig. 3.1 verifies

$$\begin{bmatrix} y_0(m) \\ y_1(m) \\ \ldots \\ y_{M-1}(m) \end{bmatrix} = \begin{bmatrix} h_0(M-1) & \ldots & h_0(1) & h_0(0) \\ h_1(M-1) & \ldots & h_1(1) & h_1(0) \\ \ldots & \ldots & \ldots & \ldots \\ h_{M-1}(M-1) & \ldots & h_{M-1}(1) & h_{M-1}(0) \end{bmatrix} \begin{bmatrix} x(mM-M+1) \\ \ldots \\ x(mM-1) \\ x(mM) \end{bmatrix} \tag{3.20}$$

Downsampling is straightforward; since the output of the analysis filters must be downsampled by M, one only needs to compute one value of $\mathbf{y}$ every M samples (by using non-overlapping frames for $\mathbf{x}$). Similarly, computing $\tilde{\mathbf{x}}=\mathbf{G}^{\mathrm{T}}\mathbf{y}$, in which $\mathbf{G}$ is an $M \times M$ matrix whose rows are the time-reversed impulse responses of the analysis filters:

$$\begin{bmatrix} x(mM-M+1) \\ \ldots \\ x(mM-1) \\ x(mM) \end{bmatrix} = \begin{bmatrix} g_0(M-1) & g_1(M-1) & \ldots & g_{M-1}(M-1) \\ \ldots & \ldots & \ldots & \ldots \\ g_0(1) & g_1(1) & \ldots & \ldots \\ g_0(0) & g_1(0) & \ldots & g_{M-1}(0) \end{bmatrix} \begin{bmatrix} y_0(m) \\ y_1(m) \\ \ldots \\ y_{M-1}(m) \end{bmatrix} \tag{3.21}$$

computes the output of the synthesis filters, but it also sums all sub-band signals to produce the output frame. Obviously, PR is achieved in this case if $\mathbf{G}^{\mathrm{T}}$ is the inverse of H, since $\mathbf{x}=\mathbf{H}^{-1}\mathbf{y}$.

Having input and output frames that do not overlap is a problem when sub-band signals are modified in a time-varying way (as it happens typically when they are quantized as a function of their instantaneous energy); the effect of each modification applies to one synthesis frame only, with no possibility of interframe smoothing. This effect is known as the *blocking effect.*[4] Hence practical filter banks make use of longer analysis and synthesis filters (as in the previous section on PQMF filters).

However, when the length N of the analysis and synthesis filters is longer than the number of channels, the filtering operations can still be implemented as the multiplication of N sample frames with $N \times M$ or $M \times N$ matrices. Analysis filtering is still achieved by **y=Hx**, in which **H** is an $M \times N$ matrix whose rows are the time-reversed impulse responses of the analysis filters and **x** is an N sample frame. Downsampling is achieved by shifting successive input frames by M samples (i.e., they *overlap* by $N–M$ samples).

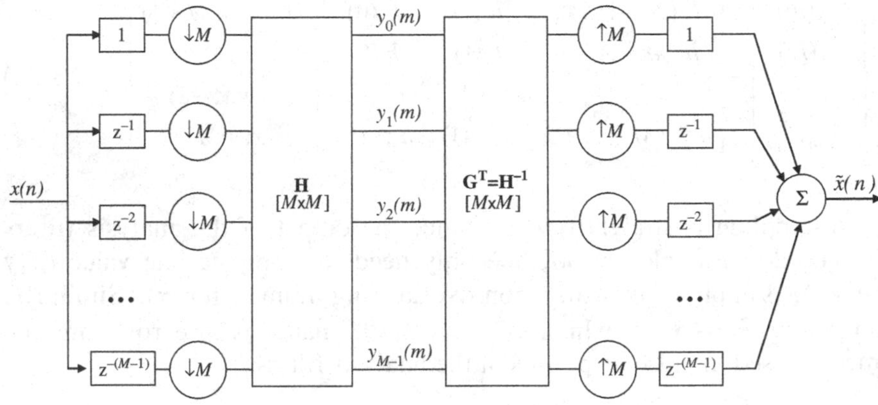

Fig. 3.6 An M-channel transform-based implementation of a filter bank using filters of length N=M

The upsampling, synthesis filtering, and summation operations are still performed as $\mathbf{G}^T\mathbf{y}$, in which **G** is an $M \times N$ matrix whose rows are the (not time-reversed) impulse responses of the analysis filters. This computes the sum of the responses of the synthesis filters each excited by the single sample stored in **y** for each sub-band. Just as the input frames, the output frames overlap in time (and should therefore be accumulated; see Fig. 3.7 for a simple example with $N = 2M$). This leads to the more general implementation of Fig. 3.8, which differs only slightly from Fig. 3.6 and still clearly shows the underlying sub-band processing interpretation.

[4] The same concept will be for the case of image coding in Chapter 8.

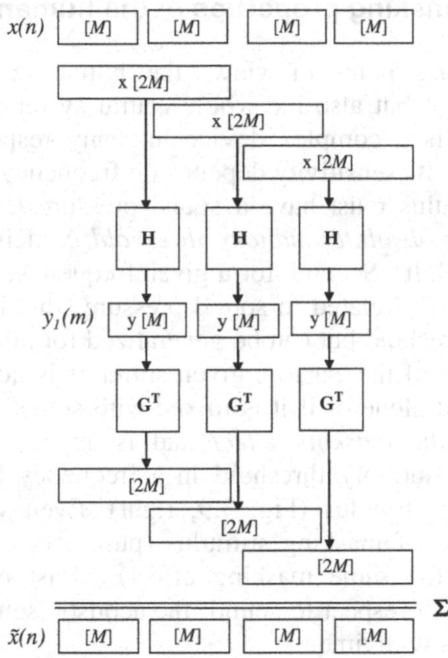

Fig. 3.7 Frame-based view of an M-channel transform-based implementation of a filter bank using filters of length N=2M

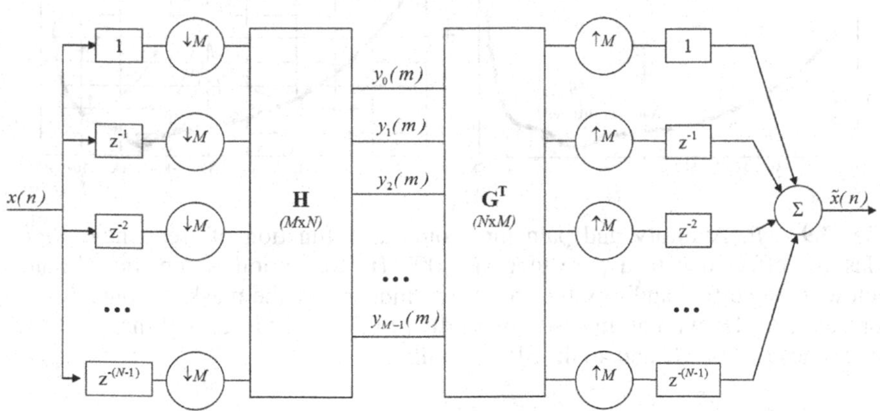

Fig. 3.8 An M-channel transform-based implementation of a filter bank using filters of length N>M

3.1.3 Using the masking properties of the human ear

From an engineering point of view, the human ear (not only the mechanical parts of it but also the whole neural system that produces an acoustic sensation) is a complex device in many respects (Painter and Spanias 2000). First, its sensitivity depends on frequency: to be just heard, a narrow-band stimulus must have a sound pressure level higher than a threshold called the *absolute auditory threshold,* which is a function of frequency (Fig. 3.9, left). Second, for a given frequency, the impression of loudness is not linearly related to sound pressure (this is obvious for the absolute auditory threshold but can be generalized for all loudness values). Third, the sensitivity of the ear to a given stimulus is not the same if this stimulus is presented alone or if it is mixed with some masking stimulus. This is termed as the *masking effect* and is approximated as a local modification of the auditory threshold in a frequency band surrounding that of the masking stimulus (Fig. 3.9, right). Even worse, this effect depends on the type of masking stimulus (pure tones and narrow-band noise do not have the same masking effect). Last but not least, not everyone is equal with respect to sound: the acoustic sensation is listener-dependent and varies with time.

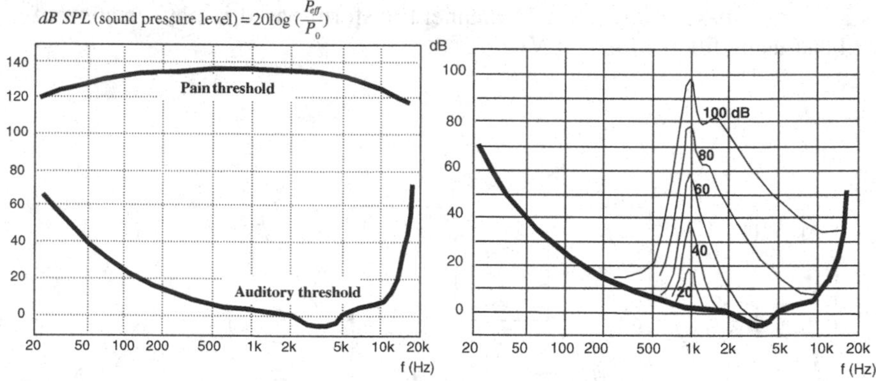

Fig. 3.9 *Left*: Auditory and pain thresholds as a function of frequency. *Right*: Masking effect due to a pure tone at 1,000 Hz for various intensities. Sounds below the modified auditory threshold are "hidden" by the masking tone. A pure tone at 1,000 Hz with an intensity of 80 dB SPL,[5] for instance, will make another pure tone at 2,000 Hz and 40 dB SPL inaudible

[5] Sound pressure level (SPL) L_p is a logarithmic measure of the root-mean-square sound pressure of a sound relative to a reference value, usually measured in decibel:

One of the most practical features of the masking effect can be modeled as operating within a discrete set of 25 auditory "channels" termed as *critical bands* (with idealized band edges equal to 20, 100, 200, 300, 400, 510, 630, 770, 920, 1,080, 1,270, 1,480, 1,720, 2,000, 2,320, 2,700, 3,150, 3,700, 4,400, 5,300, 6,400, 7,700, 9,500, 12,000, and 15,500 Hz). As an example, if a signal and a masker are presented simultaneously, then only the masker frequencies falling within the critical bandwidth contribute to masking the signal. In other words, the ear acts as a sub-band processing system, in which separate nonlinear masking effects mainly appear in separate sub-bands.

The great interest of sub-band coding is to establish a master–slave relationship (Fig. 3.10), in each sub-band, between (i) the local *masking threshold* computed by the so-called *psychoacoustic model* from an estimation of the masking stimuli present in the auditory spectrum and (ii) quantization noise, i.e., the number of bits used to quantize the corresponding sub-band signal.

Note that although sub-band coders are based on the features of the auditory process, they do not completely mimic it. As an example, although the edges of the critical bands are not uniformly spaced in frequency, most sub-band coders use equal bandwidth channels. More importantly, since the masking effect depends on the actual loudness of the masking signal, while the coder is not aware of the actual volume level set by the listener, the psychoacoustic model used in sub-band coding a priori assumes that the user sets the volume level such that the least significant bit of the original 16-bit PCM code can just be heard.

3.1.4 Audio coders

MPEG-1 audio

The MPEG-1 audio norm actually defines three complexity layers of processing, each with its own sub-band coding scheme, psychoacoustic model, and quantizer (see Pan 1995 or Painter and Spanias 2000 for a tutorial). Starting from 1.4 Mbps for a stereo audio signal, Layer-I achieves a 1:4 compression level (down to 384 kbps). With the addition of Layer-II (formerly known as Musicam: Masking Pattern adapted Universal Subband Coding And Multiplexing), a 1:6–8 ratio is reached (between 256 and 192 kbps). This coder is very much used in digital video broadcasting (DVB). MPEG-1 Layer-III, commercially known as MP3,

---

$$L_p = 20\log_{10}\left(p_{rms} / p_{ref}\right)$$

where p_{ref} is the reference sound pressure, usually set to 20 μPa (rms), although 1 μPa (rms) is used for underwater acoustics.

further compresses the data stream down to 128–112 kbps (i.e., a compression ratio of 1:10–12). In this chapter, we examine only Layer-I, which is far enough for a proof of concept.

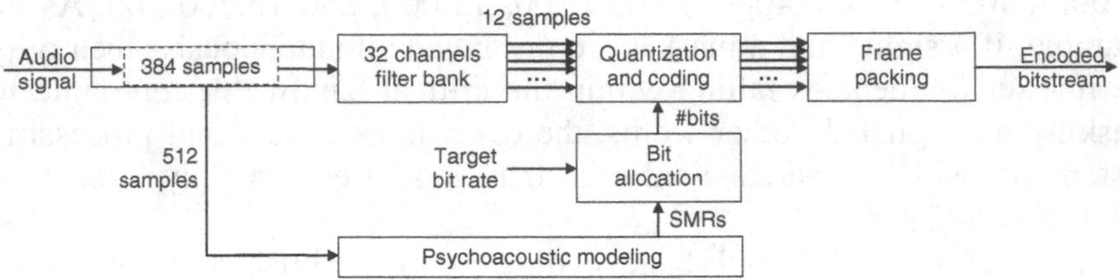

Fig. 3.10 Psychoacoustic model controlling sub-band quantizers in the MPEG-1 Layer-I audio coder

The MPEG-1 Layer-I coder (Fig. 3.10) implements a 32-channel sub-band coder. Samples are processed by frames of 384 samples (i.e., 12 samples at the downsampled rate). The psychoacoustic model is based on a 512-sample FFT estimation of the local power spectral density of the signal, $S_{xx}(f)$. Local maxima with dominant peaks are detected as tonal maskers and extracted from the FFT. A single noise masker is then computed from the remaining energy in each critical band. Maskers are then combined, accounting again for the existence of critical bands (typically, several maskers in a single band are merged into the strongest of them). Individual masking thresholds are finally estimated for each tonal and noise masker and summed to produce the overall masking threshold, $\Phi(f)$. This leads to the determination of a *signal-to-mask ratio* (SMR) in each band:

$$SMR(k) = 10\log_{10}\left(\frac{\max\limits_{\text{band } k} S_{xx}(f)}{\min\limits_{\text{band } k} \Phi(f)}\right) \tag{3.22}$$

Sub-band signals are quantized uniformly, and the number of bits for each sub-band is chosen so as to make resulting signal-to-noise ratio lower than the SMR. Higher-frequency sub-bands typically require less bits. In particular, sub-bands 27–32 are assigned 0 bits by design. Quantization is made adaptive by normalizing sub-band signals before quantizing them. Normalization is performed by blocks of 12 sub-band samples (i.e., by blocks of 384 samples at the original sampling frequency) by estimating a local scale factor and dividing samples by its value (Fig. 3.11).

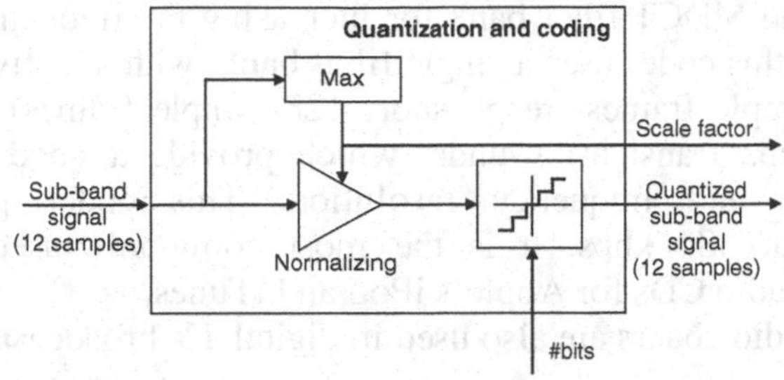

Fig. 3.11 Adaptive quantization in the MPEG-1 Layer-I audio coder

MPEG-1 Layer-II is based on the same principle but uses 1024 samples for FFT computation in its psychoacoustic model and processes frames that are three times longer than in Layer-I.[6] It also implements nonsimultaneous (sometimes called "temporal") masking, which denotes the fact that masking can occur both before masker onset and after masker removal.

MPEG-1 Layer-III further splits each of the 32 original sub-bands into 18 channels by using a modified discrete cosine transform (MDCT, see Section 3.1.2), which provides a better frequency resolution[7] when required (i.e., in stationary parts of the audio signal). Its coding scheme is dynamic; more bits are allocated to frames that require it.

MPEG-2 Audio – Advanced audio coding

MPEG-2 was normalized in 1994. The MPEG-2 BC audio coder adds a series of features to the MPEG-1 audio coder: an extension to lower sampling frequencies (down to 8 kHz) and multichannel processing for 5.1 surround sound. It remains *backward compatible* (BC) with the MPEG-1 coder, i.e., MPEG-1 audio coders can decode MPEG-2 BC, ignoring the additional information.

MPEG-2 AAC (advanced audio coding[8]) appeared in 1997 and provides a great improvement over MPEG-2 BC (hence, over MP3), especially for low bit rates (below 100 kbps). Contrary to the MPEG-1 Layer-III coder,

[6] More specifically, scale factors are computed as before but quantized by groups of three.

[7] MPEG-1 Layers I and II have a good time resolution, since sub-band samples are updated every 32 audio samples, which enables them to track audio transients. They exhibit a bad frequency resolution though, as their sub-band filter bandwidth is $F_s/2/32 \approx 700$ Hz.

[8] Not to be confused with AC3, the Dolby Digital 5.1 audio coding scheme developed by Dolby Labs.

which added an MDCT filter bank for increasing the frequency resolution when needed, this coder uses a single filter bank, with adaptive block size: large 1024-sample frames (resp., short 128-sample frames) are used for stationary (resp., transient) sounds, which provide a good compromise between time and frequency resolution. This coder provides 5.1 transparency at 384 kbps. It is the most commonly used format for compressing audio CDs for Apple's iPod and iTunes.

MPEG-2 audio coders are also used in digital TV broadcasting.

MPEG-4 audio

The AAC coder has been enhanced in the framework of MPEG-4 by including additional tools (such as long-term prediction) for intermediate bit rates. A low-delay version of the coder is also available for teleconference applications. The resulting MPEG-4 AAC coder, normalized in 1999, provides stereo transparency with a bit rate between 96 and 128 kbps and 5.1 transparency at 256 kbps.

In 2003, a technique known as *sub-band replication* (SBR) has been added to the MPEG-4 AAC coder, leading to the MPEG-4 HE AAC (high efficiency). The idea is to use the strong correlation between low-frequency (up to 4–8 KHz) and higher frequency spectral components of audio signals and to avoid sending redundant information. HE AAC provides stereo transparency at 48 kbps and 5.1 transparency at 128 kbps.

A new version of it, HE AAC v2 (sometimes called AAC+), appeared in 2004, which further uses psychoacoustic effects to better encode stereo channels as one main channel and side information for spatialization. Intermediate stereo quality is obtained at 24 kbps. This coder is used in the emerging digital radio mondiale and in 3G mobile telephony (the 3GPP standard).

The latest development has been reached in July 2006 with the definition of the HE AAC surround (or MPEG surround), which adapts the spatialization coding principles of HE AAC v2 to multichannel audio. As a result, 5.1 transparency is now available at 64 kbps.

3.2 MATLAB proof of concept: ASP_mp3.m

In this section, we provide a proof of concept of the MPEG-1Layer-I audio coding norm. We start by examining a two-channel filter bank using conventional filters (Section 3.2.1) and QMF filters (Section 3.2.2). We then extend our study to a 32-channel PQMF filter bank (Section 3.2.3) and show how it can be efficiently implemented using block-based lapped transforms (Section 3.2.4). We conclude by providing our filter bank with a perceptual quantizer for sub-band signals (Section 3.2.5).

3.2.1 Two-channel filter bank

Before examining a complete *M*-channel filter bank, we first discuss the two-channel case and the effect of a decimation and interpolation in each branch.

Let us first generate a 4-seconds chirp signal (from 0 to 4 kHz) with a sampling rate of 8 kHz, which we will use as a reference throughout this section, and plot its spectrogram (Fig. 3.12, left).

```
Fs=8000;
input_signal=chirp((0:4*Fs)/Fs,0,4,4000);
specgram(input_signal,1024,Fs,256);
```

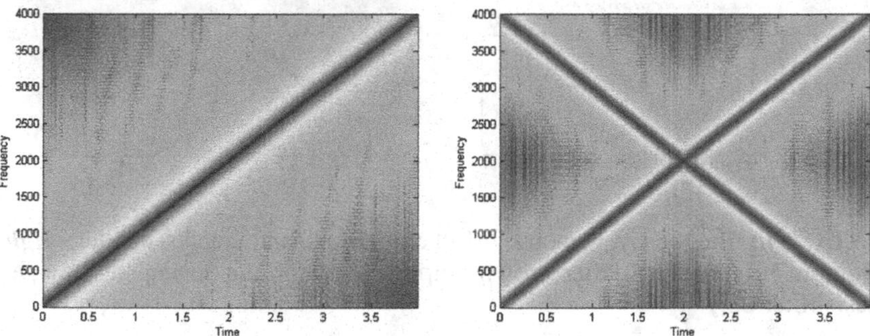

Fig. 3.12 Spectrogram of a chirp. *Left*: Original signal; *right*: after downsampling–upsampling without using analysis and synthesis filters

Applying direct downsampling–upsampling by 2 results in replacing every other sample by zero. The result is that an artifact signal is added, whose spectrum is the quadrature mirror image of that of the chirp. Two sinusoids can be heard at any time (Fig. 3.12, right).[9]

```
downsampled=input_signal(1:2:end);
upsampled(1:2:2*length(downsampled))=2*downsampled;
```

[9] Note that we multiply the signal by 2 when upsampling, so that the power of the signal remains unchanged.

Adding a quarter-band filter after upsampling eliminates the image distortion, making sure that only one sinusoid appears at any time (Fig. 3.13). During the first half of the chirp, that sinusoid is the chirp itself; during the second half, the sinusoid is an alias due to the quadrature mirror image of the chirp. We design the synthesis filter as a symmetric FIR, so as to avoid phase distortion, and we set its order to 1,000, so as to obtain high stop-band rejection (close to 80 dB). Note that the first 1,000 samples are a transient.

```
G0=fir1(1000,1/2);
freqz(G0,1);
G0_output=filter(G0,1,upsampled);

specgram(G0_output(1001:end),1024,Fs);
```

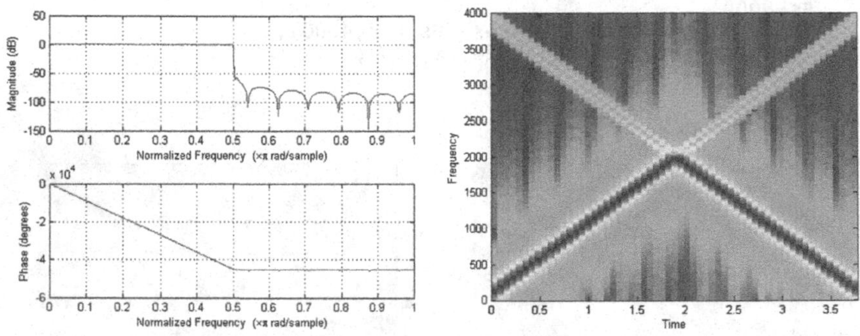

Fig. 3.13 *Left*: Frequency response of the quarter-band analysis filter; *right*: spectrogram of a chirp after downsampling–upsampling using a low-pass synthesis filter only

Adding another quarter-band filter before downsampling removes the aliasing distortion, i.e., the alias in the second half of the chirp (Fig. 3.14, left). This synthesis filter can a priori be identical to the analysis filter.

```
H0=G0;
H0_output=filter(H0,1,input_signal);
downsampled=H0_output(1:2:end);
upsampled(1:2:2*length(downsampled))=2*downsampled;
G0_output=filter(G0,1,upsampled);

specgram(G0_output(2001:end),1024,Fs, 256);
```

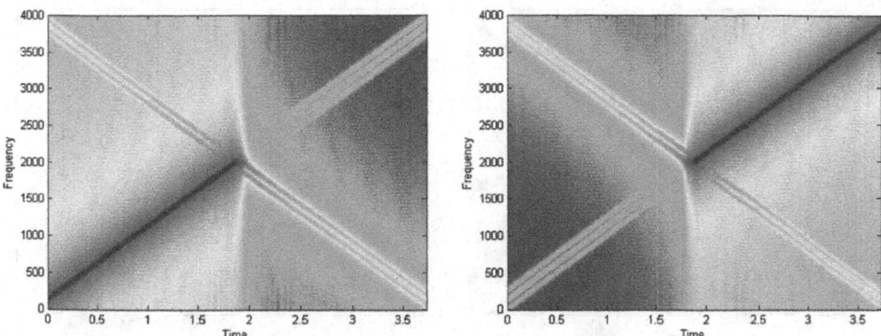

Fig. 3.14 *Left*: Spectrogram of a chirp after downsampling–upsampling using low-pass analysis and synthesis filters; *right*: same thing with high-pass analysis and synthesis filters

Using HF quarter-band filters instead of LF ones selects the second half of the chirp (Fig. 3.14, right).

```
G1=fir1(1000,1/2,'high');
H1=G1;
H1_output=filter(H1,1,input_signal);
downsampled=H1_output(1:2:end);
upsampled(1:2:2*length(downsampled))=2*downsampled;
G1_output=filter(G1,1,upsampled);
```

We now examine the output of the two-channel sub-band filter bank by adding the two signals obtained above (Fig. 3.15, left). Perfect reconstruction is not achieved, as shown in the previous spectrogram. Closer examination of the error waveform (Fig. 3.15, right) shows that most of the error lies in the center of the signal (this is not shown in the spectrogram), i.e., for frequencies for which the `H0` and `H1` filters overlap; more important, aliasing cannot be avoided in each band for these frequencies.

```
synt_signal=G0_output+G1_output;
% Shift synt_signal to account for the filters delay.
% A symmetric FIR filter of order N (length N+1) brings
% a delay of N/2 samples.
error=synt_signal(1001:end)-input_signal(1:end-1000);
```

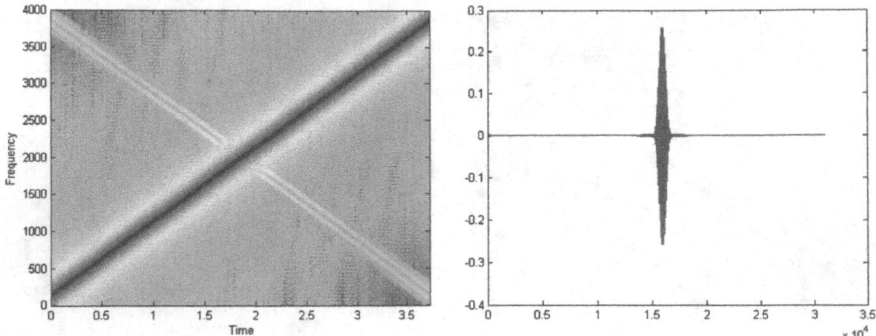

Fig. 3.15 *Left*: Spectrogram of the output of the two-channel filter bank; *right*: error signal waveform

3.2.2 Two-channel QMF filter bank

Perfect reconstruction is possible even with nonideal filters (i.e., even with some aliasing in each band) provided the overall aliasing is canceled when adding the low-pass and high-pass sub-band signals. This, as we shall see, is mainly the responsibility of the analysis and synthesis filters. Johnson's type QMF filters provide a solution for nearly perfect reconstruction. In this example, we use Johnson's "B-12" QMF (Fig. 3.16).

```
H0_QMF=[-0.006443977 0.02745539 -0.00758164 ...
    -0.0913825  0.09808522 0.4807962];
H0_QMF=[H0_QMF fliplr(H0_QMF)];
[H0,W]=freqz(H0_QMF,1);

H1_QMF=H0_QMF.*[1 -1 1 -1 1 -1 1 -1 1 -1 1 -1];
[H1,W]=freqz(H1_QMF,1);
```

These filters are not very frequency selective (because their order is low) and will therefore allow important aliasing in each band. This is confirmed by passing our chirp signal through the corresponding filter bank (Fig. 3.17).

```
% LF band
H0_output=filter(H0_QMF,1,input_signal);
subband_0=H0_output(1:2:end);
upsampled(1:2:2*length(subband_0))=2*subband_0;
G0_QMF=H0_QMF;
G0_output=filter(G0_QMF,1,upsampled);

% HF band
H1_output=filter(H1_QMF,1,input_signal);
subband_1=H1_output(1:2:end);
upsampled(1:2:2*length(subband_0))=2*subband_1;

G1_QMF=-H1_QMF;
```

```
G1_output=filter(G1_QMF,1,upsampled);
synt_signal=G0_output+G1_output;
```

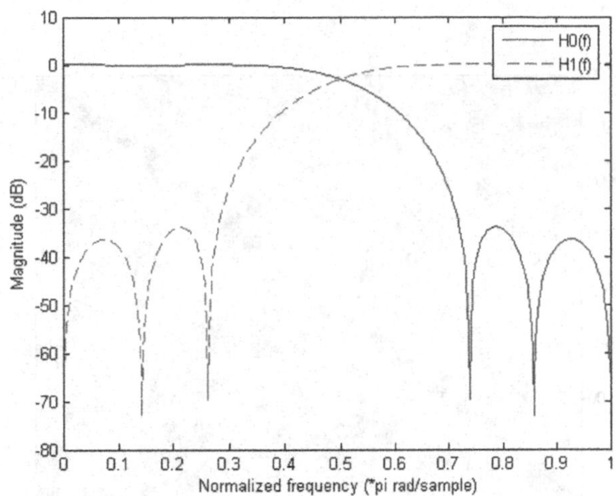

Fig. 3.16 *Left*: Spectrogram of the output of the two-channel filter bank; *right*: error signal waveform

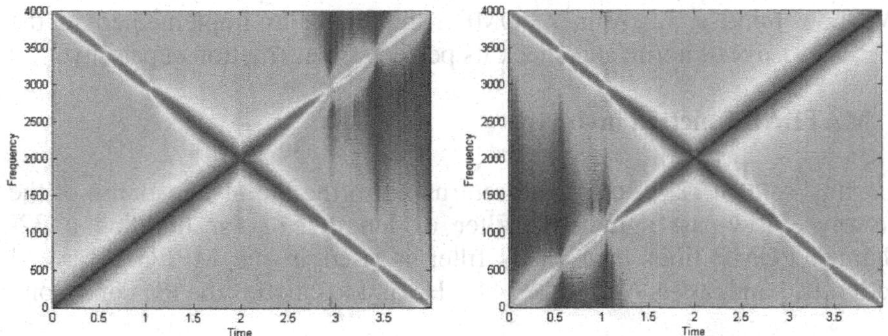

Fig. 3.17 *Left*: Spectrogram of the output of the LF channel of the QMF filter bank; *right*: spectrogram of the output of the HF channel. Important aliasing can be observed

However, perfect reconstruction is now achieved, because the QMF analysis and synthesis filters are such that the aliasing in each band sum up to zero[10] (Fig. 3.18).

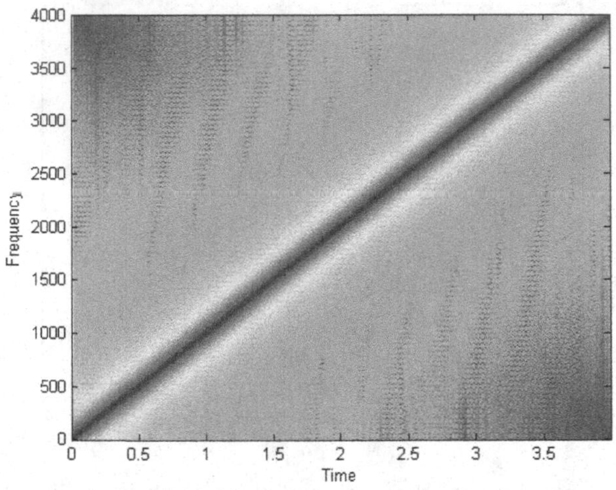

Fig. 3.18 Spectrogram of the output of the two-channel QMF filter bank

3.2.3 32-channel pseudo-QMF filter bank

We now build a 32-channel PQMF filter bank, as implemented in the MPEG-1 Layer-I norm, and check its perfect reconstruction capability.

MATLAB function involved:

- `hn = PQMF32_prototype` returns in `hn` the impulse response of the prototype low-pass symmetric filter of length 512 for building a 32-channel PQMF filter bank. This filter is used in the MPEG-1 Layer-I coder. Its bandpass is $F_s/2/32/2$=344.5 Hz and it satisfies the PR condition.

We first build the filter bank from the prototype filter and examine the magnitude frequency responses of all filters (Fig. 3.19, left). Clearly, PQMF filters are not ideal bandpass filters with bandwidth=689 Hz, but their response overlaps only with their two neighboring filters (hence their name of "pseudo"-QMF filters). The total bandwidth of each filter is

[10] More precisely, very close to 0.

indeed about 2 × 689 Hz. Note that the filter gain is 15 dB, i.e., 20*log10(32)/2. As a result, passing twice through the filter produces a gain of 32, which compensates for the decimation by 32. In the next lines, it is thus not necessary to multiply upsampled sub-band signals by 32.

```
% Load the prototype lowpass filter
hn=PQMF32_prototype;

% Build 32 cosine modulated filters centered on normalized
% frequencies Fi=(2*i+1)/64 *1/2
PQMF32_Gfilters = zeros(32, 512);
for i = 0:31
    t2 = ((2*i+1)*pi/(2*32))*((0:511)+16);
    PQMF32_Gfilters(i+1,:) = hn.*cos(t2);
end
PQMF32_Hfilters=fliplr(PQMF32_Gfilters);

for i = 1:32
    [H,W]=freqz(PQMF32_Hfilters(i,:),1,512,44100);
    plot(W,20*log10(abs(H))); hold on
end
xlabel('Frequency (Hz)'); ylabel('Magnitude (dB)');
set(gca,'xlim',[0 22050]);
ylabel('Magnitude (dB)');
hold off;
```

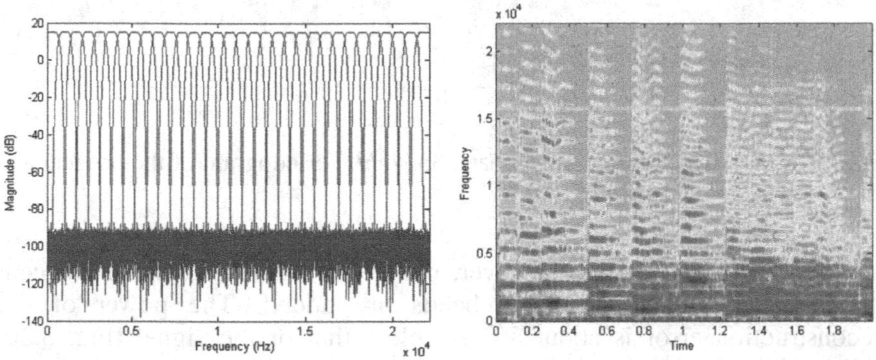

Fig. 3.19 *Left*: Magnitude frequency response of the 32 analysis filters of a PQMF filter bank; *right*: spectrogram of an audio signal containing a few notes played on the violin

Let us now check the output of the PQMF filter bank when fed with 2 seconds of violin monophonic signal sampled at 44.100 Hz (Fig. 3.19, right). As revealed by listening to sub-band 3, isolated sub-band signals are very much aliased, because each PQMF filter is not ideal (Fig. 3.20, left).

```
[input_signal,Fs]=wavread('violin.wav');
output_signal=zeros(size(input_signal));
```

```
for i=1:32

    Hi_output=filter(PQMF32_Hfilters(i,:),1,input_signal);
    subband_i=Hi_output(1:32:end);
    upsampled_i(1:32:32*length(subband_i))=subband_i;

    % synthesis filters are the symmetric of the analysis
    % filters, which ensures linear phase in each sub-band
    Gi_output=filter(PQMF32_Gfilters(i,:),1,upsampled_i');
    output_signal=output_signal+Gi_output;

    if i==3
        G3_output=Gi_output;
    end;

end;
```

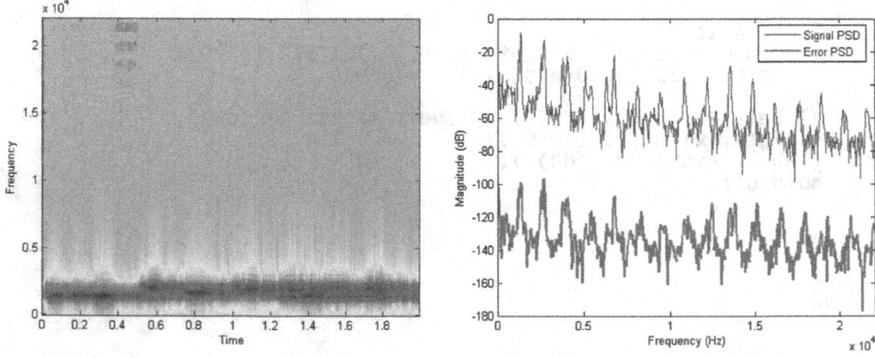

Fig. 3.20 *Left*: Spectrogram of sub-band 3; *Right*: periodogram of the original and error signals, around sample 11,000.

The PQMF filter bank, however, makes sure that aliasing in adjacent bands cancels itself when sub-bands are added. The power of the reconstruction error is about 85 dB below that of the signal (Fig. 3.20, right). Note that the output is delayed by 511 samples (since analysis and synthesis filters have a delay of 511/2 samples).

MATLAB function involved:

- `snr=snr(signal,signal_plus_noise,max_shift,showplot)` returns the signal-to-noise ratio computed from the input signals. `Max_shift` gives the maximum time shift (in samples) between `signal` and `signal_plus_noise`. The actual time shift (obtained from the maximum of the cross-correlation between both signals) is taken into account to estimate the noise. If `showplot` is specified, then the signal,

signal_plus_noise, and error are plotted, and the SNR is printed on the plot.

```
error=output_signal(512:end)-input_signal(1:end-511);

[signal_psd,w]=periodogram(input_signal(11001:12024),...
                           hamming(1024));
[error_psd,w]=periodogram(error(11001:12024),...
                           hamming(1024));
plot(w/pi*22050,10*log10(signal_psd));
hold on;
plot(w/pi*22050,10*log10(error_psd),'r','linewidth',2);
hold off;

snr_PQMF=snr(input_signal(1:end-511),output_signal(512:end),0)
```

```
snr_PQMF =   84.2443
```

3.2.4 Filter banks and lapped transforms

We now show how analysis and synthesis filters can be implemented using block transforms.

When the length of the filters used in *M*-channel sub-band filters are exactly *M* samples, these operations reduce to a linear transform of successive non-overlapping *M* sample frames of the input signal. A four-sample DFT, for instance, can implement a four-channel filter bank whose sub-band filter impulse responses are the time-reversed of the lines of the 4×4 DFT matrix. Applying it to our chirp is straightforward.

```
Fs=8000;
input_signal=chirp((1:4*Fs)/Fs,0,4,4000);

for i=1:length(input_signal)/4

    % creating a column vector with 4 samples
    input_frame=input_signal(4*(i-1)+1:4*(i-1)+4)';

    % producing one sample in each downsampled sub-band, i.e.
    % band-pass filtering and downsampling all sub-bands in
    % one operation.
    subbands=fft(input_frame);

    % producing four samples of the filter bank output, i.e.
    % upsampling, band-pass filtering all sub-bands, and
    % summing them in one operation.
    output_frame=ifft(subbands);

    % storing the output column vector in the output signal
    output_signal(4*(i-1)+1:4*(i-1)+4)=output_frame';

end;

%soundsc(output_signal,Fs);
```

Obviously, this will return the original chirp. Since the underlying filters have complex coefficients, however, each sub-band signal is complex. Moreover, this type of filter bank is not very frequency selective, as shown in Fig. 3.21. The frequency overlap between adjacent bands is about half the main lobe bandpass (as in the previous section on PQMF), but the side lobes are very high. This does not make it a good candidate for sub-band coding.

```
tmp=[1 ; exp(-j*pi/2) ; exp(-j*pi) ; exp(-j*3*pi/2)];
DFT_matrix_4x4=vander(tmp);

for i=1:4
    [H,W]=freqz(fliplr(DFT_matrix_4x4(i,:)),1,'whole');
    plot(W/pi,max(20*log10(abs(H)), -50)); hold on
end
```

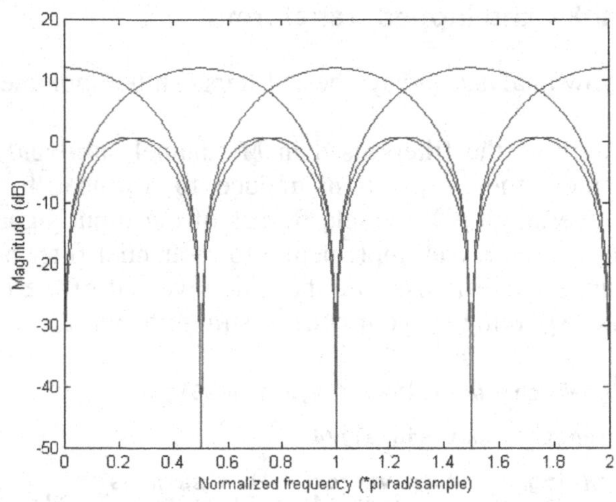

Fig. 3.21 Magnitude frequency response of the four channels of a DFT filter bank. Note that we show the responses in the full [0,1] frequency range, as these frequency responses are not two-sided

In general, the length of the impulse responses of the analysis and synthesis filters used in sub-band coders is higher than the number M of channels. The filtering operations, however, can still be implemented as the multiplication of L-sample frames with $L\times M$ or $M\times L$ matrices. For example, the 32-channel PQMF filter bank introduced in the previous section, in which the length N of the impulse response of each filter is 512 samples, can be efficiently implemented as follows (which is very much similar to the implementation of our previous DFT-based filter bank, with the addition of overlap). Obviously, we get the same results as before (Fig. 3.22).

```
[input_signal,Fs]=wavread('violin.wav');

% Block-based sub-band filtering
input_frame=zeros(512,1);
output_signal=zeros(size(input_signal));

for i=1:(length(input_signal)-512+32)/32

    % Overlap input_frames (column vectors)
    input_frame=input_signal((i-1)*32+1:(i-1)*32+512);

    % Analysis filters and downsampling
    % Since PQMF H filters are the time-reversed G filters,
    % we use the G filters matrix to simulate analysis
    % filtering
    subbands_frame_i = PQMF32_Gfilters*input_frame;

    % Synthesis filters
    output_frame = PQMF32_Gfilters'*subbands_frame_i;

    % Overlap output_frames (with delay of 511 samples)
    output_signal((i-1)*32+1:(i-1)*32+512)= ...
        output_signal((i-1)*32+1:(i-1)*32+512)+output_frame;

end

error=output_signal(512:end)-input_signal(1:end-511);
```

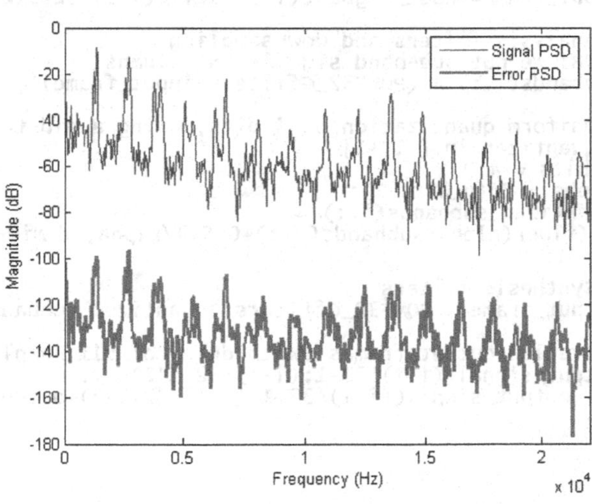

Fig. 3.22 Periodogram of the original and error signals, around sample 11,000

3.2.5 Perceptual audio coding

The sub-band filtering process developed in the previous sections transforms the original stream of samples at sampling frequency F_s into 32 parallel sub-bands sampled at $F_s/32$. It still needs a quantization and coding step to produce overall bit rate compression.

Quantizing sub-band samples uniformly[11] does not allow much transparency at low bit rates, as shown in the four-bits per sub-band sample trial below (compression factor = 4). The resulting output signal is degraded; its spectrogram is given in Fig. 3.23.

```
% Build the PQMF H and G filters
hn=PQMF32_prototype;
PQMF32_Gfilters = zeros(32, 512);
for i = 0:31
    t2 = ((2*i+1)*pi/(2*32))*((0:511)+16);
    PQMF32_Gfilters(i+1,:) = hn.*cos(t2);
end

[input_signal,Fs]=wavread('violin.wav');

% Block-based sub-band analysis filtering
input_frame=zeros(512,1);
output_signal=zeros(size(input_signal));

n_frames=(length(input_signal)-512+32)/32;
for i=1:n_frames

        % Overlap input_frames (column vectors)
        input_frame=input_signal((i-1)*32+1:(i-1)*32+512);

        % Analysis filters and downsampling
        % NB: we put sub-band signals in columns
        subbands(i,:) = (PQMF32_Gfilters*input_frame)';

        % Uniform quantization on 4 bits, using a mid-thread
          quantizer in [-1,+1]
         n_bits = 4;
         alpha = 2^(n_bits-1);
         quantized_subbands(i,:) =…
           (floor(alpha*subbands(i,:)+0.5))/alpha; % mid-thread

        % Synthesis filters
        output_frame = PQMF32_Gfilters'*quantized_subbands(i,:)';

        % Overlap output_frames (with delay of 511 samples)
        output_signal((i-1)*32+1:(i-1)*32+512)= ...
            output_signal((i-1)*32+1:(i-1)*32+512)+output_frame;

end
```

[11] We use a *mid-thread* quantizer here, as opposed to the more classical *mid-rise* quantizer (see Section 2.1.1), so as to encode low-level signals to 0. Failing to do so quickly creates *musical noise*, caused by high-frequency changes in the LSB for low-level signals. Note that we do not use the `uencode` and `udecode` functions from MATLAB, which do not implement a true mid-thread quantizer.

The output signal is strongly distorted. The SNR falls down to 10.3 dB.

```
% NB: no delay compensation required, as the first sub-band
% samples produced by the lapped transform correspond to the
% 512th original samples.

error=output_signal-input_signal;

[signal_psd,w]=periodogram(input_signal(11001:12024),...
    hamming(1024));
[error_psd,w]=periodogram(error(11001:12024),...
    hamming(1024));
plot(w/pi*22050,10*log10(signal_psd));
hold on;
plot(w/pi*22050,10*log10(error_psd),'r','linewidth',2);
hold off;

snr_4bits=snr(input_signal(512:end-512),...
              output_signal(512:end-512),0)
```

snr_4bits = 10.3338

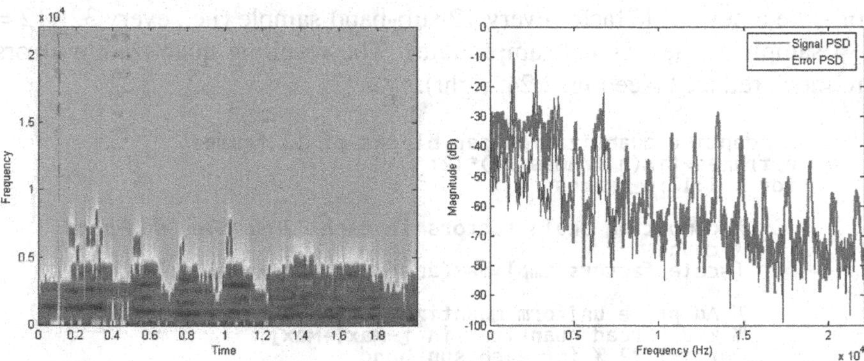

Fig. 3.23 Sub-band coding using fixed four-bit uniform quantization. *Left*: spectrogram; *right*: the periodograms of the original and error signals (computed here around sample 11,000) are almost identical, except for low frequencies

One can see that this fixed [–1,+1] quantizer range does not adequately account for the variation of sub-band signal level across sub-bands, as well as in time (Fig. 3.24, left).

```
plot(quantized_subbands(100:200,2)); hold on;
plot(subbands(100:200,2),'--r'); hold off;
```

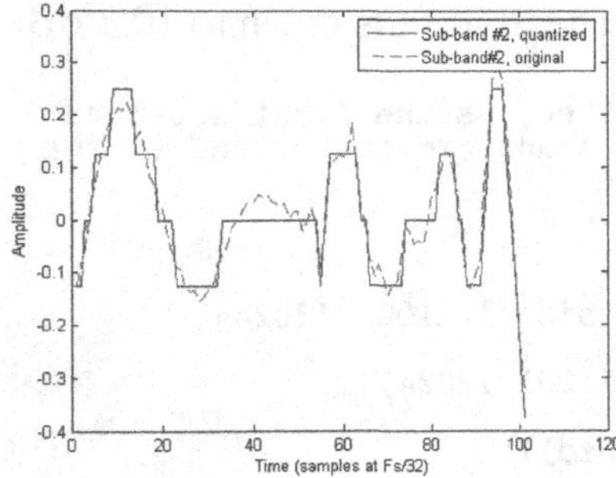

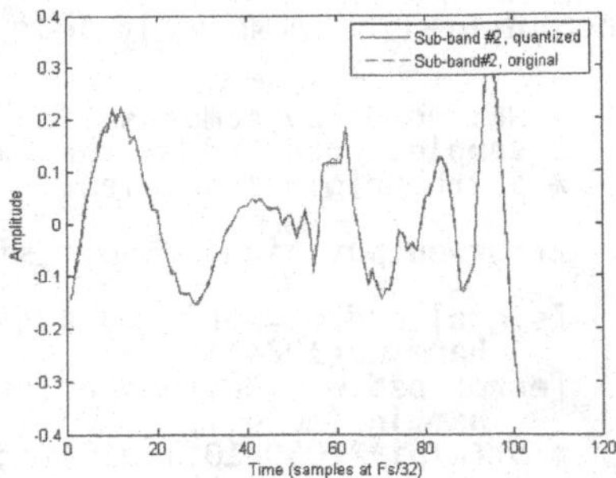

Fig. 3.24 Excerpt from sub-band #2 before and after quantization. *Left*: using fixed four-bit uniform quantization; *right*: using level-adaptive four-bit uniform quantization

An obvious means of enhancing its quality is therefore to apply a scale factor to each sub-band quantizer. As in the MPEG-1 Layer-I coder, we compute a new scale factor every 12 sub-band sample (i.e., every $32 \times 12 = 384$ sample at the original sample rate). The resulting quantization errors are much reduced (see Fig. 3.24, right).

```
% Adaptive quantization per blocks of 12 frames
n_frames=fix(n_frames/12)*12;
for k=1:12:n_frames

    % Computing scale factors in each 12 samples sub-band
    % chunk
    [scale_factors,tmp]=max(abs(subbands(k:k+11,:)));

    % Adaptive uniform quantization on 4 bits, using a
    % mid-thread quantizer in [-Max,+Max]
    for j=1:32 % for each sub-band

        n_bits = 4;
        alpha = 2^(n_bits-1)/scale_factors(j);
        quantized_subbands(k:k+11,j) = ... % mid-thread
            (floor(alpha*subbands(k:k+11,j)+0.5))/alpha;

    end;
end;
```

The resulting signal is now of much higher quality, as shown in Fig. 3.25. The overall SNR has increased to 25 dB.

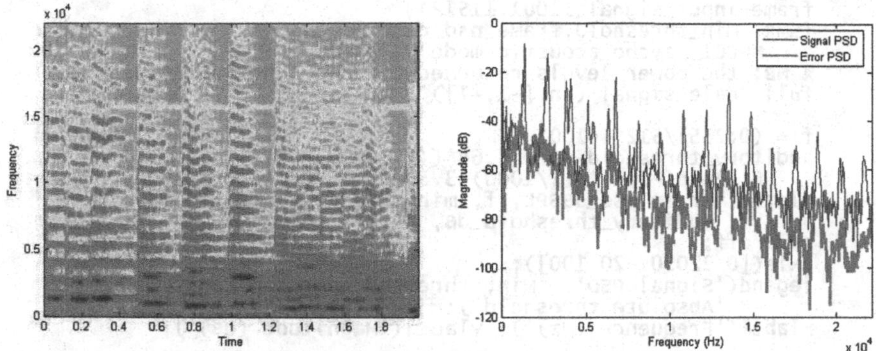

Fig. 3.25 Sub-band coding using adaptive four-bit uniform quantization. *Left*: spectrogram; *right*: periodogram of the original and error signals, around sample 11,000.

```
snr_4bits_scaled=snr(input_signal(512:end-512),…
                     output_signal(512:end-512),0)
snr_4bits_scaled =   25.0629
```

The ultimate refinement, which is by far the most effective and results from years of audio research, consists in accepting more quantization noise (by allocating less bits and thereby accepting a lower SNR) in frequency bands where it will not be heard and using these extra bits for more perceptually prominent bands. The required perceptual information is provided by a psychoacoustic model.

For any 512-sample frame taken from the input signal, the MPEG-1 Layer-I psychoacoustic model computes a global masking threshold, obtained by first detecting prominent tonal and noise maskers separately and combining their individual thresholds. The maximum of this global threshold and the absolute auditory threshold is then taken as the final threshold (Fig. 3.26, top).

MATLAB function involved:

- `function [SMR, min_threshold_subband, masking_threshold] = MPEG1_psycho_acoustic_model1(frame)`

computes the masking threshold (in dB) corresponding to psychoacoustic model #1 used in MPEG-1 audio (cf. ISO/CEI norm 11172-3:1993 (F), pp. 122–128). Input `frame` length should be 512 samples. `min_threshold_subband` returns the minimum of `masking threshold` in each of the 32 sub-bands. `SMR` returns 27 signal-to-mask ratios (in dB); `SMR`(28–32) are not used.

```
frame=input_signal(11001:11512);
[SMR, min_threshold,frame_psd_dBSPL]= ...
    MPEG1_psycho_acoustic_model1(frame);
% NB: the power levels returned by this function assume that a
full-scale signal (in [-1,+1]) corresponds to 96 DB SPL

f = (0:255)/512*44100;
auditory_threshold_dB = 3.64*((f/1000).^-0.8) - ...
    6.5*exp(-0.6.*((f/1000)-3.3).^2) + 0.001*((f/1000).^4);
plot(f, frame_psd_dBSPL, f, min_threshold, 'r', ...
    f, auditory_threshold_dB, 'k');
hold off;
axis([0 22050 -20 100]);
legend('Signal PSD', 'Min. threshold per sub-band',…
       'Absolute threshold');
xlabel('Frequency (Hz)'); ylabel('Magnitude (dB)');
```

Signal-to-mask ratios (SMR) are computed for each band in a very conservative way, as the ratio of the maximum of the signal PSD to the minimum of the masking threshold in each band. Bit allocation is performed by an iterative algorithm, which gives priority to sub-bands with higher SMR. The resulting SNR in each sub-band should be greater than or equal to the SMR so as to push the noise level below the masking threshold (Fig. 3.26, bottom).

MATLAB function involved:

- `function [N_bits,SNR]= MPEG1_bit_allocation(SMR,bit_rate)` implements a simplified bit allocation greedy algorithm. `SMR` is the signal-to-mask ratios in each sub-band, as defined by the MPEG1 psycho-acoustic model. `bit_rate` is in kbps. `N_bits` is the number of bits in each sub-band. `SNR` is the maximum SNR in each sub-band after quantization, i.e., the SNR assuming each sub-band contains a full-range sinusoid. `N_bits` and `SNR` are set to zero for sub-bands 28–32.

```
% Allocating bits for a target bit rate of 192 kbits/s
% (compression ratio = 4)
[N_bits, SNR] = MPEG1_bit_allocation(SMR, 192000);

stairs((0:32)/32*22050,[SMR SMR(32)]);
hold on;
stairs((0:32)/32*22050,[SNR SNR(32)],'--');
axis([0 22050 -20 100]);
legend('SMR', 'SNR');
xlabel('Frequency (Hz)'); ylabel('Magnitude (dB)');
```

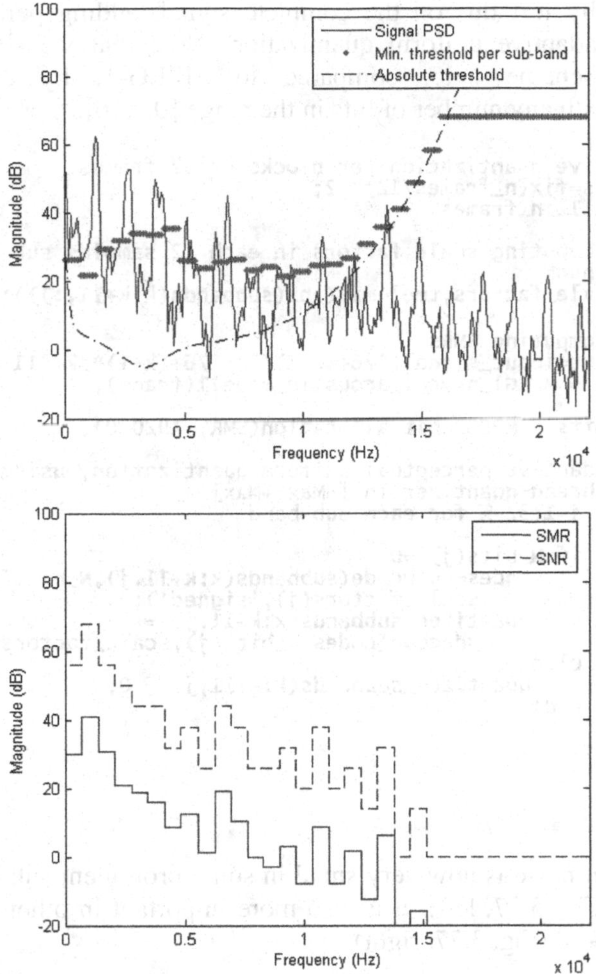

Fig. 3.26 The MPEG-1 psychoacoustic model #1. *Top*: Masking threshold per sub-band; *bottom*: signal-to-mask ratios (SMR) and the maximum (theoretical) SNR obtained after bit allocation. SNR is systematically above SMR so as to make quantization noise inaudible[12].

[12] Note that this is made possible only by the fact that we have allocated enough bits for quantizing frames. When the allowed bit rate decreases, SNR can be lower than SMR.

Let us finally test this on the complete signal, adding perceptual bit allocation to adaptive uniform quantization. Note that we simplify the quantization scheme here, compared to MPEG-1, by considering quantization with any number of bits in the range [0…16].

```
% Adaptive quantization per blocks of 12 frames
n_frames=fix(n_frames/12)*12;
for k=1:12:n_frames

    % Computing scale factors in each 12 samples sub-band
    % chunk
    [scale_factors,tmp]=max(abs(subbands(k:k+11,:)));

    % Computing SMRs 13
    frame=input_signal(176+(k-1)*32:176+(k-1)*32+511);
    SMR = MPEG1_psycho_acoustic_model1(frame);

    N_bits = MPEG1_bit_allocation(SMR, 192000);

    % Adaptive perceptual uniform quantization, using a mid-
    % thread quantizer in [-Max,+Max]
    for j=1:32 % for each sub-band

        if N_bits(j)~=0
            codes= uencode(subbands(k:k+11,j),N_bits(j),...
                scale_factors(j),'signed');
            quantized_subbands(k:k+11,j) = ...
                udecode(codes,N_bits(j),scale_factors(j));
        else
            quantized_subbands(k:k+11,j) = 0;
        end;

    end;

end;
```

Quantization noise is now very small in some prominent sub-bands like sub-band #2 (Fig. 3.27, left). It is also more important in other sub-bands like sub-band #20 (Fig. 3.27, right).

[13] The input frame for the psychoacoustic model is delayed by 176 samples, as it should be centered in the middle of the local block of 12 sub-band samples, and the first sub-band sample corresponds to original sample 256 (actually to sample 512, but the analysis filter introduces a delay of 256 samples). So the center of the first frame should be on the original sample 256+11*32/2=432, and the beginning of this frame falls on sample 11*32/2=176.

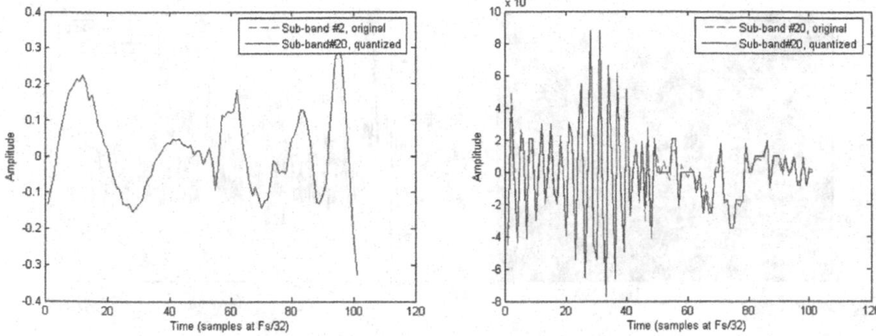

Fig. 3.27 Excerpts from sub-bands #2 and #20

The overall SNR has increased to 36.6 dB. The perceptual SNR is actually much higher, since most of the noise cannot be heard. The resulting signal is high-quality audio (Fig. 3.28).

```
snr_scaled_perceptual=snr(input_signal(512:end-512),...
    output_signal(512:end-512),0)
```

```
snr_scaled_perceptual =   36.6457
```

The price to pay is that the scale factors and the number of bits per sub-band must now be stored every 12 frames (i.e., every 12*32 sub-band samples). In the MPEG-1 norm, scale factors are expressed in decibels and quantized on 6 bits each. This comes from the fact that the ear perceives loudness as the log of the energy and has about 96 dB of hearing dynamics with a sensitivity threshold of about 1 dB. With 6 bits, the quantization step is 96/64 dB and the error lies in [−96/64/2, +96/64/2], which is below the 1 dB threshold. Assuming 4 bits are required for encoding the number of bits used in each of the 27 first sub-bands (sub-bands 28–32 are ignored in MPEG-1), this leads to 27*10=270 bits every 12 frames.[14] It is easy to obtain the number of bits used for the last block of 12 frames in our audio test and the related bit rate.

```
bits_per_block=sum(N_bits)*12+270
bit_rate=bits_per_block*44100/384
```

```
bits_per_block =    1674
bit_rate =   1.9225e+005
```

[14] In practice, MPEG-1 does not even allow all integer values in [1,16] for the number of bits in each band, which makes it possible to quantize it to less than 4 bits. The total number of bits used for bit allocation can then be reduced to 88 instead of 4*27=108.

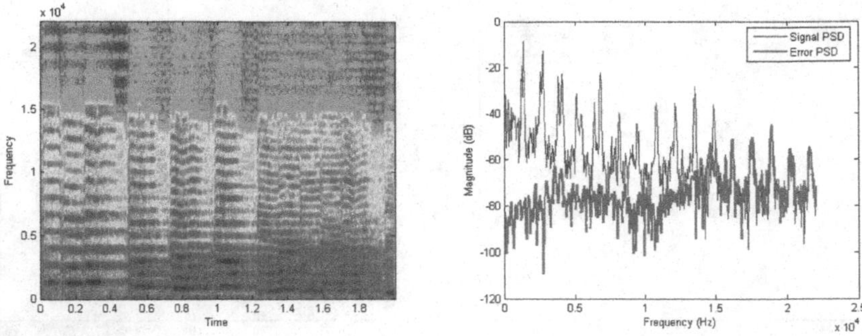

Fig. 3.28 Sub-band coding using adaptive perceptual uniform quantization. *Left*: spectrogram; *right*: periodogram of the original and error signals, around sample 11,000.

3.3 Going further

Unfortunately, MATLAB simulation code for audio coders is not widespread, with the remarkable exception of the toolbox designed by F. Petitcolas at Cambridge University (Petitcolas 2003).

Several other implementations (not in MATLAB) are available from the MPEG homepage (Chariglione 2007) and the MPEG.ORG website (MPEG TV 1998).

3.4 Conclusion

Although a bit tricky to implement correctly, perceptual audio coders are based on simple principles: perfect reconstruction filter banks and sub-band quantization driven by the masking effect.

If audio compression is one of the most prominent applications of sub-band (or transform) coding, it is by far not the only one. As early as in the 1950s, the idea of using the frequency-dependent sensitivity of the human eye to color (the eye is mostly sensitive to luminance, much less to chrominance) led to the definition of the NTSC encoding system in which most of the blue signal is discarded, keeping most of the green and only some of the red. More recently, the JPEG and MPEG/video coding standards were also based on transform coding, using the discrete cosine transform (see Chapters 8 and 9).

References

Chariglione L (2007) The MPEG Home Page [online] Available: http://www.chiariglione.org/mpeg/ [17/04/07]

Crochiere RE, Rabiner LR (1983) Multirate Digital Signal Processing. Englewood Cliffs, NJ: Prentice Hall.

Esteban D, Galand C (1977) Application of quadrature mirror filters to split-band voice coding schemes. Proc. IEEE Int. Conf. Acoustics, Speech and Signal Processing. Hartford, Connecticut, pp 191–195

Johnston JD (1980) A filter family designed for use in quadrature mirror filter banks. Proc. IEEE Int. Conf. Acoustics, Speech and Signal Processing, pp. 291–294

Malvar H (1990) Lapped transforms for efficient transform/subband coding. IEEE Trans. Acoust., Speech, Signal Processing, 38: 969–978

MPEG TV (1998) MPEG pointers and resources [online] Available: http://www.mpeg.org [17/04/07]

Nussbaumer HJ, Vetterli M (1984) Pseudo quadrature mirror filters. Digital Signal Processing, pp 8–12

Painter T, Spanias A (2000) Perceptual coding of digital audio. Proc. IEEE 88-4: 451–512

Pan D (1995) A tutorial on MPEG audio compression. IEEE Multimedia Magazine 2–2:60–74.

Petitcolas N (2003) MPEG for Matlab [online] Available: http://www.petitcolas.net/fabien/software/mpeg/index.html [17/04/07]

Princen J, Bradley A (1986) Analysis/synthesis filter bank design based on time domain aliasing cancellation. IEEE Trans. Acoust. Speech, Signal Processing 34: 1153–1161

Vaidyanathan PP (1993) Multirate Systems and Filter Banks. Englewood Cliffs, NJ: Prentice-Hall

Chapter 4

How does a dictation machine recognize speech?

T. Dutoit (°), L. Couvreur (°), H. Bourlard (*)

(°) Faculté Polytechnique de Mons, Belgium
(*) Ecole Polytechnique Fédérale de Lausanne, Switzerland

This chapter is not about how to wreck a nice beach[1]

There is magic (or is it witchcraft?) in a speech recognizer that transcribes continuous radio speech into text with a word accuracy of even not more than 50%. The extreme difficulty of this task, though, is usually not perceived by the general public. This is because we are almost deaf to the infinite acoustic variations that accompany the production of vocal sounds, which arise not only from physiological constraints (coarticulation) but also from the acoustic environment (additive or convolutional noise, Lombard effect) or from the emotional state of the speaker (voice quality, speaking rate, hesitations, etc.)[2]. Our consciousness of speech is indeed not stimulated until after it has been processed by our brain to make it appear as a sequence of meaningful units: phonemes and words.

In this chapter, we will see how statistical pattern recognition and statistical sequence recognition techniques are currently used for trying to mimic this extraordinary faculty of our mind (section 4.1). We will follow, in Section 4.2, with a MATLAB-based proof of concept of word-based automatic speech recognition (ASR) based on hidden Markov models (HMM), using a bigram model for modeling (syntactic–semantic) language constraints.

---

[1] It is, indeed, about *how to recognize speech.*
[2] Not to mention interspeaker variability or regional dialects.

T. Dutoit, F. Marqués (eds.), *Applied Signal Processing*,
DOI 10.1007/978-0-387-74535-0_4,

4.1 Background – Statistical pattern recognition

Most modern ASR systems have a pipe-line block architecture (Fig. 4.1).

The acoustic wave is first digitized, usually with a sampling frequency of 8 kHz for telephone applications and 16 kHz for multimedia applications. A *speech detection* module then detects segments of speech activity in the digital signal: only those segments that compose the speech signal are transmitted to the following block. The purpose of speech detection is to reduce the computational cost and the probability of ASR error when unexpected acoustic events happen. Doing this automatically, however, is by itself a difficult problem. Speech detection is sometimes implemented manually: the speaker is asked to push a button while speaking in order to activate the ASR system (*push-to-talk* mode).

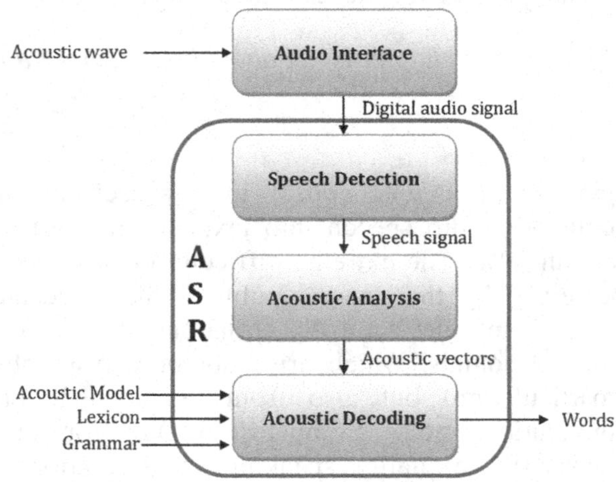

Fig. 4.1 Classical architecture of an automatic speech recognition system

The *acoustic analysis* module processes the speech signal in order to reduce its variability while preserving its linguistic information. A time–frequency analysis is typically performed (using frame-based analysis, with 30 ms frames shifted every 10 ms), which transforms the continuous input waveform into a sequence $X = [x(1), x(2), \ldots, x(N)]$ of *acoustic feature vectors* $x(n)$.[3] The performances of ASR systems (in particular, their *robustness*, i.e., their resistance to noise) are very much dependent on this formatting of the acoustic observations. Various types of feature vectors can be used, such as the LPC coefficients described in Chapter 1, although specific feature vectors, such as the linear prediction cepstral coefficients (LPCC) or the mel frequency cepstral coefficients (MFCC; Picone 1993), have been developed in practice for speech recognition, which are somehow related to LPC coefficients.

The *acoustic decoding* module is the heart of the ASR system. During a training phase, the ASR system is presented with several examples of every possible word, as defined by the *lexicon*. A statistical model (Section 4.1.1) is then computed for every word such that it models the distribution of the acoustic vectors. Repeating the estimation for all the words, we finally obtain a set of statistical models, the so-called *acoustic model*, which is stored in the ASR system. At runtime, the acoustic decoding module searches the sequence of words whose corresponding sequence of models is the "closest" to the observed sequence of acoustic feature vectors. This search is complex since neither the number of words nor their segmentation are known in advance. Efficient decoding algorithms constrain the search for the best sequence of words by a *grammar*, which defines the authorized, or at least the most likely, sequence of words. It is usually described in terms of a statistical model: the *language model*.

In large vocabulary ASR systems, it is hard if not impossible to train separate statistical models for all words (and even to gather the speech data that would be required to properly train a word-based acoustic model). In such systems, words are described as sequences of phonemes in a *pronunciation lexicon*, and statistical modeling is applied to phonemic units. Word-based models are then obtained by concatenating the phoneme-based models. Small vocabulary systems (<50 words), on the contrary, usually consider words as basic acoustic units and therefore do not require a pronunciation lexicon.

4.1.1 The statistical formalism of ASR

The most common statistical formalism of ASR,[4] which we will use throughout this chapter, aims to produce the *most probable* word sequence

---

[3] Although $x(n)$ is a vector, it will not be written with a bold font in this chapter to avoid overloading all equations.

[4] There are numerous textbooks that explain these notions in detail. See, for instance, Gold and Morgan (2000), Bourlard and Morgan (1994) or Bourlard (2007). For a more general introduction to pattern recognition, see also Polikar (2006) or the more complete Duda et al. (2000).

W^*, given the acoustic observation sequence X. This can be expressed mathematically by the so-called *Bayesian* or *maximum a posteriori* (MAP) decision rule as

$$W^* = \arg\max_{W_i} P(W_i \mid X, \Theta) \tag{4.1}$$[5]

where W_i represents the ith possible word sequence and the conditional probability is evaluated over all possible word sequences,[6] and Θ represents the set of parameters used to estimate the probability distribution.

Since each word sequence W_i may be realized as an infinite number of possible acoustic realizations, it is represented by its model $M(W_i)$, also written M_i for the sake of simplicity, which is assumed to be able to produce all such possible acoustic realizations. This yields

$$M^* = \arg\max_{M_i} P(M_i \mid X, \Theta) \tag{4.2}$$

where M^* is (the model of) the sequence of words representing the linguistic message in input speech X, M_i is (the model of) a possible word sequence, W_i, $P(M_i \mid X,\Theta)$ is the posterior probability of (the model of) a word sequence, given the acoustic input X, and the maximum is evaluated over all possible models (i.e., all possible word sequences).

Bayes' rule can be applied to Equation (4.2), yielding

$$P(M_i \mid X, \Theta) = \frac{P(X \mid M_i, \Theta) P(M_i \mid \Theta)}{P(X \mid \Theta)} \tag{4.3}$$

---

[5] In Equation (4.1), W_i and X are not random variables; they are values taken by their respective random variables. As a matter of fact, we will often use in this chapter a shortcut notation for probabilities, when this does not bring confusion. The probability $P(A{=}a|B{=}b)$ that a discrete random variable A takes value a given the fact that random variable B takes value b will simply be written $P(a|b)$. What is more, we will use the same notation when A is a *continuous* random variable for referring to *probability density* $p_{A|B=b}(a)$.

[6] It is assumed here that the number of possible word sequences is finite, which is not true for natural languages. In practice, a specific component of the ASR, the *decoder*, takes care of this problem by restricting the computation of Equation (4.1) for a limited number of most probable sequences.

where $P(X \mid M_i, \Theta)$ represents the contribution of the so-called *acoustic model* (i.e., the likelihood that a specific model M_i has produced the acoustic observation X), $P(M_i \mid \Theta)$ represents the contribution of the so-called *language model* (i.e., the a priori probability of the corresponding word sequence), and $P(X|\Theta)$ stands for the a priori probability of the acoustic observation. For the sake of simplicity (and tractability of the parameter estimation process), state-of-the-art ASR systems always assume independence between the acoustic model parameters, which will now be denoted Θ_A, and the parameters of the language model, which will be denoted Θ_L.

Based on the above, we thus have to address the following problems:

- *Decoding (recognition)*: Given an unknown utterance X, find the most probable word sequence W^* (i.e., the most probable word sequence model M^*) such that

$$M^* = \arg\max_{M_i} \frac{P(X \mid M_i, \Theta_A) P(M_i \mid \Theta_L)}{P(X \mid \Theta_A, \Theta_L)} \tag{4.4}$$

  Since during recognition all parameters Θ_A and Θ_L are frozen, probability $P(X \mid \Theta_A, \Theta_L)$ is *constant for all hypotheses of word sequences* (i.e., for all choices of i) and can thus be ignored so that Equation (4.4) simplifies to

$$M^* = \arg\max_{M_i} P(X \mid M_i, \Theta_A) P(M_i \mid \Theta_L) \tag{4.5}$$

- *Acoustic modeling*: Given (the model of) a word sequence, M_i, estimate the probability $P(X \mid M_i, \Theta_A)$ of the unknown utterance X.

  This is typically carried out using hidden Markov models (HMM; see Section 4.1.3). It requires estimating the acoustic model Θ_A. At training time, a large amount of training utterances X_j (j = 1,... , J) with their associated models M_j are used to estimate the optimal acoustic parameter set Θ_A^* such that

$$\Theta_A^* = \underset{\Theta_A}{\arg\max} \prod_{j=1}^{J} P(X \mid M_i, \Theta_A)$$
$$= \underset{\Theta_A}{\arg\max} \sum_{j=1}^{J} \log(P(X \mid M_i, \Theta_A)) \quad (4.6)$$

which is referred to as the *maximum likelihood* (ML) or *maximum log likelihood criterion.*[7]

- *Language modeling*: The goal of the language model is to estimate prior probabilities of sentence models $P(M_i \mid \Theta_L)$.

  At training time, the language model parameters Θ_L are commonly estimated from large text corpora. The language model is most often formalized as word-based Markov models (see Section 4.1.2), in which case Θ_L is the set of transition probabilities of these chains, also known as *n-grams*.

4.1.2 Markov models

A *Markov model* is the simplest form of a *stochastic finite state automaton* (SFSA). It describes a sequence of observations $X = [x(1), x(2), \dots, x(N)]$ as the output of a finite state automaton (Fig. 4.2) whose internal states $\{q_1, q_2, \dots, q_K\}$ are univocally associated with possible observations $\{x_1, x_2, \dots, x_K\}$ and whose state-to-state transitions are associated with probabilities a given state q_k always outputs the same observation x_k, except initial and final states (q_I and q_F, which output nothing); the *transition probabilities* from any state sum to one. The most important constraint imposed by a (first-order) Markov model is known as follows:
the probability of a state (or that of the associated observation) only depends on the previous state (or that of the associated observation).

---

[7] Although both criteria are equivalent, it is usually more convenient to work with the sum of *log* likelihoods. As a matter of fact, computing products of probabilities (which are often significantly lower than 1) quickly exceeds the floating point arithmetic precision. Even the log of a sum of probabilities can be estimated, when needed, using log likelihoods (i.e., without having to compute likelihoods at any time), using

$$\log(a+b) = \log(a) + \log\left(1 + e^{(\log b - \log a)}\right)$$

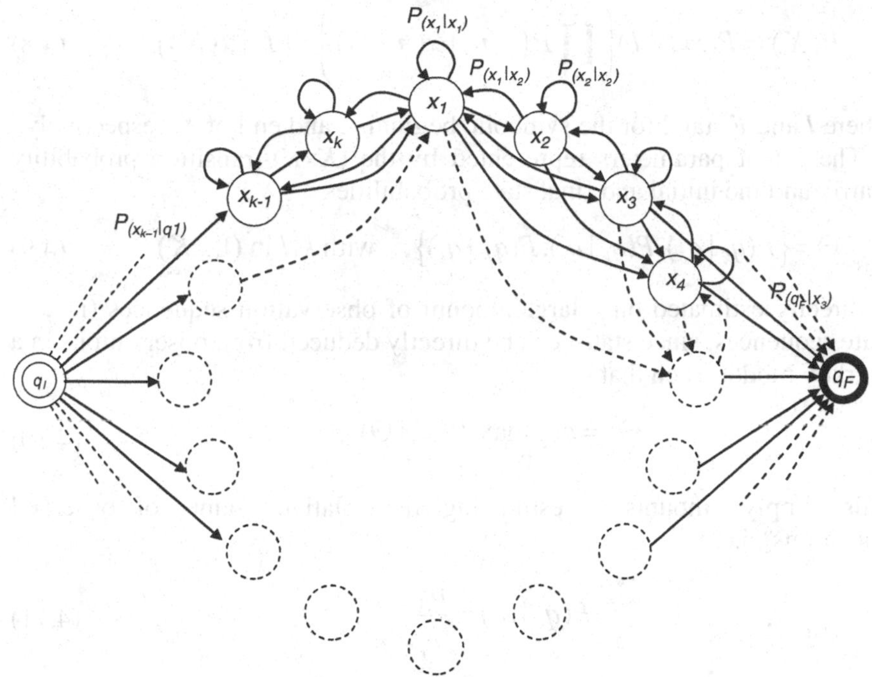

Fig. 4.2 A typical Markov model[8]. The leftmost and rightmost states are the initial and final states 5 respectively. Each internal state q_k in the center of the figure is associated with a specific observation x_k (and is labeled as such). Transition probabilities are associated with arcs (only a few transition probabilities are shown)

The probability of X given such a model reduces to the probability of the sequence of states $[q_I, q(1), q(2), \ldots, q(N), q_F]$ corresponding to X, i.e., to a product of transition probabilities (including transitions from state q_I and transitions to state q_F):

$$P(X) = P(q(1) \mid q_I)\left(\prod_{n=2}^{N} P(q(n) \mid q(n-1))\right) P(q_F \mid q(N)) \qquad (4.7)$$

where $q(n)$ stands for the state associated with observation $x(n)$. Given the one-to-one relationship between states and observations, this can also be written as

[8] The model shown here is *ergodic*: transitions are possible from each state to all other states. In practical applications (such as in ASR, for language modeling), some transitions may be prohibited.

$$P(X) = P(x(1)|I)\left(\prod_{n=2}^{N} P\left(x(n)\,|\,x(n-1)\right)\right)P\left(F\,|\,x(N)\right) \tag{4.8}$$

where I and F stand for the symbolic beginning and end of X, respectively.

The set of parameters represented by the ($K \times K$)-transition probability matrix and the initial and final state probabilities

$$\Theta = \left\{P(q_k\,|\,q_I), P(q_k\,|\,q_l), P(q_F\,|\,q_l)\right\}, \quad \text{with } k,l \text{ in } (1,\ldots K) \tag{4.9}$$

is directly estimated on a large amount of observation sequences (i.e., of state sequences, since states can be directly deduced from observations in a Markov model) such that

$$\Theta^* = \arg\max_{\Theta} P(X\,|\,\Theta) \tag{4.10}$$

This simply amounts to estimating the relative counts of observed transitions[9], i.e.,

$$P(q_k\,|\,q_l) = \frac{n_{lk}}{n_l} \tag{4.11}$$

where n_{lk} stands for the number of times a transition from state q_l to state q_k occurred, while n_l represents the number of times state q_l was visited.

Markov models are intensively used in ASR for language modeling, in the form of *n-grams*, to estimate the probability of a word sequence $W = [w(1), w(2), \ldots, w(L)]$ as

$$P(W) = P\left(w(1)\,|\,I\right)\left(\prod_{l=2}^{L} P\left(w(l)\,|\,w(l-1), w(l-2), \ldots w(l-n+1)\right)\right). \\ P\left(F\,|\,w(L)\right) \tag{4.12}$$ [10]

In particular, *bigrams* further reduce this estimation to

$$P(W) = P\left(w(1)\,|\,I\right)\left(\prod_{l=2}^{L} P\left(w(l)\,|\,w(l-1)\right)\right)P\left(F\,|\,w(L)\right) \tag{4.13}$$

[9] This estimate is possibly smoothed in case there is not enough training data so as to avoid forbidding state sequences not found in the data (those that are rare but not impossible).

[10] In this case, states are not associated with words but rather to sequences of $n-1$ words. Such models are called *Nth* order Markov models.

In this case, each observation is a word from the input word sequence *W*, and each state of the model (except *I* and *F*) is characterized by a single word, which an observed word could possibly follow with a given probability.
As Jelinek (1991) pointed out, "That this simple approach is so successful is a source of considerable irritation to me and to some of my colleagues. We have evidence that better language models are obtainable, we think we know many weaknesses of the trigram model, and yet, when we devise more or less subtle methods of improvement, we come up short."

Markov models cannot be used for acoustic modeling, as the number of possible observations is infinite.

4.1.3 Hidden Markov models

Modifying a Markov model by allowing several states (if not all) to output the same observations with state-dependent *emission probabilities* (Fig. 4.3) turns it into a hidden Markov model (HMM, Rabiner 1989). In such a model, the sequence of states cannot be univocally determined from the sequence of observations (such a SFSA is called *ambiguous*). The HMM is thus called "hidden" because there is an underlying stochastic process (i.e., the sequence of states) that is not observable but affects the sequence of observations.

Although Fig. 4.3 shows a *discrete* HMM, in which the number of possible observations is finite, *continuous* HMMs are also very much used, in which the output space is a continuous variable (often even multivariate). Emission probabilities are then estimated by assuming they follow a particular functional distribution: $P(x_m|q_k)$ is computed analytically (it can no longer be obtained by counting). In order to keep the number of HMM parameters as low as possible, this distribution often takes the classical form of a (multivariate, *d*-dimensional) Gaussian[11]:

$$\begin{aligned} P(x\,|\,q_k) &= N(x,\mu_k,\Sigma_k) \\ &= \frac{1}{(2\pi)^{d/2}\left|\Sigma_k\right|^{1/2}}\exp\left(-\frac{1}{2}(x-\mu_k)^T\Sigma_k^{-1}(x-\mu_k)\right) \end{aligned} \tag{4.14}$$

where μ_k and Σ_k denote the mean vector and the covariance matrix associated with state q_k, respectively. When this model is not accurate

[11] Gaussian PDFs have many practical advantages; they are entirely defined by their first two moments and are linear once derivated.

enough, mixtures of (multivariate) Gaussians (*Gaussian mixture model*, GMM) are also used, which allow for multiple modes[12]:

$$P(x \mid q_k) = \sum_{g=1}^{G} c_{kg} N(x, \mu_{kg}, \Sigma_{kg}) \tag{4.15}$$

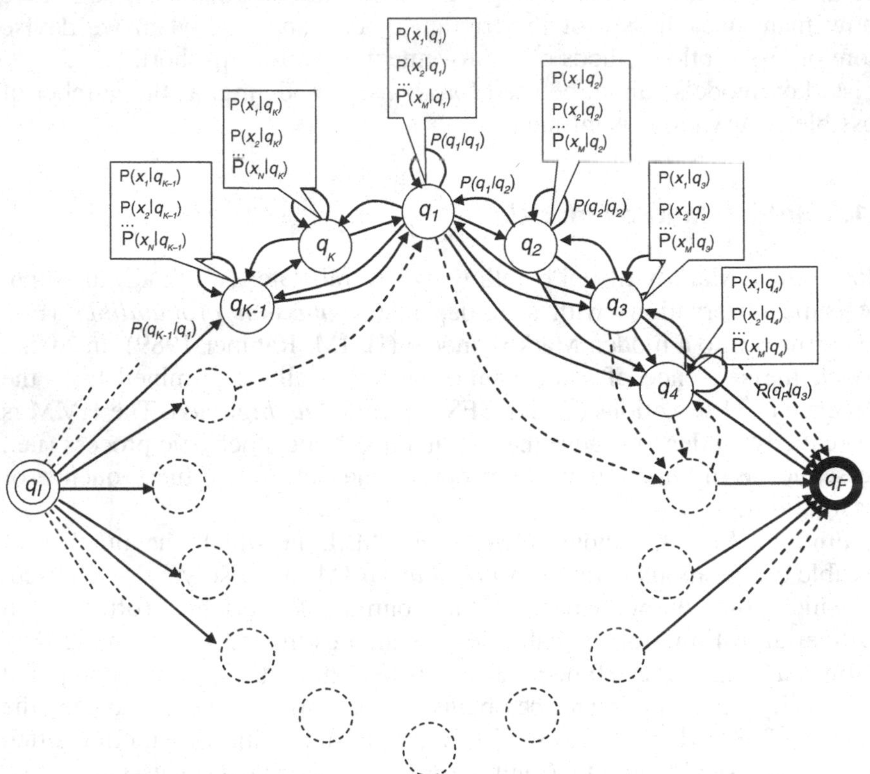

Fig. 4.3 A typical (discrete) hidden Markov model. The leftmost and rightmost states are the initial and final states, respectively. Each state q_k in the center of the figure is associated with several possible observations (here, to all observations $\{x_1, x_2, \ldots, x_M\}$) with the corresponding emission probability. Transition probabilities are associated with arcs (only a few transition probabilities are shown). The HMM is termed as *discrete* because the number of possible observations is finite.

[12] It is also possible (and has proved very efficient in ASR) to use artificial neural networks (ANN) to estimate emission probabilities (Bourlard and Wellekens 1990, Bourlard and Morgan 1994). We do not examine this option here.

where G is the total number of Gaussian densities and c_{kg} are the mixture gain coefficients (thus representing the prior probabilities of Gaussian mixture components). These gains must verify the following constraint:

$$\sum_{g=1}^{G} c_{kg} = 1 \quad \forall k = 1,...,K \tag{4.16}$$

Assuming that the total number of states K is fixed, the set of parameters Θ of the model comprises all the Gaussian means and variances, gains, and transition probabilities.

Two approaches can be used for estimating $P(X \mid M, \Theta)$.

In the *full likelihood approach*, this probability is computed as a sum on all possible paths of length N. The probability of each path is itself computed as in Equation (4.7)

$$P(X \mid M,\Theta) = \sum_{paths\ j} P\big(q_j(1) \mid q_I\big) P\big(x(1) \mid q_j(1)\big). \tag{4.17}$$

$$\left(\prod_{n=2}^{N} P\big(q_j(n) \mid q_j(n-1)\big) P\big(x(n) \mid q_j(n)\big) \right) P\big(q_F \mid q_j(N)\big)$$

where $q_j(n)$ stands for the state in $\{q_1, q_2, \ldots, q_K\}$, which is associated with $x(n)$ in path j. In practice, estimating the likelihood according to Equation (4.17) involves a very large number of computations, namely $O(NK^N)$, which can be avoided by the so-called *forward recurrence formula* with a lower complexity, namely $O(K^2N)$. This formula is based on the recursive estimation of an intermediate variable $\alpha_n(l)$:

$$\alpha_n(l) = P\big(x(1), x(2), ..., x(n), q(n) = q_l\big) \tag{4.18}$$

$\alpha_n(l)$ stands for the probability that a partial sequence $[x(1), x(2), \ldots, x(n)]$ is produced by the model in such a way that $x(n)$ is produced by state q_l. It can be obtained by using (Fig. 4.4) the following equation:

$$\begin{aligned}
&\alpha_1(l) = P(x(1) \mid q_l) P\big(q_l \mid q_I\big) \quad (l = 1,...,K) \\
&for\ n = 2,...,N\ (and\ l = 1,...,K) \\
&\quad \alpha_n(l) = P(x(n) \mid q_l) \sum_{k=1}^{K} \alpha_{n-1}(k) P\big(q_l \mid q_k\big) \\
&P(X \mid M,\Theta) = \alpha_{N+1}(F) = \sum_{k=1}^{K} \alpha_N(k) P\big(q_F \mid q_k\big)
\end{aligned} \tag{4.19}$$

In the *Viterbi approximation approach,* the estimation of the data likelihood is restricted to the most probable path of length N generating the sequence X

$$P(X \mid M,\Theta) \cong \max_{paths\ j} P\big(q_j(1) \mid I\big)\, P\big(x_1 \mid q_j(1)\big).$$
$$\left(\prod_{n=2}^{N} P\big(q_j(n) \mid q_j(n-1)\big) P\big(x_n \mid q_j(n)\big)\right) P\big(q_F \mid q_j(N)\big) \quad (4.20)$$

and the sums in Equation (4.19) are replaced by the max operator. Note that it is also easy to memorize the most probable path given some input sequence by using Equation (4.19) and additionally keeping in memory, for each $n=(1,\ldots,N)$ and for each $l=(1,\ldots,K)$, the value of k producing the highest term of $\alpha_{n+1}(l)$ in Equation (4.19). Starting from the final state (i.e., the one leading to the highest term for $\alpha_{N+1}(F)$), it is then easy to trace back the best path, thereby associating one "best" state to each feature vector.

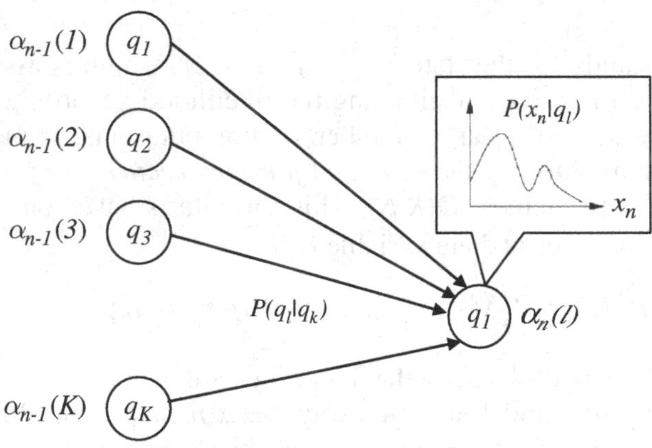

Fig. 4.4 Illustration of the sequence of operations required to compute the intermediate variable $\alpha_n(l)$

HMMs are intensively used in ASR acoustic models where every sentence model M_i is represented as a HMM. Since such a representation is not tractable due to the infinite number of possible sentences, sentence HMMs are obtained by compositing subsentence HMMs such as word HMMs, syllable HMMs, or more generally phoneme HMMs. Words, syllables, or phonemes are then generally described using a specific HMM topology (i.e., allowed state connectivity) known as *left-to-right* HMMs

(Fig. 4.5), as opposed to the general ergodic topology shown in Fig. 4.3. Although sequential signals, such as speech, are nonstationary processes, left-to-right HMMs assume that the sequence of observation vectors is a piecewise stationary process. That is, a sequence $X = [x(1), x(2), \ldots, x(N)]$ is modeled as a sequence of discrete stationary states with instantaneous transitions between these states.

4.1.4 Training HMMs

HMM training is classically based on the maximum likelihood criterion; the goal is to estimate the parameters of the model that maximize the likelihood of a large number of training sequences X_j ($j = 1,\ldots, J$). For Gaussian HMMs (which we will examine here, as they are used in most ASR systems), the set of parameters to estimate comprises all the Gaussian means and variances, gains (if GMMs are used), and transition probabilities.

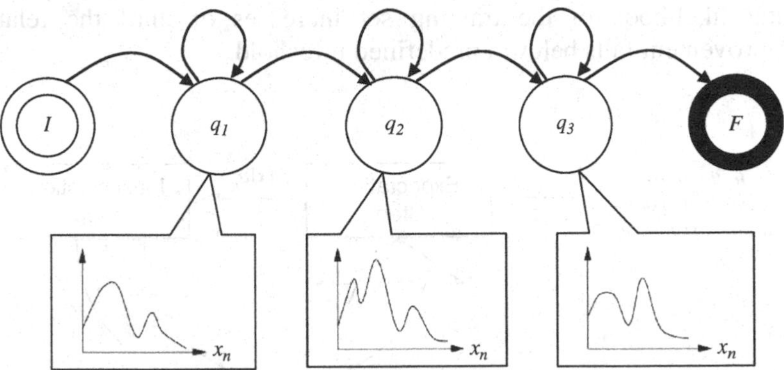

Fig. 4.5 A left-to-right continuous HMM, shown here with (univariate) continuous emission probabilities (which look like mixtures of Gaussians). In speech recognition, this could be the model of a word or of a phoneme, which is assumed to be composed of three stationary parts

Training algorithms

A solution to this problem is a particular case of the *expectation–maximization* (EM) algorithm (Moon 1996). Again, two approaches are possible.

In the *Viterbi approach* (Fig. 4.6), the following steps are taken:

1. Start from an initial set of parameters $\Theta^{(0)}$. With a left-to-right topology, one way of obtaining such a set is by estimating the parameters from a linear segmentation of feature vector sequences,

i.e., by assuming that each training sequence X_j ($j = 1,\dots, J$) is produced by visiting each state of its associated model M_j the same amount of times. Then apply the expectation step to this initial linear segmentation.

2. (*Expectation* step) Compute transition probabilities as in Equation (4.11). Obtain emission probabilities for state k by estimating the Gaussian parameters in Equation (4.14) or the GMM parameters in Equation (4.15) and (4.16) from all feature vectors associated with state k in the training sequences (see below).
3. (*Maximization* step) For all training utterances X_j and their associated models M_j, find the maximum likelihood paths ("best" paths), maximizing $P(X_j|M_j)$ using the Viterbi recursion, thus yielding a new segmentation of the training data. This step is often referred to as "forced alignment," since we are forcing the matching of utterances X_j on given models M_j.
4. Given this new segmentation, collect all the vectors (over all utterances X_j) associated with states q_k and reestimate emission and transition probabilities as in the expectation step. Iterate as long as the total likelihood of the training set increases or until the relative improvement falls below a predefined threshold.

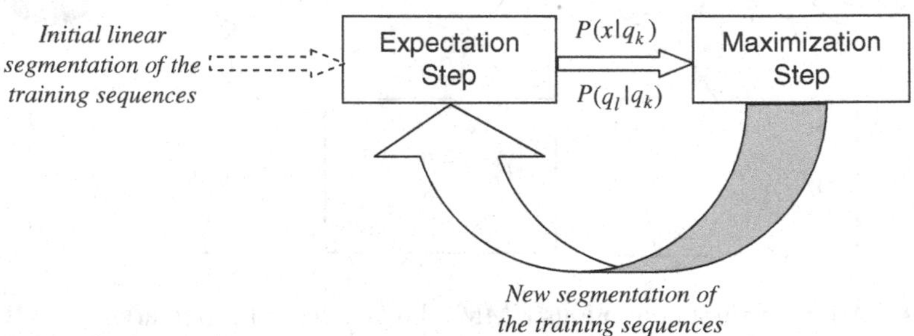

Fig. 4.6 The *expectation–maximization* (EM) algorithm using the Viterbi approach

In the *forward–backward* or *Baum–Welch approach*, all paths are considered. Feature vectors are thus no longer univocally associated with states when reestimating the emission and transitions probabilities: each of them counts for some weight in the reestimation of the whole set of parameters.

The convergence of the iterative processes involved in both approaches can be proved to converge to a local optimum (whose quality will depend on the quality of the initialization).

Estimating emission probabilities

In the Viterbi approach, one needs to estimate the emission probabilities of each state q_k, given a number of feature vectors $\{x_{1k}, x_{2k}, \ldots, x_{Mk}\}$ associated with it. The same problem is encountered in the Baum–Welch approach, with feature vector partially associated with each state. We will explore the Viterbi case here, as it is easier to follow.[13]

When a multivariate Gaussian distribution $N(\mu_k, \Sigma_k)$ is assumed for some state q_k, the classical estimation formulas for the mean and covariance matrix, given samples x_{ik} stored as column vectors, are the following:

$$\tilde{\mu}_k = \frac{1}{M}\sum_{i=1}^{M} x_{ik}$$
$$\tilde{\Sigma}_k = \frac{1}{M-1}\sum_{i=1}^{M}(x_{ik} - \tilde{\mu}_k)(x_{ik} - \tilde{\mu}_k)^T \tag{4.21}$$

It is easy to show that $\tilde{\mu}_k$ is the maximum likelihood estimate of μ_k. The ML estimator of Σ_k, though, is not exactly the one given by Equation (4.21); the ML estimator normalizes by M instead of $(M-1)$. However, it is shown to be biased when the exact value of μ_k is not known, while Equation (4.21) is unbiased.

When a multivariate GMM distribution is assumed for some state q_k, estimating its weights c_{kg}, means μ_{kg}, and covariance matrices Σ_{kg} for $g=1, \ldots, G$ as defined in Equation (4.15) cannot be done analytically. The EM algorithm is used again for obtaining the maximum likelihood estimate of the parameters, although in a more straightforward way than above (there is not such a thing as transition probabilities in this problem). As before, two approaches are possible: the *Viterbi–EM approach*, in which each feature vector is associated with one of the underlying Gaussians, and the *EM approach*, in which each vector is associated with all Gaussians, with some weight (for a tutorial on the EM algorithm, see Moon 1996, Bilmes 1998).

The Viterbi–EM and EM algorithms are very sensitive to the initial values chosen for their parameters. In order to maximize their chances to converge to a global maximum of the likelihood of the training data, the *K-means* algorithm is sometimes used for providing a first estimate of the

[13] Details on the Baum–Welch algorithm can be found in Bourlard (2007).

parameters. Starting from an initial set of G prototype vectors, this algorithm iterates on the following steps:

1. For each feature vector x_{ik} (i=1, …,M), compute the squared Euclidian distance from the kth prototype and assign x_{ik} to its closest prototype.
2. Replace each prototype with the mean of the feature vectors assigned to it in step 1.

Iterations are stopped when no further assignment changes occur.

4.2 MATLAB proof of concept: ASP_dictation_machine.m

Although speech is in essence a nonstationary signal, and therefore calls for dynamic modeling, it is convenient to start this script by examining static modeling and classification of signals, seen as a statistical pattern recognition problem. We do this by using Gaussian multivariate models in Section 4.2.1 and extending it to Gaussian mixture models (GMM) in Section 4.2.2. We then examine, in Section 4.2.3, the more general dynamic modeling using Hidden markov models (HMM) for isolated word classification. We follow in Section 4.2.4 by adding a simple bigram-based language model, implemented as a Markov model, to obtain a connected word classification system. We end the chapter in Section 4.2.5 by implementing a word-based speech recognition system,[14] in which the system does not know in advance how many words each utterance contains.

4.2.1 Gaussian modeling and Bayesian classification of vowels

We will examine here how Gaussian multivariate models can be used for the classification of signals.

A good example is that of the classification of sustained vowels, i.e., of the classification of incoming acoustic feature vectors into the corresponding phonemic classes. Acoustic feature vectors are generally highly multidimensional (as we shall see later), but we will work in a 2D space so as to be able to plot our results.

In this chapter, we will work on a hypothetic language, whose phoneme set is composed of only four vowels {/a/, /e/, /i/, /u/} and whose lexicon reduces to {"why" /uai/, "you" /iu/, "we" /ui/, "are" /ae/, "hear" /ie/, "here" /ie/}.

[14] Note that we will not use the words *classification* and *recognition* indifferently. Recognition is indeed more complex than classification, as it involves the additional task of segmenting an input stream into segments for further classification.

Every speech frame can then be represented as a 2D vector of speech features in the form of pairs of formant values (the first and the second spectral formants, F_1 and F_2; see Chapter 1, Section 1.1).

Our first task will be to classify vowels by using Gaussian probability density functions (PDF) for class models and Bayesian (MAP) decision. Let us load a database of features extracted from the vowels and words of this language.[15] Vowel samples are grouped in matrices of size $N \times 2$, where each of the N rows is a training example and each example is characterized by a formant frequency pair $[F_1, F_2]$. Supposing that the whole database adequately covers our imaginary language, it is easy to compute the prior probability $P(q_k)$ of each class q_k (q_k in {/a/,/e/,/i/,/u/}). The most common phoneme in our hypothetic language is /e/.

```
load data; % vowels={a,e,i,u};

N_samples=0;
for j=1:4
    N_samples = N_samples+size(vowels{j}.training,1);
end;
for j=1:4
    prior(j) = size(vowels{j}.training,1)/N_samples;
end;
prior
```

prior = 0.1500 0.4000 0.1500 0.3000

As can be seen in Fig. 4.7 (left), our four vowel classes have serious overlap in the 2D vector space.

```
plot(vowels{1}.training(:,1),vowels{1}.training(:,2),'k+');
hold on;
plot(vowels{2}.training(:,1),vowels{2}.training(:,2),'r*');
plot(vowels{3}.training(:,1),vowels{3}.training(:,2),'gp');
plot(vowels{4}.training(:,1),vowels{4}.training(:,2),'bs');
```

Let us now assume that we are asked to identify an unknown vowel from its (F_1, F_2) features. One way of solving this problem is by performing multivariate Gaussian modeling of each class, i.e., finding the mean and covariance matrices of the data in each class.

---

[15] These samples were actually generated from statistical models of the vowels, which we chose for tutorial purposes. See Appendix 1 in the **ASP_dictation_machine.m** script, and the **gendata.m** file.

MATLAB function involved:

- `plot_2Dgauss_pdf(mu,sigma)` plots the mean and standard deviation ellipsis of the 2D Gaussian process that has mean `mu` and covariance matrix `sigma` in a 2D plot.

```
for j=1:4
    mu{j}=mean(vowels{j}.training)';
    sigma{j}=cov(vowels{j}.training);
    plot_gauss2D_pdf(mu{j},sigma{j})
end;
```

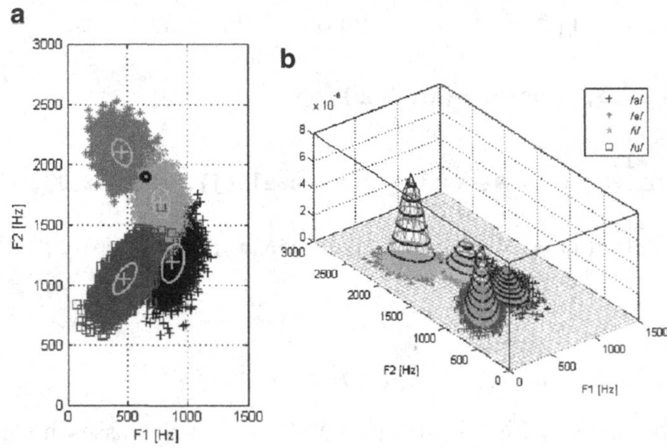

Fig. 4.7 *Left*: Samples of the four vowels {/a/, /e/, /i/, /u/} of our imaginary language in the (F_1,F_2) plane superimposed with the standard deviation ellipsis of their 2D Gaussian model. *Right*: 2D Gaussian estimates of the PDF of these vowels in the (F_1,F_2) plane

Figure 4.7 shows that /i/, for instance, has its mean F_1 at 780 Hz and its mean F_2 at 1,680 Hz.[16] The covariance matrix for the /i/ class is almost diagonal (the scatter plot for the class has its principal axes almost parallel to the coordinate axes, which implies that F_1 and F_2 are almost uncorrelated; see Appendix 1 of the **`ASP_dictation_machine.m`** file). Its diagonal elements are thus close to the square of the length of the half-major and half-minor axes of the standard deviation ellipsis: 76 and 130 Hz, respectively.

[16] These values are the ones fixed in our imaginary language; they do not correspond to those of English vowels at all.

```
mu{3}
sqrtm(sigma{3})
```

```
ans =   1.0e+003 *
    0.7814    1.6827

ans =
   75.3491    -4.4051
   -4.4051   125.5608
```

Let us estimate the likelihood of a test feature vector, given the Gaussian model of class /e/, using the classical Gaussian PDF formula. The feature vector is shown as a black dot in Fig. 4.7.

```
sample=[650 1903];
x = sample-mu{2};
likelihood = exp(-0.5* x* inv(sigma{2}) *x') / sqrt((2*pi)^2 ...
    * det(sigma{2}))

plot(sample(1),sample(2),'ko','linewidth',3)
```

```
likelihood =   7.1333e-007
```

The likelihood of this vector is higher in class /i/ than in any other class (this is also intuitively obvious from the scatter plots shown previously), as shown below.

MATLAB function involved:

- `gauss_pdf(x,mu,sigma)` returns the likelihood of sample x (N×D) with respect to a Gaussian process with mean `mu` (1×D) and covariance `sigma` (D×D). When a set of samples is provided as input, a set of likelihoods is returned.

```
for j=1:4
    likelihood(j) = gauss_pdf(sample,mu{j},sigma{j});
end;
likelihood
```

```
likelihood =   1.0e-005 *

    0.0000    0.0713    0.1032    0.0000
```

Likelihood values are generally very small. Since we will use products of them in the next paragraphs, we will systematically prefer their log likelihood estimates.

```
log(likelihood)
```

```
ans =   -29.3766  -14.1533  -13.7837  -36.9803
```

Since not all phonemes have the same prior probability, Bayesian (MAP) classification of our test sample is not equivalent to finding the class with maximum likelihood. Posterior probabilities *P(class|sample)* must be estimated by multiplying the likelihood of the sample by the prior of each class and dividing by the marginal likelihood of the sample (obtained by summing its likelihood for all classes). Again, for convenience, we compute the log of posterior probabilities. The result is that our sample gets classified as /e/ rather than as /i/, because the prior probability of /e/ is much higher than that of /i/ in our imaginary language.

```
marginal=sum(likelihood); % is a constant
log_posterior=log(likelihood)+log(prior)-log(marginal)
```

```
log_posterior =  -18.0153   -1.8112   -2.4224  -24.9258
```

Note that the marginal likelihood of the sample is not required for classifying it, as it is a subtractive constant for all log-posterior probabilities. We will not compute it in the sequel.

Multiplying likelihoods by priors can be seen as a weighting, which accounts for the intrinsic frequency of occurrence of each class. Plotting the posterior probability of classes in the (F_1, F_2) plane gives a rough idea of how classes are delimited (Fig. 4.7, right).

MATLAB function involved:

- `mesh_2Dgauss_pdf(mu,sigma,prior,gridx,gridy,ratioz)` plots the PDF of a 2D-Gaussian PDF in a 3D plot. `mu` (1×2) is the mean of the density, `sigma` (2×2) is the covariance matrix of the density. `prior` is a scalar used as a multiplicative factor on the value of the PSD. `gridx` and `gridy` must be vectors of the type (x:y:z). `ratioz` is the (scalar) aspect ratio on the Z-axis.

```
hold on;
for j=1:4
    mesh_gauss2D_pdf(mu{j},sigma{j},prior(j),0:50:1500, ...
        0:50:3000, 7e-9);
    hold on;
end;
```

One can easily compare the performance of max likelihood vs. max posterior classifiers on test data sets taken from our four vowels (and having the same prior distribution as from the training set). The error rate is smaller for Bayesian classification: 2.4 vs. 2.2%.

MATLAB function involved:

- `gauss_classify(x,mus,sigmas,priors)` returns the class of the point `x` (1×D) with respect to Gaussian classes using Bayesian classification. `mus` is a cell array of the (1×D) means, `sigmas` is a cell array of the (D×D) covariance matrices. `priors` is a vector of Gaussian priors. When a set of points (NxD) is provided as input, a set of classes is returned.

```
total=0;
errors_likelihood=0;
errors_bayesian=0;
for i=1:4
    n_test=size(vowels{i}.test,1);
    class_likelihood=gauss_classify(vowels{i}.test,mu,...
        sigma,[1 1 1 1]);
    errors_likelihood=errors_likelihood...
        +sum(class_likelihood'~=i);
    class_bayesian=gauss_classify(vowels{i}.test,mu,...
        sigma,prior);
    errors_bayesian=errors_bayesian...
        +sum(class_bayesian'~=i);
    total=total+n_test;
end;
likelihood_error_rate=errors_likelihood/total
bayesian_error_rate=errors_bayesian/total
```

```
likelihood_error_rate = 0.0240
bayesian_error_rate   = 0.0220
```

4.2.2 Gaussian Mixture Models (GMM)

In the previous section, we have seen that Bayesian classification is based on the estimation of class PDFs. Up to now, we have modeled the PDF for each class /a/, /e/, /i/, /u/ as a Gaussian multivariate (one per class). This implicitly assumes that the feature vectors in each class have a (unimodal) normal distribution, as we used the `mean` and `cov` functions, which return the estimates of the mean and covariance matrix of supposedly Gaussian multivariate data samples. It turns out that the vowel data we used had actually been sampled according to Gaussian distributions so that this hypothesis was satisfied.

Let us now try to classify the words of our imaginary language using the same kind of approach as above. We will use 100 samples of the six words {"why" /uai/, "you" /iu/, "we" /ui/, "are" /ae/, "hear" /ie/, "here" /ie/}[17] in our imaginary language, for which each speech frame is again characterized by an $[F_1, F_2]$ feature vector. These samples (Fig. 4.8) are stored in the `words` variable.

[17] Again, the phonetic transcriptions of these words are not those of English (while they remain easy to remember for tutorial purposes).

```
for i=1:6
    subplot(2,3,i)
    plot(words{i}.training_all(:,1),…
        words{i}.training_all(:,2),'+');
    title(words{i}.word);
hold on;
end;
```

Note that "you" and "we" have the same statistical distribution because of their phonemic content (in our imaginary language): /iu/ and /ui/. Also note that "hear" and "here" also have the same distribution, because they have exactly the same phonemic transcription: /ie/. We will come back to this later.

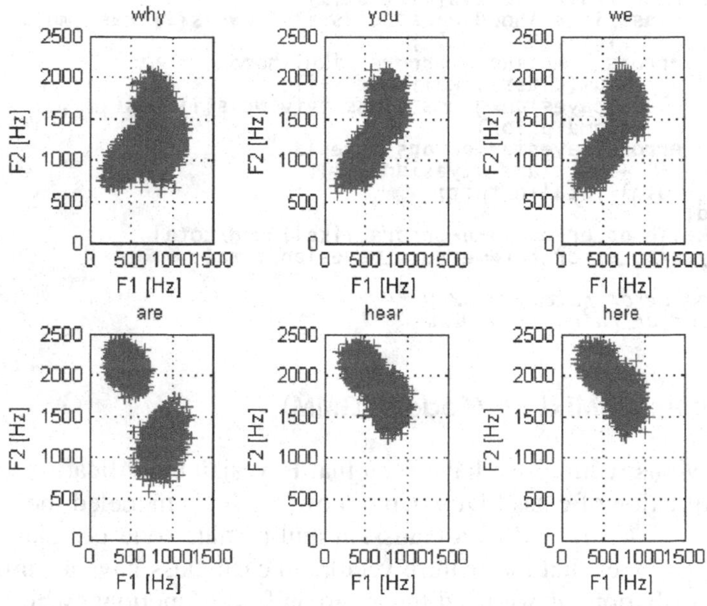

Fig. 4.8 Samples of the six words {"why" /uai/, "you" /iu/, "we" /ui/, "are" /ae/, "hear" /ie/, "here" /ie/} of our imaginary language in the (F_1,F_2) plane

We are now facing a PDF estimation problem; the PDF of the data in each class is no longer Gaussian. This is typical of practical ASR systems; in word-based ASR, each class accounts for the realization of several phonemes and is thus better described as a multimodal distribution, i.e., a distribution with several maxima. The same holds for phoneme-based ASR as well. As a matter of fact, speech is very much submitted to coarticulation, which often results in several modes for the acoustic realization of each phoneme, as a function of the phonetic context in which it appears.

If we apply a unimodal Gaussian model to word "why," for instance, we get a gross estimation of the PDF (Fig. 4.9, left). This estimation does not correctly account for the fact that several areas in the (F_1, F_2) plane are more densely crowded. The maximum value of the Gaussian PDF is very low, since it spans more of the (F_1, F_2) space than it should (and the integral is constrained to one).

```
training_set=words{1}.training_all;
test_set=words{1}.test_all;
mu_all=mean(training_set);
sigma_all=cov(training_set);

plot(training_set(:,1),training_set(:,2),'+');
hold on;
mesh_gauss2D_pdf(mu_all,sigma_all,...
        1, 0:50:1500, 0:50:2500,7e-9);
```

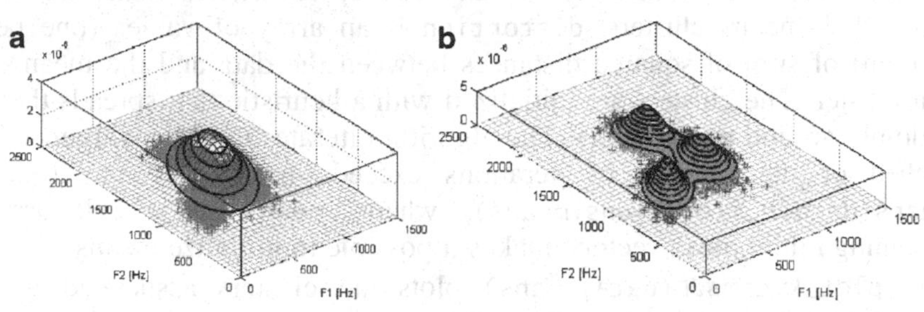

Fig. 4.9 *Left*: 2D Gaussian estimation of the PDF of the word "why" in the (F_1,F_2) plane. *Right*: the same PDF estimated from a GMM estimate (using a mixture of three Gaussians)

The total log likelihoods of the training and test data given this Gaussian model are obtained as sum of the log likelihoods of all feature vectors.

```
log_likelihood_training=...
   sum(log(gauss_pdf(training_set,mu_all,sigma_all)))
log_likelihood_test=...
   sum(log(gauss_pdf(test_set,mu_all,sigma_all)))
```

```
log_likelihood_training =   -7.4102e+004
log_likelihood_test = -8.3021e+004
```

One way of estimating a multimodal PDF is by clustering data and then estimating a unimodal PDF in each cluster. An efficient way to do this (using a limited number of clusters) is by using *K*-means clustering. Starting with *k* prototype vectors or *centroids,* this algorithm first associates each feature vector in the training set to its closest centroid. It then replaces every centroid by the mean of all feature vectors that have been associated with it. The algorithm iterates by reassociating each feature vector to one of the newly found centroids and so on until no further change occurs.

MATLAB function involved:

- `[new_means,new_covs,new_priors,distortion]= ... kmeans(data,n_iterations,n_clusters)` , where `data` is the matrix of observations (one observation per row) and `n_clusters` is the desired number of clusters, returns the mean vectors, covariance matrices, and priors of *K*-means clusters. `distortion` is an array of values (one per iteration) of sum of squared distances between the data and the mean of their cluster. The clusters are initialized with a heuristic that spreads them randomly around `mean(data)`. The algorithm iterates until convergence is reached or the number of iterations exceeds `n_iterations`. Using `kmeans(data,n_iterations,means)`, where `means` is a cell array containing initial mean vectors, makes it possible to initialize means.
- `plot_kmeans2D(data,means)` plots the clusters associated with `means` in `data` samples using a Euclidian distance.

```
% Initializing prototypes "randomly" around the mean
initial_means{1} = [0,1] * sqrtm(sigma_all) + mu_all;
initial_means{2} = [0,0] * sqrtm(sigma_all) + mu_all;
initial_means{3} = [1,2] * sqrtm(sigma_all) + mu_all;

[k_means,k_covs,k_priors,totalDist]=kmeans(training_set,...
   1000,initial_means);

plot_kmeans2D(training_set, k_means);
```

The *K*-means algorithm converges monotonically, in 14 iterations, to a (local) minimum of the global distortion defined as the sum of all distances between feature vectors and their associated centroids (Fig. 4.10, left).

```
plot(totalDist,'.-');
xlabel('Iteration'); ylabel('Global LS criterion'); grid on;
```

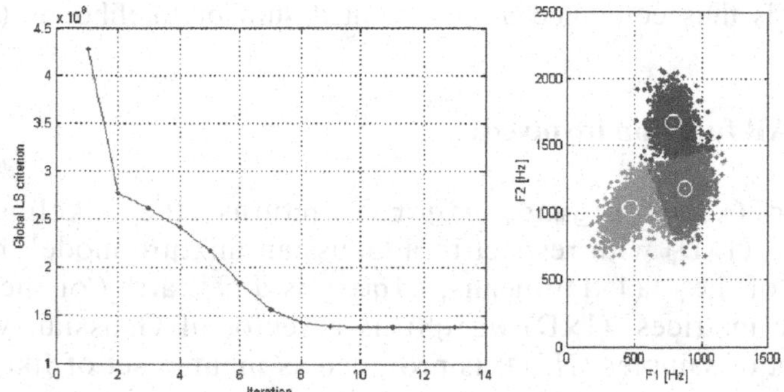

Fig. 4.10 Applying the *K*-means algorithm (with *k*=3) to the sample feature vectors for word "why." *Left*: evolution of the total distortion; *right*: final clusters

The resulting subclasses, though, do not strictly correspond to the phonemes of "why" (Fig. 4.10, right). This is because the global criterion that is minimized by the algorithm is purely geometric. It would actually be very astonishing in these conditions to find the initial vowel subclasses. This is not a problem, as what we are trying to do is to estimate the PDF of the data, not to classify it into "meaningful" subclasses. Once clusters have been created, it is easy to compute the corresponding (supposedly unimodal) Gaussian means and covariance matrices for each cluster (this is actually done inside our `kmeans` function), and to plot the sum of their PDFs, weighted by their priors. This produces an estimate of the PDF of our speech unit (Fig. 4.9, right).

MATLAB function involved:

- `mesh_GMM2D_pdf(mus,sigmas,weights,gridx,gridy,ratioz)` plots the PDF of a 2D Gaussian mixture model PDF in a 3D plot. `mus` is a cell array of the (1×2) means, `sigmas` is a cell array of the (2×2) covariance matrices. `weights` is a vector of Gaussian weights. `gridx` and `gridy` must be vectors of the type (x:y:z). `ratioz` is the (scalar) aspect ratio on the Z-axis.

```
plot(training_set(:,1),training_set(:,2),'+');
hold on;
mesh_GMM2D_pdf(k_means,k_covs,k_priors, ...
     0:50:1500, 0:50:2500,2e-8);
hold off;
```

The total log likelihoods of the training and test data are obtained as above, except we now consider that each feature vector "belongs" to each cluster with some weight equal to the prior probability of the cluster. Its likelihood is thus computed as a weighted sum of likelihoods (one per Gaussian).

MATLAB function involved:

- `GMM_pdf(x,mus,sigmas,weights)` returns the likelihood of sample `x` (1×D) with respect to a Gaussian mixture model. `mus` is a cell array of the (1×D) means, `sigmas` is a cell array of the (D×D) covariance matrices. (1×D) `weight` is a vector of Gaussian weights. When a set of samples (N×D) is provided as input, a set of likelihoods is returned.

```
log_likelihood_training=...
   sum(log(GMM_pdf(training_set,k_means,k_covs,k_priors)))
log_likelihood_test=...
   sum(log(GMM_pdf(test_set,k_means, k_covs, k_priors)))
```

```
log_likelihood_training =
 -7.1310e+004

log_likelihood_test =
 -7.9917e+004
```

The *K*-means approach used above is not optimal in the sense that it is based on a purely geometric convergence criterion. The central algorithm for training GMMs is based on the EM (expectation–maximization) algorithm. As opposed to *K*-means, EM truly maximizes the likelihood of the data, given the GMM parameters (means, covariance matrices, and weights). Starting with *k* initial unimodal Gaussians (one for each subclass), it first estimates, for each feature vector, the probability of each subclass, given that vector. This is the estimation step, which is based on soft classification; each feature vector belongs to all sub-classes, with some weights. In the maximization step, the mean and covariance of each subclass is updated using all feature vectors and taking those weights into account. The algorithm iterates on the E and M steps until the total likelihood increase for the training data falls under some threshold.

The final estimate obtained by EM, however, corresponds only to a local maximum of the total likelihood of the data, whose value may be very sensitive to the initial unimodal Gaussian estimates provided as input. A frequently used value for these initial estimates is precisely the one provided by the *K*-means algorithm.

Applied to the sample feature vectors of "why," the EM algorithm converges monotonically, in seven steps, from the *K*-means solution to a (local) maximum of the total likelihood of the sample data (Fig. 4.11).

MATLAB function involved:

- `[new_means,new_sigmas,new_priors,total_loglike]= ...`
`GMM_train(data,n_iterations,n_gaussians)`, where `data` is the matrix of observations (one observation per row) and `n_gaussians` is the desired number of clusters, returns the mean vectors, covariance matrices, and priors of GMM Gaussian components. `total_loglike` is an array of values (one per iteration) of the total likelihood of the data, given the GMM model. GMMs are initialized with a heuristic that spreads them randomly around `mean(data)`. The algorithm iterates until convergence is reached or the number of iterations exceeds `n_iterations`.
`GMM_train(data,n_iterations,means,covs,priors)` makes it possible to initialize the means, covariance matrices, and priors of the GMM components.
- `plot_GMM2D(data, means, covs)` shows the standard deviation ellipsis of the Gaussian components of a GMM defined by `means` and `covs`, on a 2D plot, together with `data` samples.

```
[means,covs,priors,total_loglike]=GMM_train(training_set,…
   100,k_means,k_covs,k_priors);

plot(training_set(:,1),training_set(:,2),'+');
plot_GMM2D(training_set,means,covs);

plot(total_loglike,'.-');
xlabel('Iteration'); ylabel('Global Log Likelihood'); grid on;
```

The total log likelihoods of the training and test data given this GMM model are obtained as above. The increase compared to estimating the GMM parameters from *K*-means clustering is small, but this is due to the oversimplified PDF we are dealing with. GMMs are very much used in speech recognition for acoustic modeling.

```
log_likelihood_training=…
   sum(log(GMM_pdf(training_set,means,covs,priors)))
log_likelihood_test=…
   sum(log(GMM_pdf(test_set,means,covs,priors)))
```

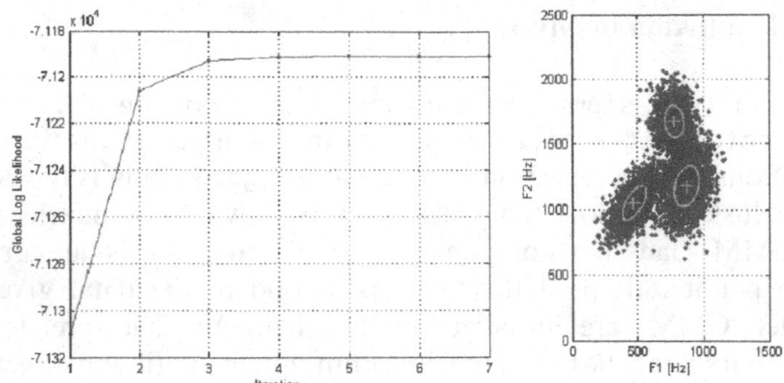

Fig. 4.11 Applying the EM algorithm (with three Gaussians) to the sample feature vectors for word "why". *Left*: evolution of the total log likelihood of the data; *right*: standard deviation ellipses of the three Gaussian components. Note that we do not set colors to feature vectors, as EM precisely does not strictly assign Gaussians to feature vectors

```
log_likelihood_training =
 -7.1191e+004

log_likelihood_test =
 -7.9822e+004
```

Now let us try to recognize sample words in {"why", "you", "we", "are", "hear", "here"}. We now use the sequence of feature vectors from our unknown signal (instead of a single vector as before), estimate the joint likelihood of all vectors in this sequence given each class, and obtain the posterior probabilities in the same way as above. If we assume that each sample in our sequence is independent of the others (which is in practice a rather bold claim, even for stationary signals; we will come back to this in the next section when introducing dynamic models), then the joint likelihood of the sequence is simply the product of the likelihoods of each sample.

We first estimate a GMM for each word using three Gaussians per word.[18] The estimated GMMs are plotted in Fig. 4.12.

```
for i=1:6

   [GMMs{i}.means,GMMs{i}.covs,GMMs{i}.priors,total_loglike]=...
         GMM_train(words{i}.training_all,100,3);
end;

for i=1:6
   subplot(2,3,i)
```

[18] When two Gaussians are enough, one of the three ends up having very small weight.

```
    plot(words{i}.training_all(:,1),...
         words{i}.training_all(:,2),'+');
    title(words{i}.word);
    hold on;
    mesh_GMM2D_pdf(GMMs{i}.means,GMMs{i}.covs,GMMs{i}.priors, ...
            0:50:1500, 0:50:2500,8e-9);
end;
```

Let us then try to recognize the first test sequence taken from "why" (Fig. 4.13). Since we do not know the priors of words in our imaginary language, we will set them all to 1/6. As expected, the maximum log likelihood is encountered for word "why"; our first test word is correctly recognized.

```
word_priors=ones(1,6)*1/6;
test_sequence=words{1}.test{1};
for i=1:6
    log_likelihood(i) = sum(log(GMM_pdf(test_sequence,...
        GMMs{i}.means,GMMs{i}.covs,GMMs{i}.priors)));
end;
log_posterior=log_likelihood+log(word_priors)
[maxlp,index]=max(log_posterior);
recognized=words{index}.word
```

```
log_posterior =
 -617.4004 -682.0656 -691.2229 -765.6281 -902.7732 -883.7884
recognized = why
```

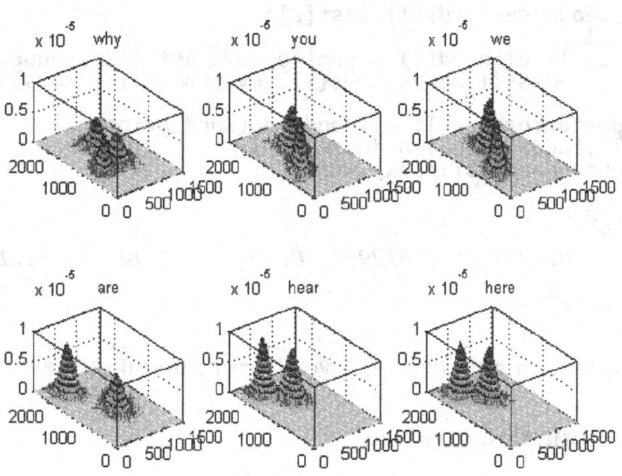

Fig. 4.12 GMMs estimated by the EM algorithm from the sample feature vectors of our six words: "why," "you," "we," "are," "hear," "here"

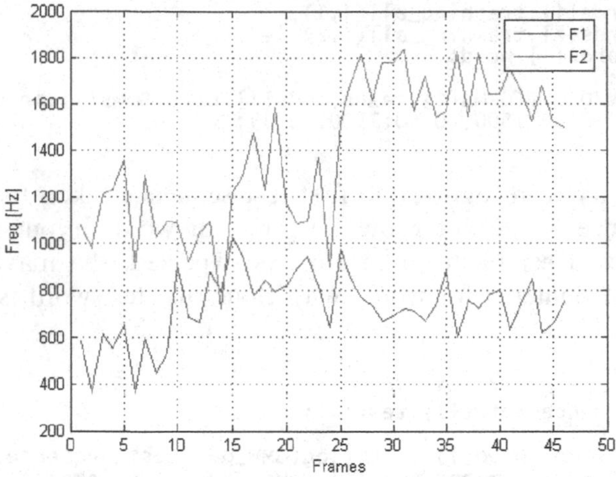

Fig. 4.13 Sequence of feature vectors of the first sample of "why." The three phonemes (each corresponding to a Gaussian in the GMM) are quite apparent

Not all sequences are correctly classified, though. Sequence 2 is recognized as a "we."

```
test_sequence=words{1}.test{2};
for i=1:6
    log_likelihood(i) = sum(log(GMM_pdf(test_sequence,...
        GMMs{i}.means,GMMs{i}.covs,GMMs{i}.priors)));
end;
log_posterior=log_likelihood+log(word_priors)
[maxlp,index]=max(log_posterior);
recognized=words{index}.word
```

```
log_posterior =
  1.0e+003 *
   -0.6844   -0.6741   -0.6729   -0.9963   -1.1437   -1.1181

recognized = we
```

We may now compute the total word error rate on our test database.

MATLAB function involved:

- `GMM_classify(x,GMMs,priors)` returns the class of sample `x` with respect to GMM classes using Bayesian classification. `x` {(N×D)} is a cell array of test sequences. `priors` is a vector of class priors. The function returns a vector of classes.

```
total=0;
errors=0;
```

```
for i=1:6
    n_test=length(words{i}.test);
    class=GMM_classify(words{i}.test,GMMs,word_priors);
    errors=errors+sum(class'~=i);
    class_error_rate(i)=sum(class'~=i)/n_test;
    total=total+n_test;

    subplot(2,3,i);
    hist(class,1:6);
    title(words{i}.word);
    set(gca,'xlim',[0 7]);
end;
overall_error_rate=errors/total
class_error_rate
```

```
overall_error_rate =
    0.3000

class_error_rate =
    0.0800    0.4300    0.4400    0.0100    0.3900    0.4500
```

Obviously, our static approach to word classification is not a success. Only 70% of the words are recognized. The rather high error rates we obtain are not astonishing. Except for "why" and "are," which have fairly specific distributions, "hear" and "here" have identical PDFs, as well as "you" and "we." These pairs of words are thus frequently mistaken for one another (Fig. 4.14).

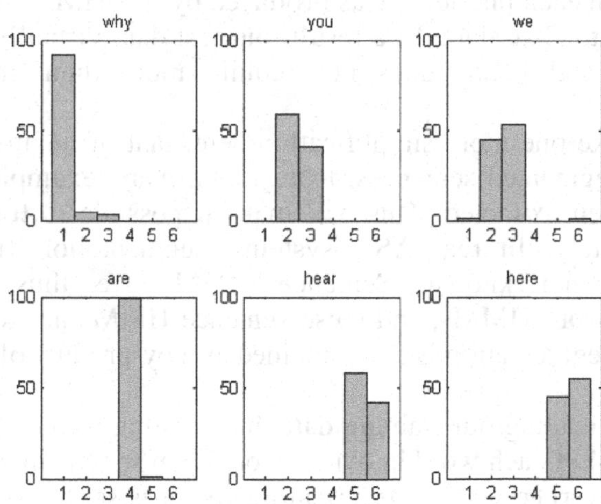

Fig. 4.14 Histograms of the outputs of the GMM-based word recognizer for samples of each of the six possible input words. The integer values on the x, axes refer to the index of the output word, in {"why," "you," "we," "are," "hear," "here"}

4.2.3 Hidden Markov models (HMM)

In the previous sections, we have seen how to create a model, either Gaussian or GMM, for estimating the PDF of speech feature vectors, even with complicated distribution shapes, and have applied it to the classification of isolated words. The main drawback of such a static classification, as it stands, is that it does not take time into account. For instance, the posterior probability of a sequence of feature vectors does not change when the sequence is time-reversed, as in words "you" /iu/ and "we" /ui/. This is due to the fact that our Bayesian classifier implicitly assumed that successive feature vectors are statistically independent.

In this section, we will model each word in our imaginary language using a two-state HMM (plus their initial and final states), except for "why," which will be modeled as a three-state HMM. One should not conclude that word-based ASR systems set the number of internal HMM states for each word to the number of phonemes they contain. The number of states is usually higher than the number of phonemes, as phonemes themselves are produced in several articulatory steps, which may each require a specific state. The reason for our choice is directly dictated by the fact that the test data we are using throughout this script were randomly generated by HMMs (see Appendix 1 in the MATLAB script) in which each phoneme was produced by one HMM state modeled as a multivariate Gaussian. As a result, our test data virtually exhibits no coarticulation and hence does not require more than one state per phoneme.

We will make one more simplification here: that of having access to a corpus of presegmented sentences, from which many examples of our six words have been extracted. This will make it possible to train our word HMMs separately. In real ASR systems, segmentation (in words or phonemes) is not known. Sentence HMMs are thus created by concatenating word HMMs, and these sentence HMMs are trained. Word (or phoneme) segmentation is then obtained as a by-product of this training stage.

We start by loading our training data and creating initial values for the left–right HMM of each word in our lexicon. Each state is modeled using a Gaussian multivariate whose mean feature vector is set to a random value close to the mean of all feature vectors in the word. The elements *trans*(*i*,*j*) of the transition matrix give the probability of going from state q_i to q_j (state 1 being the initial state). Transition probabilities between internal (emitting) states are set to a constant value of 0.8 for staying in the same state and 0.2 for leaving to the next state.

```
% Initializing HMM parameters
% "why" is a special case: it has 3 states
```

```
mu=mean(words{1}.training_all);
sigma=cov(words{1}.training_all);
HMMs{1}.means = {[],mu,mu,mu,[]};
HMMs{1}.covs  = {[],sigma,sigma,sigma,[]};
HMMs{1}.trans = [ 0.0 1.0  0.0  0.0  0.0
                  0.0 0.8  0.2  0.0  0.0
                  0.0 0.0  0.8  0.2  0.0
                  0.0 0.0  0.0  0.8  0.2
                  0.0 0.0  0.0  0.0  1.0 ];
for i=2:6
    mu=mean(words{i}.training_all);
    sigma=cov(words{i}.training_all);
    HMMs{i}.means = {[],mu,mu,[]};
    HMMs{i}.covs  = {[],sigma,sigma,[]};
    HMMs{i}.trans = [0.0 1.0  0.0  0.0
                     0.0 0.8  0.2  0.0
                     0.0 0.0  0.8  0.2
                     0.0 0.0  0.0   1  ];
end
```

Let us train our HMM models using the Baum–Welch (or forward–backward) algorithm, which is a particular implementation of the EM algorithm we already used for training our GMMs in the previous section. This algorithm will adapt the parameters of our word HMMs so as to maximize the likelihood of each training set, given each HMM model.

MATLAB function involved:

- `new_hmm = HMM_train_FB(data,old_hmm,dmin,qmax)|` returns the maximum likelihood reestimation of a Gaussian hidden Markov model (i.e., a single, possibly multivariate, Gaussian probability density function per state) based on the forward–backward algorithm (aka. Baum–Welch reestimation formulas). Note that most operations are performed in the log domain for accuracy[19]. `dmin` and `qmax` are the minimum log likelihood relative improvement and the maximum number of iterations until convergence, respectively.

```
for i=1:6
  HMMs{i}=HMM_gauss_train(words{i}.training,HMMs{i},0.001,50);
end;
```

The word "why" is now correctly modeled as a sequence of three states, each with a Gaussian multivariate PDF, which matches those of the underlying phonemes in the word: /uai/ (Fig. 4.15).

```
for i=2:4
    subplot(1,3,i-1)
    plot(words{1}.training_all(:,1),...
        words{1}.training_all(:,2),'+');
    title(['state ' num2str(i-1)]); % emiting states only
    hold on;
```

[19] This function uses a homemade `logsum.m` function, which computes the log of a sum of likelihoods from log likelihoods, as mentioned in Section 4.1.1.

```
    mesh_gauss2D_pdf(HMMs{1}.means{i},HMMs{1}.covs{i},1, ...
        0:50:1500, 0:50:2500,1e-8);
end;
```

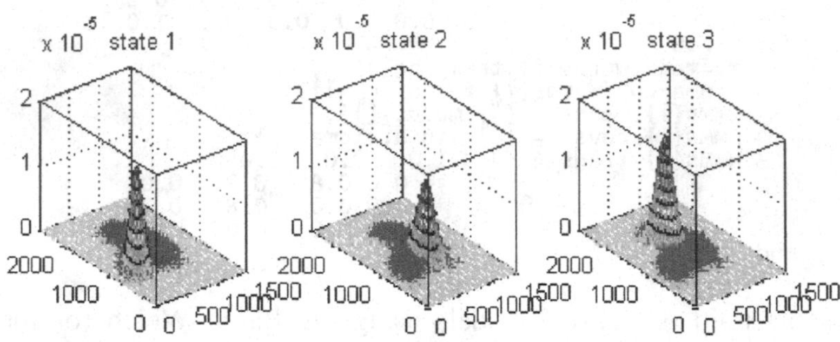

Fig. 4.15 PDF of the three Gaussian HMM states obtained from samples of "why"

The transition probabilities between the states of "why" have been updated by the Baum–Welch algorithm.

```
HMMs{1}.trans

ans =
         0    1.0000         0         0         0
         0    0.9970    0.0030         0         0
         0         0    0.9951    0.0049         0
         0         0         0    0.9387    0.0613
         0         0         0         0    1.0000
```

As a result of this better modeling, the total likelihood of the data for word "why" is higher than with our previous static GMM model. The previous model can actually be seen as a single-state HMM whose emission probabilities are modeled by a GMM.

```
log_likelihood_training=0;
for i=1:length(words{1}.training)
    training_sequence=words{1}.training{i};
    log_likelihood_training=log_likelihood_training+...
        HMM_gauss_loglikelihood(training_sequence,HMMs{1});
end;

log_likelihood_test=0;
for i=1:length(words{1}.test)
    test_sequence=words{1}.test{i};
    log_likelihood_test=log_likelihood_test+...
        HMM_gauss_loglikelihood(test_sequence,HMMs{1});
end;
```

```
log_likelihood_training
log_likelihood_test
```

```
log_likelihood_training = -6.7144e+004
log_likelihood_test = -7.5204e+004
```

HMM-based isolated word classification can now be achieved by finding the maximum of the posteriori probability of a sequence of feature vectors, given all six HMM models. The second test sequence for "why" (which was not correctly recognized using GMMs and a single state) now passes our classification test.

```
word_priors=ones(1,6)*1/6;

test_sequence=words{1}.test{2};
for i=1:6
   log_posterior(i) = HMM_gauss_loglikelihood(...
      test_sequence, HMMs{i})+log(word_priors(i));
end
log_posterior
[tmp,index]=max(log_posterior);
recognized=words{index}.word
```

```
log_posterior =
  1.0e+003 *
   -0.6425   -1.0390   -0.6471   -1.0427   -1.1199   -1.1057

recognized = why
```

The HMM model does not strictly assign states to feature vectors; each feature vector can be emitted by any state with a given probability. It is possible, though, to estimate the best path through the HMM, given the data by using the Viterbi algorithm (Fig. 4.16).

MATLAB function involved:

- `plot_HMM2D_timeseries(x,stateSeq)` plots a 2D sequence `x` (one observation per row) as two separate figures, one per dimension. It superimposes the corresponding state sequence `stateSeq` as colored dots on the observations. `x` and `stateSeq` must have the same length.

```
best_path=HMM_gauss_viterbi(test_sequence,HMMs{index});
plot_HMM2D_timeseries(test_sequence,best_path);
```

We may now compute the total word error rate again.

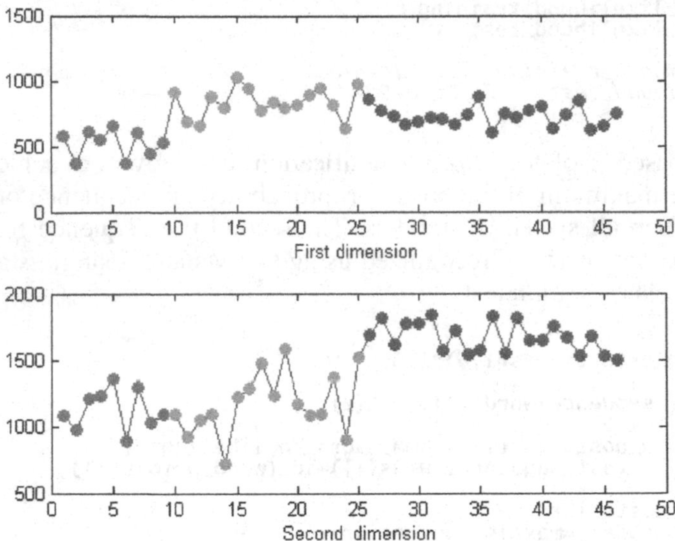

Fig. 4.16 Best path obtained by the Viterbi algorithm from the sequence of feature vectors of "why" in Fig. 4.13

MATLAB function involved:

- `HMM_gauss_classify(x,HMMs,priors)` returns the class of sample `x` with respect to HMM classes using Bayesian classification. HMM states are modeled by a Gaussian multivariate. `x` (N×D) is a cell array of test sequences. `priors` is a vector of class priors. The function returns a vector of classes.

```
total=0;
errors=0;
for i=1:6
    n_test=length(words{i}.test);
    class=HMM_gauss_classify(words{i}.test,HMMs,word_priors);
    errors=errors+sum(class'~=i);
    class_error_rate(i)=sum(class'~=i)/n_test;
    total=total+n_test;

    subplot(2,3,i);
    hist(class,1:6);
    title(words{i}.word);
    set(gca,'xlim',[0 7]);

end;
overall_error_rate=errors/total
class_error_rate
```

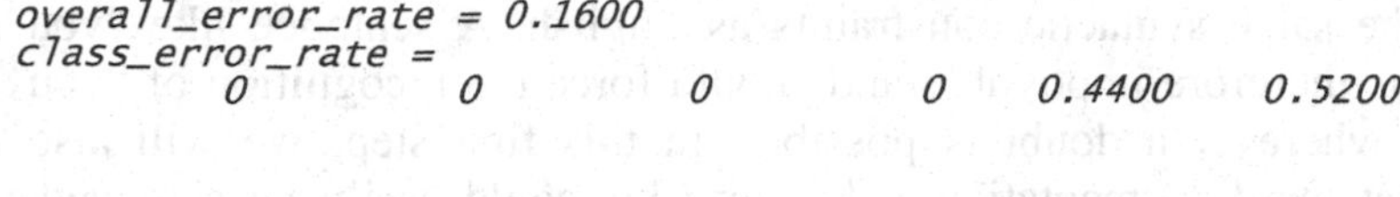

```
overall_error_rate = 0.1600
class_error_rate =
        0         0         0         0    0.4400    0.5200
```

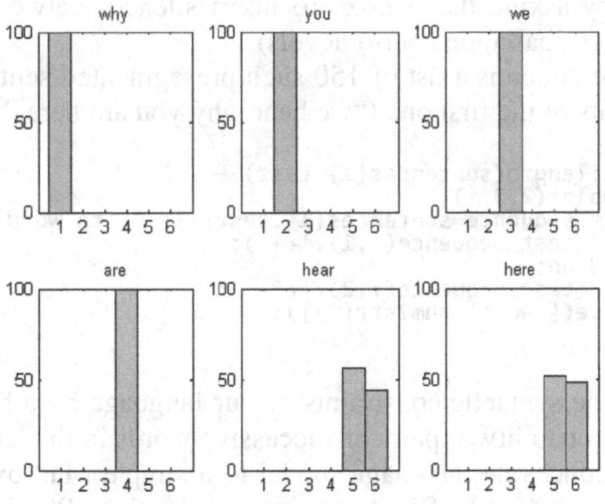

Fig. 4.17 Histograms of the outputs of the HMM-based word classifier for samples of each of the six possible input words

Note the important improvement in the classification of "you" and "we" (Fig. 4.17), which are now modeled as HMMs with distinctive parameters. Eighty-four percent of the (isolated) words are now recognized. The remaining errors are due to the confusion between "here" and "hear."

4.2.4 *N*-grams

In the previous section, we have used HMM models for the words of our imaginary language, which led to a great improvement in isolated word classification. It remains that "hear" and "here," having strictly identical PDFs, cannot be adequately distinguished. This kind of ambiguity can be resolved only when words are embedded in a sentence by using constraints imposed by the language on word sequences, i.e., by modeling the syntax of the language.

We will now examine the more general problem of connected word classification in which words are embedded in sentences. This task requires adding a language model on top of our isolated word classification system. For convenience, we will assume that our imaginary language imposes the same syntactic constraints as English. A sentence like "you are hear" is therefore impossible and should force the recognition of "you are here" wherever a doubt is possible. In this first step, we will also assume that word segmentation is known (this could easily be achieved, for instance, by asking the speaker to insert silences between words and detecting silences based on energy levels).

Our data file contains a list of 150 such presegmented sentences. Let us plot the contents of the first one ("we hear why you are here," Fig. 4.18).

```
for i=1:length(sentences{1}.test)
    subplot(2,3,i);
    test_sequence=sentences{1}.test{i}; % ith word
    plot(test_sequence(:,1),'+-');
    hold on;
    plot(test_sequence(:,2),'r*-');
    title(['Word' num2str(i)]);
end;
```

We model the syntactic constraints of our language by a bigram model based on the probability of pairs of successive words in the language. Such an approach reduces the language model to a simple Markov model. The component `bigram(i,j)` of its transition matrix gives P($word_i$|$word_j$): the probability that the *j*th word in the lexicon is followed by the *i*th word. Clearly, "You are hear" is made impossible by `bigrams(5,6)=0`.

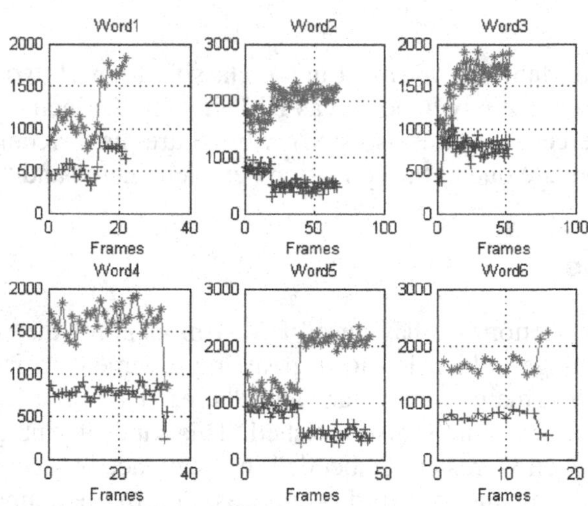

Fig. 4.18 Sequences of feature vectors for the six (presegmented) words in the first test sentence

```
% states = I U {why,you,we,are,hear,here} U F
% where I and F stand for the begining and the end of a
sentence

bigrams = ...
    [0  1/6  1/6  1/6  1/6  1/6  1/6  0  ; % P(word|I)
     0  0    1/6  1/6  1/6  1/6  1/6  1/6; % P(word|"why")
     0  1/5  0    0    1/5  1/5  1/5  1/5; % P(word|"you")
     0  0    0    0    1/4  1/4  1/4  1/4; % P(word|"we")
     0  0    1/4  1/4  0    0    1/4  1/4; % P(word|"are")
     0  1/4  1/4  0    0    0    1/4  1/4; % P(word|"hear")
     0  0    1/4  1/4  1/4  0    0    1/4; % P(word|"here")
     0  0    0    0    0    0    0      1]; % P(word|F)
```

Let us now try to classify a sequence of words taken from the test set. We start by computing the log likelihood of each unknown word, given the HMM model for each word in the lexicon. Each column of the log likelihood matrix stands for a word in the sequence; each line stands for a word in the lexicon {why,you,we,are,hear,here}.

```
n_words=length(sentences{1}.test);
log_likelihoods=zeros(6,n_words);

for j=1:n_words
    for k=1:6 % for each possible word HMM model
        unknown_word=sentences{1}.test{j};
        log_likelihoods(j,k) = HMM_gauss_loglikelihood(...
            unknown_word,HMMs{k});
    end;
end;

log_likelihoods
```

```
log_likelihoods =

  1.0e+003 *

  -0.2754    -0.3909    -0.2707    -0.4219    -0.6079    -0.5973
  -1.4351    -1.4067    -1.3986    -0.9186    -0.7952    -0.7977
  -0.6511    -0.8062    -0.7147    -0.9689    -0.8049    -0.8024
  -0.5203    -0.4208    -0.5043    -0.6925    -0.5306    -0.5284
  -0.9230    -1.0715    -1.0504    -0.5506    -0.6912    -0.6935
  -0.2510    -0.2772    -0.2400    -0.2851    -0.1952    -0.1953
```

With the approach we used in the previous section, we would classify this sentence as “we hear why you are hear” (by choosing the max likelihood candidate for each word independently of its neighbors).

```
[tmp,indices]=max(log_likelihoods);
for j=1:n_words
    recognized_sequence{j}=words{indices(j)}.word;
end;
recognized_sequence
```

```
recognized_sequence =
    'we'     'hear'     'why'     'you'     'are'     'hear'
```

We implement our language model as a Markov model on top of our word HMMs. The resulting model for the sequence to recognize is a discrete HMM in which there are as many internal states as the number of words in the lexicon (six in our case). Each state can emit any of the n_words input words (which we will label as '1', '2', ... 'n_words'), with emission probabilities equal to the likelihoods computed above. Bigrams are used as transition probabilities. Finding the best sequence of words from the lexicon, given the sequence of observations [1, 2, ..., n_words], is obtained by looking for the best path in this model, using the Viterbi algorithm again.

As shown below, we now correctly classify our test sequence as "we hear why you are here."

MATLAB function involved:

• [state,likelihood] = HMM_viterbi(transition,emission) performs the Viterbi search (log version) of the best state sequence for a discrete hidden Markov model.

transition: (K+2)x(K+2) matrix of transition probabilities, first and last rows correspond to initial and final (non-emitting) states.
emission: NxK matrix of state-conditional emission probabilities corresponding to a given sequence of observations of length N.
state: (Nx1) vector of state-related indexes of best sequence.
likelihood: best sequence likelihood.

```
best_path=HMM_viterbi(log(bigrams),log_likelihoods);
for j=1:n_words
    recognized_sequence{j}=words{best_path(j)}.word;
end;
recognized_sequence
```

```
recognized_sequence =
    'we'    'hear'    'why'    'you'    'are'    'here'
```

We may finally compute the word error rate on our complete test data.

```
n_sentences=length(sentences);

total=0;
errors=0;
class_error_rate=zeros(1,6);
class=cell(6); % empty cells

for i=1:n_sentences

    n_words=length(sentences{i}.test);
    log_likelihoods=zeros(6,n_words);
```

```
        for j=1:n_words
            unknown_word=sentences{i}.test{j};
            for k=1:6 % for each possible word HMM model
               log_likelihoods(j,k) = HMM_gauss_loglikelihood(...
                  unknown_word,HMMs{k});
            end;
        end;

        best_path=HMM_viterbi(log(bigrams),log_likelihoods);

        for j=1:n_words
            recognized_word=best_path(j);
            actual_word=sentences{i}.wordindex{j};
            class{actual_word}= [class{actual_word}, ...
                                 recognized_word];

            if (recognized_word~=actual_word)
                errors=errors+1;
                class_error_rate(actual_word)=...
                    class_error_rate(actual_word)+1;
            end;
        end;

        total=total+n_words;

    end;

    overall_error_rate=errors/total
    class_error_rate

overall_error_rate = 0.1079

class_error_rate =
     0     0     0     0    37    31
```

We now have an efficient connected word classification system for our imaginary language. The final recognition rate is now 89.2%. Errors are still mainly due to “here” being confused with “hear” (Fig. 4.19). As a matter of fact, our bigram model is not constrictive enough. It still allows nonadmissible sentences such as in sentence #3: “why are you hear.” Bigrams cannot solve all “hear” vs. “here” ambiguities because of the weaknesses of this poor language model. Trigrams could do a much better job (“are you hear,” for instance, will be forbidden by a trigram language model) at the expense of additional complexity.

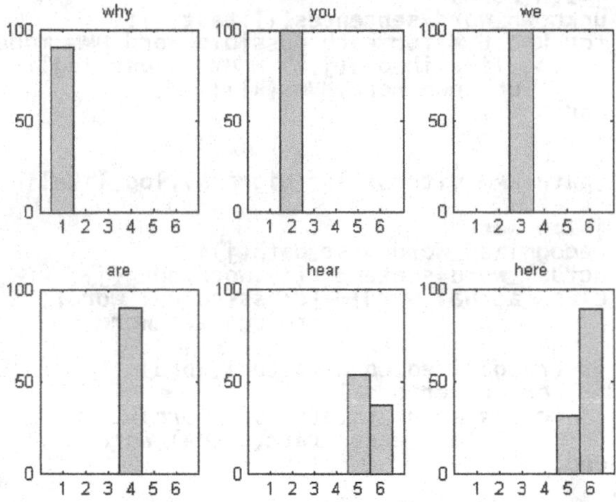

Fig. 4.19 Histograms of the outputs of the HMM-based word classifier after adding a bigram language model

4.2.5 Word-based continuous speech recognition

In this section, we will relax the presegmentation constraint, which will turn our classification system into a true word-based speech recognition system (albeit still in our imaginary language).

The discrete sentence HMM we used previously implicitly imposed initial and final states of word HMMs to fall after some specific feature vectors[20]. When word segmentation is not known in advance, the initial and final states of all word HMMs must be erased for the input feature vector sequence to be properly decoded into a sequence of words.

The resulting sentence HMM is a Gaussian HMM (as each word HMM state is modeled as a Gaussian) composed of all the word HMM states connected in a left–right topology inside word HMMs and connected in an ergodic topology between word HMMs. For the six words of our language, this makes 13 internal states plus the sentence-initial and sentence-final states. The transition probabilities between word-internal states are taken from the previously trained word HMMs, while the transition probabilities between word-final and word-initial states are taken from our bigram model.

[20] The sentence HMM therefore had to be changed for each new incoming sentence.

```
sentence_HMM.trans=zeros(15,15);

% word-initial states, including  sentence-final state;
word_i=[2 5 7 9 11 13 15];
word_f=[4 6 8 10 12 14]; % word-final states;

% P(word in sentence-initial position)
sentence_HMM.trans(1,word_i)=bigrams(1,2:8);

% copying trans. prob. for the 3 internal states of "why"
sentence_HMM.trans(2:4,2:4)=HMMs{1}.trans(2:4,2:4);

% distributing P(new word|state3,"why") to the first states of
% other word models, weighted by bigram probabilities.
sentence_HMM.trans(4,word_i)=...
    HMMs{1}.trans(4,5)*bigrams(2,2:8);

% same thing for the 2-state words
for i=2:6
   sentence_HMM.trans(word_i(i):word_f(i),word_i(i):word_f(i))=...
       HMMs{i}.trans(2:3,2:3);
   sentence_HMM.trans(word_f(i),word_i)=...
       HMMs{i}.trans(3,4)*bigrams(i+1,2:8);
end;
```

The emission probabilities of our sentence HMM are taken from the word-internal HMM states.

```
k=2;
sentence_HMM.means{1}=[]; % sentence-initial state
for i=1:6
    for j=2:length(HMMs{i}.means)-1
        sentence_HMM.means{k}=HMMs{i}.means{j};
        sentence_HMM.covs{k}=HMMs{i}.covs{j};
        k=k+1;
    end;
end;
sentence_HMM.means{k}=[]; % sentence-final state
```

We search for the best path in our sentence HMM[21] given the sequence of feature vectors of our test sequence, with the Viterbi algorithm, and plot the resulting sequence of states (Fig. 4.20).

MATLAB function involved:

• `[states,log_likelihood] = HMM_gauss_viterbi(x,HMM)` returns the best state sequence and the associated log likelihood of the sequence of feature vectors `x` (one observation per row) with respect to a Markov model `HMM` defined by

```
HMM.means
HMM.covs
HMM.trans
```

[21] The new sentence model is no longer sentence-dependent: the same HMM can be used to decode any incoming sequence of feature vectors into a sequence of words.

This function implements the forward recursion to estimate the likelihood on the best path.

```
n_words=length(sentences{1}.test);
complete_sentence=[];
for i=1:n_words
    complete_sentence=[complete_sentence ; ...
        sentences{1}.test{i}];
end;

best_path=HMM_gauss_viterbi(complete_sentence,sentence_HMM);
plot_HMM2D_timeseries(complete_sentence,best_path);

state_sequence=best_path(diff([ 0 best_path])~=0)+1;
word_indices=state_sequence(ismember(state_sequence,word_i));
[tf,index]=ismember(word_indices,word_i);

recognized_sentence={};
for j=1:length(index)
    recognized_sentence{j}=words{index(j)}.word;
end;
recognized_sentence
```

```
recognized_sentence =
    'we'    'hear'    'why'    'you'    'are'    'here'
```

We may finally compute again the word error rate on our complete test data (this is done in the accompanying MATLAB script). The final error rate of our word-based continuous speech recognizer is about 86.8%. This is only 2.4% less than when using presegmented words, which shows the efficiency of our sentence HMM model for both segmenting and classifying words. In practice, nonsegmented data are used for both training and testing, which could still slightly increase the word error rate.

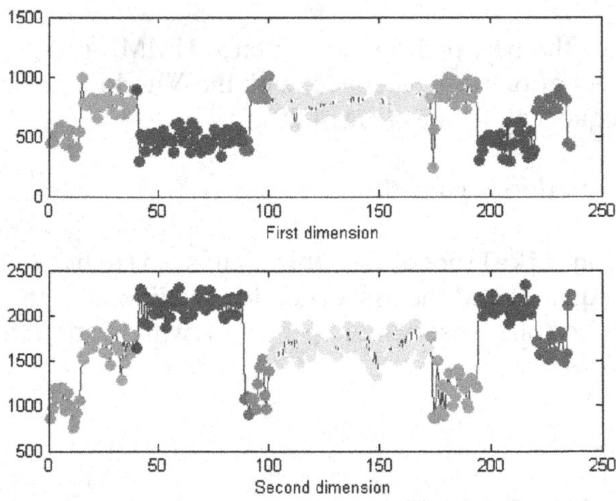

Fig. 4.20 Best path obtained by the Viterbi algorithm from the sequence of feature vectors of the first test sentence.

4.3 Going further

Dictation machines still differ from this proof of concept in several ways. Mel frequency cepstral coefficients (MFCCs) are used in place of our (F_1, F_2) formants for the acoustic model. Their first and second time-derivatives are added as features, as a simple way of accounting for the correlation between feature vectors *within* HMM states. Moreover, given the number of possible words in natural languages (several tens of thousands), ASR systems involve one additional layer in the statistical description of sentences: that of *phonemes*. The word HMMs we have trained above are replaced by phoneme HMMs. Word HMMs themselves are composed of phoneme HMMs (in the same way as we have built our sentence HMM from word HMMs), and additional pruning mechanisms are used in the decoder to constrain the search for the best sequence of words from the input feature vector sequence.

Several MATLAB-based HMM toolboxes, such as Kevin Murphy's (Murphy 2005), Steinar Thorvaldsen's (Thorvaldsen 2005, applied to biology), or Olivier Cappé's (Cappé 2001), are publicly available. MATLAB also provides its own HMM toolbox under the Statistics Toolbox. The most famous HMM toolbox, originally developed for large vocabulary speech recognition, is the HTK toolkit developed at Cambridge University (in ANSI C; Young et al. 2006).

4.4 Conclusion

In this chapter, we have seen how GMMs are used for the classification of (supposedly stationary) signals and how HMMs provide a means of modeling nonstationary signals as sequences of stationary states. We have also implemented a simple bigram model, whose coupling with our word HMMs has resulted in a unique sentence HMM, able to perform continuous speech recognition, i.e., to find how many words are present in an incoming stream of feature vectors and what those words are.

References

Bilmes JA (1998) A gentle tutorial of the EM algorithm and its applications t parameter estimation for Gaussian mixture and hidden Markov models. Technical Report 97-021, ICSI, Berkeley, CA, USA.

Bourlard (2007) Automatic speech and speaker recognition. In: Speech and Language Engineering, M. Rajman, ed., EPFL Press, pp 267–335

Bourlard H, Morgan N (1994) Connectionist Speech Recognition – A Hybrid Approach. Kluwer Academic Publishers, Dordrecht.

Bourlard H, Wellekens C (1990) Links between Markov models and multilayer perceptrons. IEEE Trans on Pattern Analysis and Machine Intelligence, 12(12)

Cappé O (2001) H2M: A Set of MATLAB/OCTAVE Functions for the EM Estimation of Mixtures and Hidden Markov Models [on line] Available: http://www.tsi.enst.fr/~cappe/h2m/h2m.html [3/06/07]

Duda RO, Hart PE, Stork DG (2000) Pattern Classification. Wiley-Interscience

Gold B, Morgan N (2000) Speech and Audio Signal Processing, Processing and Perception of Speech and Music. Wiley, Chichester.

Jelinek F (1991) Up from Trigrams! Proceedings of Eurospeech 91, Genova, vol. 3, pp 1037–1040

Moon TK (1996) The Expectation-Maximization Algorithm. IEEE Signal Processing Magazine, 13(6), 47–60

Murphy K (2005) Hidden Markov Model (HMM) Toolbox for Matlab [on line] Available: http://www.cs.ubc.ca/~murphyk/Software/HMM/hmm.html [20/05/07]

Picone JW (1993) Signal Modeling Techniques in Speech Recognition. Proceedings of the IEEE, 81(2), 1214–1247

Polikar R (2006) Pattern Recognition. In: Wiley Encyclopedia of Biomedical Engineering, M. Akay, ed., New York, Wiley

Rabiner LR (1989) A Tutorial on Hidden Markov Models and Selected Applications in Speech Recognition. Proceedings of the IEEE, 77(2), 257–286

Thorvaldsen S (2005) A tutorial on Markov models based on Mendel's classical experiments. Journal of Bioinformatics and Computational Biology, 3(6), 1441–1460. [on line] Available: http://www.math.uit.no/bi/hmm/ [3/06/07]

Young S, Evermann G, Gales M, Hain T, Kershaw D, Liu X, Moore G, Odell J, Ollason D, Povey D, Valtchev V, Woodland P (2006) The HTK Book (for HTK Version 3.4) [on line] Available: http://htk.eng.cam.ac.uk/prot-docs/htkbook.pdf [3/06/07]

Chapter 5

How does an audio effects processor perform pitch shifting?

T. Dutoit (°), J. Laroche (*)

(°) Faculté Polytechnique de Mons, Belgium
(*) Creative Labs, Inc., Scotts Valley, California

"Christmas, Christmas time is near, Time for toys and time for cheer, ..."
The chipmunk song (Bagdasarian 1958)

The old-fashioned charm of modified high-pitched voices has recently been revived by the American movie Alvin and the Chipmunks, based on the popular group and animated series of the same name which dates back to the 1950s. These voices were originally performed by recording speech spoken (or sung) slowly (typically at half the normal speed) and then accelerating ("pitching up") the playback (typically at double the speed, thereby increasing the pitch by one octave). The same effect can now be created digitally and in real time; it is widely used today in computer music.

5.1 Background – The phase vocoder

It has been shown in Chapter 3 (Sections 3.1.2 and 3.2.4) that linear transforms and their inverse can be seen as sub-band analysis/synthesis filter banks. The main idea behind the *phase vocoder*, which originally

T. Dutoit, F. Marqués (eds.), *Applied Signal Processing*,
DOI 10.1007/978-0-387-74535-0_5, © Springer Science+Business Media, LLC 2009

appeared in Flanagan and Golden (1966[1]) for speech transmission, is precisely that of using a discrete Fourier transform (DFT) as a filter bank for sub-band processing of signals (Section 5.1.1). This, as we shall see, is a particular case of the more general short-time Fourier transform (STFT) processing (Section 5.1.2) with special care to perfect reconstruction of the input signal (Section 5.1.3), which opens the way to nonlinear processing effects, such as timescale modification (Section 5.1.4), pitch shifting (Section 5.1.5), and harmonization.

5.1.1 DFT-based signal processing

The most straightforward embodiment for frequency-based processing of an audio signal $x(n)$ is given in Figs. 5.1 and 5.2. Input frames **x** containing N samples $[x(mN), x(mN+1), \ldots, x(mN+N-1)]$ are used as input for an N-bin DFT. The N resulting complex frequency-domain values $\mathbf{X}=[X_0(m), X_1(m), \ldots, X_{N-1}(m)]$ are given by the following equation:

$$X_k(m) = \sum_{n=0}^{N-1} x(n+mN)e^{-jn\Omega_k} \quad \text{with } \Omega_k = k\frac{2\pi}{N} \tag{5.1}$$

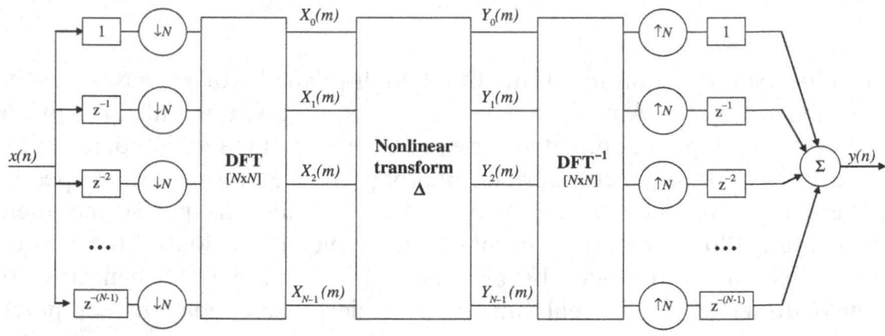

Fig. 5.1 N-bin DFT-based signal processing: block diagram

[1] Flanagan and Golden called this system the *phase vocoder* after H. Dudley's *channel vocoder*, which transmitted only the spectral envelope (i.e., a gross estimation of the amplitude spectrum) of speech, while spectral details were sent in a separate channel and modeled though a voiced/unvoiced generator. In contrast, in the phase vocoder, phase information was sent together with amplitude information and brought a lot more information on the speech excitation.

They are changed into $\mathbf{Y}$=[$Y_0(m)$, $Y_1(m)$,..., $Y_{N-1}(m)$] according to the target audio effect $\mathbf{\Delta}$, and an inverse DFT produces output audio samples $\mathbf{y}$=[$y(mM)$, $y(mM+1)$, ..., $y(mM+N-1)$].

As already explained in Chapter 3, $X_k(m)$ can also be interpreted, for a fixed value of k, as the result of passing the input signal through a sub-band filter $H_k(z)$ whose frequency response $H_k(\varphi)$ is centered on Ω_k and whose impulse response is given by

$$h_k(n) = \{1, e^{-j\Omega_k}, e^{-j2\Omega_k}, ..., e^{-j(N-1)\Omega_k}, 0, 0, ...\} \tag{5.2}$$

and further decimating its output by a factor N (Fig. 5.3). These sub-band processing and block-processing views are very complimentary (Fig. 5.4) and will be used often in this chapter.

Note that in this first simple embodiment, the input and output frames do not overlap; all sub-band signals $X_k(m)$ ($k=0,1,...N-1$) are implicitly downsampled by a factor N compared to the sampling rate of the input signal $x(n)$.

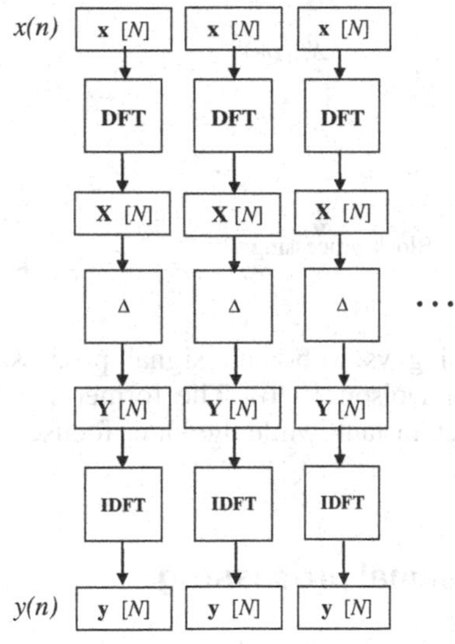

Fig. 5.2 *N*-bin DFT-based signal processing: frame-based view

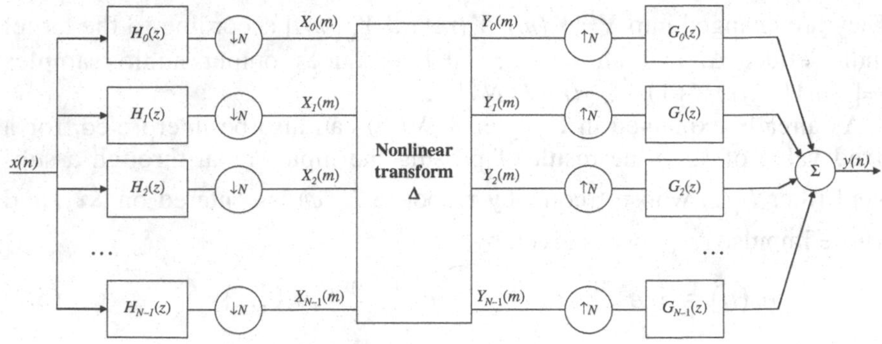

Fig. 5.3 *N*-channel DFT-based signal processing: sub-band processing view

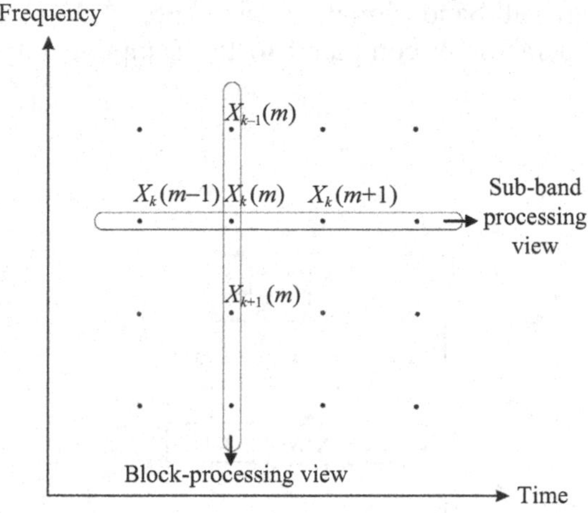

Fig. 5.4 Block processing vs. sub-band signal processing views in the time–frequency domain (after Dolson 1986). The former examines the outputs of all sub-band filters at a given instant, while the latter focuses on the output of a single sub-band filter over time

5.1.2 STFT-based signal processing

The setting depicted in Fig. 3.6 has three apparent problems for the type of nonlinear processing we will apply to sub-band signals in this chapter. These problems are related to frequency selectivity, frame shift, and to the so-called *blocking effect*, respectively. Solving them implies, as we shall see, to use analysis weighting window, overlapping, and synthesis weighting windows.

Analysis weighting windows

DFT filter banks are not frequency selective, as already shown in Fig 3.19, which reveals important side lobes in the frequency response of each filter. This is a serious problem for the design of a phase vocoder; as we will see in Section 5.1.4, the processing applied to each sub-band signal in a phase vocoder is based on the hypothesis that when the input signal is composed of sinusoidal components (called *partials*, in computer music), each sub-band signal will depend on at most one partial.

A solution is to use an analysis weighting window before the DFT.[2] As a matter of fact, the frequency response of the filter banks implemented by such a weighted DFT is the Fourier transform of the weighting window itself. Using windows such as Hanning, Hamming, Blackman, or others makes it possible to decrease the level of side lobes at the expense of enlarging the bandwidth of each sub-band filter (Fig. 5.5). This drawback can easily be compensated by increasing N, the number of samples in each input frame.

Overlapping analysis frames

In Fig. 3.7, the sampling period of each sub-band signal, i.e., the shift M between successive DFT frames $\mathbf{y}$, was explicitly set to N samples, while the normalized bandwidth of each sub-band signal $X_k(m)$ was approximately equal to $2/N$ (the width of the main lobe of the DFT of the weighting window, which is implicitly rectangular in Fig. 3.7). According to Shannon's theorem, we should have

$$\begin{gathered}\frac{F_e}{M} \geq B \\ M \leq \frac{F_e}{B}\end{gathered} \tag{5.3}$$

where B/F_e is the normalized bandwidth of the weighting window. To avoid aliasing, sub-band signals should thus be sampled with a sampling period smaller than $N/2$ samples, i.e., input DFT frames should have an overlap greater than $N/2$ (Fig. 5.6).

---

[2] Note that weighting windows are not *always* necessary in STFT-based processing. For FFT-based linear time-invariant filtering, for instance, overlapping rectangular windows do the job (Smith 2007a).

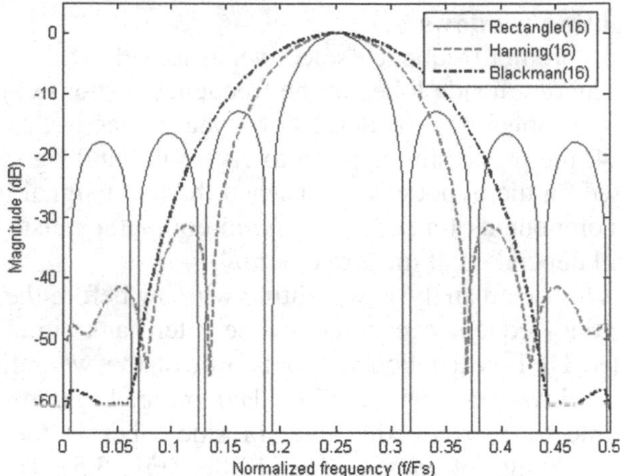

Fig. 5.5 Discrete-time Fourier transform of various weighting windows of length N=16 samples multiplied by a sine wave at 1/4th of the sampling frequency. The main lobe width of the rectangular, Hanning, and Blackman windows is equal to $2/N$, $\approx 4/N$, and $\approx 6/N$, respectively; side lobes are 13, 31, and 57 dB lower than the main lobe, respectively.

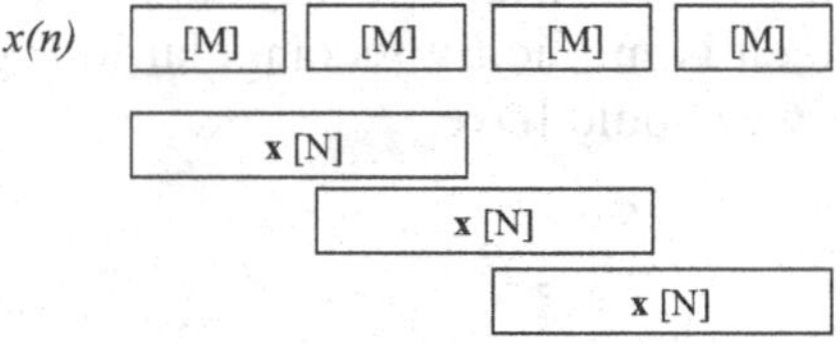

Fig. 5.6 Analysis frames with a frame shift $M=N/2$ (to be compared to the top of Fig. 3.7)

In practice though, we know that if we apply no modification to sub-band signals, i.e., if we set $Y_k(m)= X_k(m)$, the output signal $y(n)$ will be strictly equal to the input signal $x(n)$. This paradox is only apparent; in practice, each sub-band signal is indeed aliased, but the aliasing between adjacent sub-bands cancels itself, as shown in our study of perfect reconstruction filters (Section 3.1.1). This, however, strictly holds only if *no processing is applied to sub-band signals.*

In contrast, given the complex sub-band signal processing performed in the phase vocoder, it is safer to set the frame shift M in accordance with the Shannon theorem, i.e., to a maximum value of $N/2$ for a rectangular window, $N/4$ for a Hanning window, and $N/6$ for a Blackman window. Besides, we will see in Section 5.1.4 that the hypotheses underlying sub-band processing in the phase vocoder impose the same constraint.

This transforms our initial DFT (5.1) into the more general *short-time Fourier transform* (STFT):

$$X_k(m) = \sum_{n=0}^{N-1} x(n+mM)w(n)e^{-jn\Omega_k} \qquad (k=0, 1, \ldots, N-1) \tag{5.4}$$

Synthesis weighting windows

Since the analysis is based on overlapping input frames, synthesis is performed by summing overlapping output frames (as in Fig 3.7). However, when nonlinear *block-adaptive* sub-band processing is performed, as in the phase vocoder, processing discontinuities appear at synthesis frame boundaries. This has already been mentioned in Section 3.1.2.

A partial[3] solution to this problem is to use *synthesis weighting windows* before adding overlapping output frames. The resulting processing scheme is termed as *weighted overlap-add* (WOLA):

$$y(n) = \sum_{m=-\infty}^{\infty} y_m(n-mM)w(n-mM)$$
$$\text{with } y_m(n) = \frac{1}{N}\sum_{k=0}^{N-1} Y_k(m)e^{jn\Omega_k} \tag{5.5}$$

where, in practice, the number of terms in the first summation is limited by the finite length N of the weighting window $w(n)$. The association of IFFT and WOLA is sometimes referred to as *inverse short-time Fourier transform* (ISTFT).

Figure 5.7 shows the final block-processing view for STFT-based signal processing: an STFT-based analysis block, whose output (amplitudes and phases) is processed and sent to an ISTFT block. The corresponding sub-band processing view is not repeated here; it is still that of Fig. 5.3, provided the down- and upsampling ratios are set to M rather than N and in which filters $H(z)$ and $G(z)$ now also account for the effect of the nonrectangular weighting window.

[3] The use of synthesis weighting windows only spreads discontinuities around the overlap area of successive frames; it does not cancel them. As a result, special care has to be taken to avoid introducing avoidable discontinuities in modified STFTs (see Section 5.1.4).

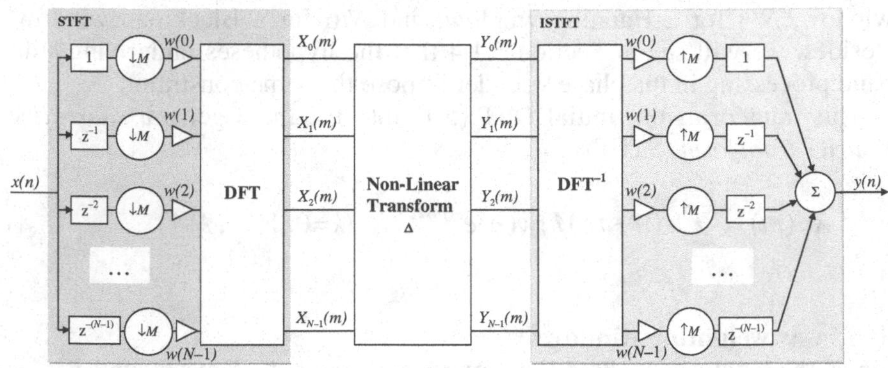

Fig. 5.7 STFT-based signal processing: block-processing view

5.1.3 Perfect reconstruction

The results obtained with an STFT-based signal processing system obviously depend on the weighting window used and on the shift between successive analysis windows.

Assuming that the analysis and synthesis windows are identical and making the further assumption that DFTs are not modified, WOLA's output is given by

$$y(n) = \sum_{m=-\infty}^{\infty} x(n) w^2(n - mM) \tag{5.6}$$

Perfect reconstruction is therefore achieved when:[4]

$$\sum_{m=-\infty}^{\infty} w^2(n - mM) = 1 \quad \forall n \in \mathbb{Z} \tag{5.7}$$

This condition is sometimes called the *constant overlap-add* (COLA) constraint, applied here on $w^2(n)$. It depends on the shape of the weighting window $w(n)$ and on the ratio of the frame shift M to the frame length N.

A simple example is that of the square root of the triangular (or *Bartlett*) window with M=N/2. The Hanning (or *Hann*) window with M=N/4 is usually preferred (Fig. 5.8), as its Fourier transform exhibits a better compromise between the level of its side lobes and the width of its

[4] Notice the similarity between this condition and Equation (3.18), in which only m=0 and 1 are taken into account, since in the modulated lapped transform the frame shift M is set to half the frame length N.

passband. The square root of the Hanning window can also be used with $M=N/2$.[5]

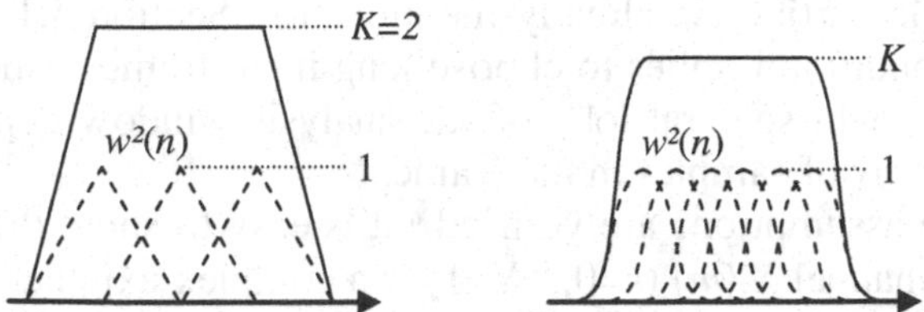

Fig. 5.8 Overlap-adding triangular windows with $M=N/2$ (left) and squared Hanning windows $M=N/4$ (right)

Usual weighting windows, when used in the COLA conditions, sum up to $K\neq 1$ (Fig. 5.8). Using the Poisson summation formula, it is easy to show that the value of K is given by (Smith 2007b) the following equation:

$$K = \frac{\sum_{0}^{N-1} w^2(n)}{M} \tag{5.8}$$

It is then easy to multiply the weighting window by 1/sqrt(K) to enforce Equation (5.7).

5.1.4 Timescale modification with the phase vocoder

The phase vocoder is a specific implementation of an STFT-based processing system in which

1. The input signal is assumed to be composed of a sum of (not necessarily harmonic) sinusoidal terms, named *partials*:

$$x(n) = \sum_{i=1}^{P} A_i \cos(n\varphi_i + \phi_i) \tag{5.9}$$

where P is the number of partials, A_i and ϕ_i are their amplitudes and initial phases, respectively, and φ_i are their normalized angular

[5] The square root of the Hanning window is nothing else than the sinusoidal window of the modulated lapped transform (MLT) used in perceptual audio coders (Section 3.1.1), with a frame shift set to half the frame length.

frequencies [$\varphi_i=\omega_i/F_s$ (rad/sample)]. The value of $n\varphi_i+\phi_i$ gives the *instantaneous phase* of the *i*th partial at sample *n*.

2. The output of each DFT bin is assumed to be mainly influenced by a single partial. As already mentioned in Section 5.1.1, satisfying this condition implies to choose long input frames, since the width of the main spectral lobe of the analysis window depends on the number N of samples in the frame.

When these assumptions are verified,[6] it is easy to show that the output of each DFT channel $X_k(m)$ (k=0,..,N−1) is a complex exponential function over *m*. For simplicity, let us assume that the input signal is a single imaginary exponential:

$$x(n) = Ae^{jn\varphi+\phi} \tag{5.10}$$

Then $X_k(m)$ is given by the following equation:

$$\begin{aligned} X_k(m) &= \sum_{n=0}^{N-1} Ae^{j(n+mM)\varphi+\phi} w(n) e^{-jn\Omega_k} \\ &= \left(A\sum_{n=0}^{N-1} e^{jn(\varphi-\Omega_k)} w(n) \right) e^{jmM\varphi+\phi} \qquad (k=0, 1, \ldots, N\text{-}1) \\ &= H_k(\varphi) e^{jmM\varphi+\phi} \end{aligned} \tag{5.11}$$

where $H_k(\varphi)$ is the frequency response of the *k*th analysis sub-band filter for frequency φ and therefore does not depend on *m*. The output of each DFT channel is thus an imaginary exponential function whose angular frequency $M\varphi$ depends only on the frequency φ of the underlying partial (and *not* on the central frequency Ω_k of the DFT bin). As a result, it is predictable from any of its past values:

$$\begin{aligned} &|X_k(m+1)| = |X_k(m)| \\ \text{and } &\angle X_k(m+1) = \angle X_k(m) + M\varphi \end{aligned} \tag{5.12}$$

where $\angle X_k(m)$ is the *instantaneous phase* of the output of the *k*th DFT channel (Fig. 5.10). Note that only a *wrapped* version of this phase is available in practice, since it is computed as an arctangent and therefore obtained in the range [−π,π].

[6] Clearly, these conditions are antagonistic. Since the partials of an audio signal are generally modulated in amplitude and frequency, condition 1 is valid only for short frames, while verifying condition 2 requires long input frames.

Time-scale modification, with horizontal phase locking
Modifying the duration of the input signal can then be achieved by setting an analysis frame shift M_a in Equation (5.5) lower (time expansion) or higher (time compression) than the synthesis frame shift M_s,[7] and adapting the DFT channel values $Y_k(m)$ accordingly:

$$\begin{aligned} |Y_k(m)| &= |X_k(m)| \\ \text{and } \angle Y_k(m+1) &= \angle Y_k(m) + M_s \varphi_k(m) \end{aligned} \tag{5.13}$$

in which $\varphi_k(m)$ is the frequency of the partial, which mainly influences $X_k(m)$.

The initial synthesis phases $\angle Y_k(0)$ are simply set to the corresponding analysis phases $\angle X_k(0)$, and the subsequent values are obtained from Equation (5.13). This operation is termed as *horizontal phase locking* (Fig. 5.9), as it ensures phase coherence across frames in each DFT channel independently of its neighbors. In other words, it guarantees that the short-time synthesis frames $y_m(n)$ in Equation (5.5) will overlap coherently.[8]

Obtaining $\varphi_k(m)$ from Equation (5.12), though, is not straightforward. Since the values of $\angle X_k(m)$ and $\angle X_k(m+1)$ are only known modulo 2π, we have

$$\angle X_k(m+1) - \angle X_k(m) = M\varphi_k(m) + l2\pi \tag{5.14}$$

where l is an a priori unknown integer constant. Let us rewrite Equation (5.14) as

$$\angle X_k(m+1) - \angle X_k(m) = M(\theta_k(m) + \Omega_k) + l2\pi \tag{5.15}$$

where $\theta_k(m)$ is the frequency offset between $\varphi_k(m)$ and Ω_k (Fig. 5.10).

If we assume that $\theta_k(m)$ is small enough for $|M\theta_k(m)|$ to be less than π, and if $[]_{2\pi}$ denotes the reduction of the phase to its principal value in $[-\pi,\pi]$, then

$$\begin{aligned} [M\theta_k(m)]_{2\pi} &= [\angle X_k(m+1) - \angle X_k(m) - M\Omega_k - l2\pi]_{2\pi} \\ M\theta_k(m) &= [\angle X_k(m+1) - \angle X_k(m) - M\Omega_k]_{2\pi} \end{aligned} \tag{5.16}$$

which gives the expected value of $\theta_k(m)$.

---

[7] The synthesis frame shift M_s is left constant, rather than the analysis frame shift M_a, so as to take the COLA constraint into account.

[8] Remember that this is only valid, though, for constant frequency partials.

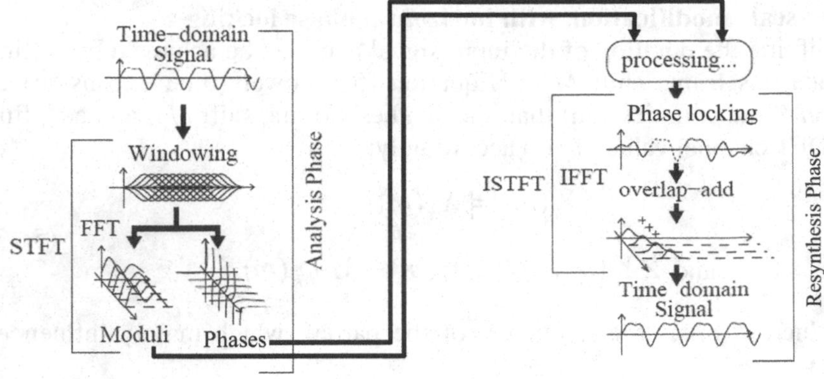

Fig. 5.9 The phase vocoder (after Arfib et al. 2002)

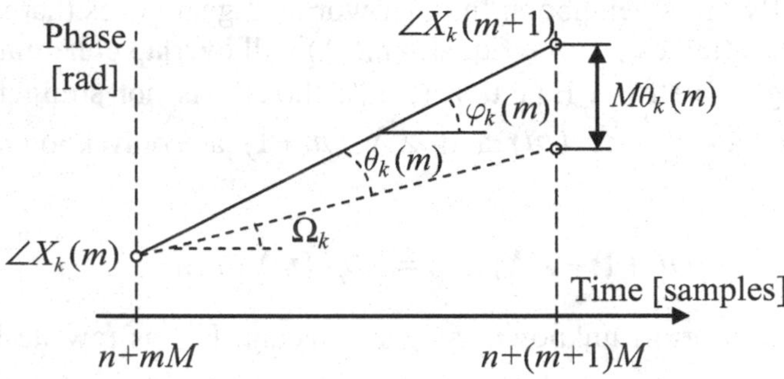

Fig. 5.10 Theoretical phase increment from frame m to frame m+1 in DFT channel k due to a main partial with frequency $\varphi_k(m)$ in its bandpass. In practice, $X_k(m)$ and $X_k(m+1)$ are only known modulo 2π (this is not shown here)

The assumption of the small value of $\theta_k(m)$ can be translated into an offset b in DFT bins whose central frequency is close enough to $\varphi_k(m)$ for Equation (5.16) to be applicable:

$$
\begin{aligned}
&| M\left(b\frac{2\pi}{N}\right)| < \pi \\
&| b | < \frac{N}{2M}
\end{aligned}
\tag{5.17}
$$

Since the half main spectral lobe width of a Hanning window covers approximately 2 bins, we conclude that if a Hanning-based vocoder is designed in such a way that horizontal phase locking should be applied to the whole main lobe around each partial, then the frame shift M must be such that $M<N/4$. This condition happens to be identical as the one derived in Section 5.1.2.

Time-scale modification, with horizontal and vertical phase locking

The two most prominent timescaling artifacts of the classical phase vocoder exposed above are *transient smearing* and *phasiness*. Both occur even with modification factors close to 1. Transient smearing is heard as a slight loss of percussiveness in the signal. Piano signals, for instance, may be perceived as having less "bite." Phasiness is heard as a characteristic coloration of the signal; in particular, time-expanded speech often sounds as if the speaker is much further from the microphone than in the original recording. Phasiness can be minimized by ensuring not only phase consistency *within* each DFT channel over time but also phase coherency across the channels in a given synthesis frame. A system implementing both horizontal and vertical phase locking is called a *phase-locked vocoder*.

Coming back to the simple case of a single imaginary exponential, and further assuming that the weighting window is rectangular, $H_k(\varphi)$ in Equation (5.11) becomes

$$\begin{aligned} H_k(\varphi) &= \sum_{n=0}^{N-1} e^{jn(\varphi-\Omega_k)} = \frac{1-e^{jN(\varphi-\Omega_k)}}{1-e^{j(\varphi-\Omega_k)}} \\ &= e^{j\frac{N-1}{2}(\varphi-\Omega_k)} \left(\frac{\sin\frac{N(\varphi-\Omega_k)}{2}}{\sin\frac{(\varphi-\Omega_k)}{2}} \right) \qquad (k=0, 1, \ldots, N\text{-}1) \end{aligned} \tag{5.18}$$

Since the bracketed expression in Equation (5.18) is positive in its main lobe, the following cross-channel phase relationship exists in the main lobe:

$$\begin{aligned} H_{k+1}(\varphi) - H_k(\varphi) &= e^{j\frac{N-1}{2}(\Omega_k-\Omega_{k+1})} \\ &= e^{-j\frac{N-1}{N}\pi} \end{aligned} \tag{5.19}$$

This relationship is even simplified if the weighted samples are first circularly shifted by $(N-1)/2$ samples[9] before the DFT, as this introduces an additional linear phase factor which cancels the influence of Ω_k in $H_k(\varphi)$. As a result, Equation (5.19) becomes:

$$H_{k+1}(\varphi) - H_k(\varphi) = 0 \tag{5.20}$$

This result can easily be generalized to any weighting window.

Puckette (1995) proposed a method called *loose phase locking,* in which the phase in each synthesis channel is implicitly influenced by the phases of the surrounding synthesis channels. The method first computes the synthesis phases in all channels $\angle Y_k(m+1)$ according to Equation (5.13) and then modifies the results as

$$\angle Y'_k(m+1) = \angle\big(Y_k(m-1) + Y_k(m) + Y_k(m+1)\big) \quad (k=0, 1, \ldots, N\text{-}1) \tag{5.21}$$

As a result, if a DFT channel is the maximum of the DFT magnitudes, its phase is basically unchanged, since $Y_k(m-1)$ and $Y_k(m+1)$ have a much lower amplitude. Conversely, the phase of a channel surrounding the maximum will roughly be set to that of the maximum: the channels around the peaks of the DFT are roughly *phase locked.*

The fundamental limitation (and also the attraction) of the loose phase-locking scheme is that it avoids any explicit determination of the signal structure; the same calculation is performed in every channel, independently of its content. Laroche and Dolson (1999a) have proposed an improved phase-locking scheme, named *rigid phase locking* and based on the explicit identification of peaks in the spectrum. Their new phase-updating technique starts with a coarse peak-picking stage where vocoder channels are searched for local maxima. In the simplest implementation, a channel whose amplitude is larger than its four nearest neighbors is said to be a peak. The series of peaks divides the frequency axis into *regions of influence* located around each peak. The basic idea is then to update the phases for the peak channels only, according to Equation (5.13); the phases of the remaining channels within each region are then locked in some way to the phase of the peak channel. In particular, in their so-called *identity phase-locking* scheme, Laroche and Dolson constrain the synthesis phases around each peak to be related in the same way as the analyses phases. If p is the index of the dominant peak

---

[9] In practice, since FFTs are used for implementing the phase vocoder, N is even, and samples are shifted by $N/2$. Consequently, (5.20) is only approximately verified.

channel, the phases of all channels k ($k \neq p$) in the peak's region of influence are set to

$$\angle Y_k(m) = \angle Y_p(m) + \angle X_k(m) - \angle X_p(m) \qquad (k=0, 1, \ldots, N\text{-}1) \qquad (5.22)$$

Another improvement is also proposed by the authors in the *scaled phase-locking* scheme, which accounts for the fact that peak channels may evolve with time.

5.1.5 Pitch shifting with the phase vocoder

The resampling-based approach

The standard pitch-scale modification technique combines time-scale modification and resampling. Assuming a pitch-scale modification by a factor β is desired (i.e., all frequencies must be multiplied by β), the first stage consists of using the phase vocoder to perform a time-scale modification of the signal (its duration is multiplied by β). In the second stage, the resulting signal is resampled at a new sampling frequency F_s/β, where F_s is the original sampling period. The output signal ends up with the same duration as the original signal, but its frequency content has been expanded by a factor β during the resampling stage, which is the desired result (Fig. 5.11).

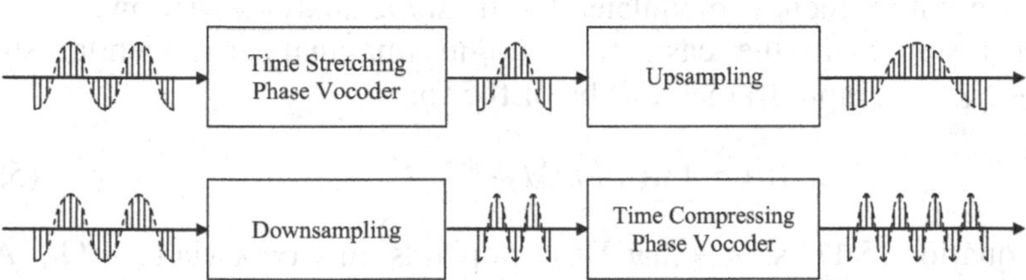

Fig. 5.11 A possible implementation of a pitch shifter using a phase vocoder. *Top*: Downward pitch scaling ($\beta < 1$); *bottom*: upward pitch scaling

Note that it is possible to reverse the order of these two stages, which yields the same result if the window size is multiplied by β in the phase vocoder time-scaling stage. However, the cost of the algorithm is a function of the modification factor and of the order in which the two stages are performed. For example, for upward pitch shifting ($\beta > 1$), it is more advantageous to resample first and then time scale, because the resampling stage yields a shorter signal. For downward pitch shifting, it is better to time scale first and then resample, because the time-scaling stage yields a signal of smaller duration (as shown in Fig. 5.11).

This standard technique has several drawbacks. Its computational cost is a function of the modification factor. If the order in which the two stages are performed is fixed, the cost becomes increasingly large for larger upward or downward modifications. An algorithm with a fixed cost is usually preferable. Another drawback of the standard technique is that only *one* "linear" pitch-scale modification is allowed, i.e., the frequencies of all the components are multiplied by the same factor. As a result, harmonizing a signal (i.e., adding several copies pitch shifted with different factors) requires repeated processing at a prohibitive cost for real-time applications.

The STFT-based approach

A more flexible algorithm has been proposed by Laroche and Dolson (1999b), which allows nonlinear frequency modifications, enabling the same kind of alterations that usually require non-real-time sinusoidal analysis/synthesis techniques.

The underlying idea behind the new techniques consists of identifying peaks in the short-term Fourier transform and then translating them to new arbitrary frequencies. If the relative amplitudes and phases of the bins around a sinusoidal peak are preserved during the translation, then the time-domain signal corresponding to the shifted peak is simply a sinusoid at a different frequency, modulated by the same analysis window.

For instance, in the case of a single imaginary exponential signal $x(n) = Ae^{jn\varphi+\phi}$, input frames will be of the form

$$x_m(n) = A\,w(n+mM)e^{j(n+mM)\varphi+\phi} \tag{5.23}$$

and Equation (5.11) shows that $X_k(m)$ depends only on k via $(\varphi - \Omega_k)$. As a result, imposing:[10]

$$Y_k(m) = X^{(*)}_{(k+S)\,\mathrm{mod}\,N}(m) \quad (k = 0,1,\ldots,N-1) \tag{5.24}$$

in which S is an integer number of DFT bins will result in output frames of the form

[10] Note the "mod N" in Equation (5.24), which makes sure that, if a shifted region of influence spills out of the frequency range [0,2π], it is simply reflected back into [0,2π] after complex conjugation (indicated by our nonstandard use of the "(*)" notation) to account for the fact that the original signal is real.

$$y_m(n) = A\, w(n+mM)e^{j(n+mM)(\varphi+\Delta\varphi)+\phi} \tag{5.25}$$

with $\Delta\varphi = S\dfrac{2\pi}{N}$. These output frames, however, will not overlap-add coherently into

$$y(n) = Ae^{jn(\varphi+S\frac{2\pi}{N})+\phi} \tag{5.26}$$

since Equation (5.24) does not ensure that the peak phases are consistent from one frame to the next. As a matter of fact, since the frequency has been changed from φ to $\varphi+\Delta\varphi$, phase coherency requires rotating all phases by $M\Delta\varphi$ and accumulating this rotation for each new frame, which results in the final pitch-scaling scheme:

$$Y_k(m) = X_{(k+S)\,\mathrm{mod}\,N}(m)\, e^{jmM\Delta\varphi} \quad (k = 0,1,\ldots,N-1) \tag{5.27}$$

Pitch shifting a more complex signal is achieved by first detecting spectral peaks and regions of influence, as in the phase-locked vocoder, and then shifting each region of influence separately by β, using Equation (5.27) with $\Delta\varphi=(\beta-1)\varphi$. Shifted areas of influence that overlap in frequency are simply added together.

When the value of the frequency shift $\Delta\varphi$ does not correspond to an integer number of bins *S* (which is the most general case), either it can be rounded to the closest integer value of *S* (which is very acceptable for large DFT sizes and low sampling rates, for which DFT channels are very narrow), or linear interpolation can be used to distribute each shifted bin into existing DFT bins (see Laroche and Dolson 1999b).

It is useful to note that because the channels around a given peak are rotated by the same angle, the differences between the phases of the channels around a peak in the input STFT are preserved in the output STFT short-term Fourier transform. This is similar to the "identity phase-locking" scheme, which dramatically minimizes the phasiness artifact often encountered in phase-vocoder time or pitch-scale modifications.

Also note, finally, that the exact value of the frequency φ of each partial is not required, since only $\Delta\varphi$ is used in Equation (5.27). This is an important savings compared to the standard phase vocoder and makes this approach more robust.

5.2 MATLAB proof of concept: ASP_audio_effects.m

In this section, we show that STFT-based signal processing provides very flexible tools for audio signal processing. We start by radically modifying the phases of the STFT to create a robotization effect (Section 5.2.1). Then we examine the MATLAB implementation of a phase vocoder, with and without vertical phase locking (Section 5.2.2). We conclude the chapter with two pitch-scale modification algorithms, which use the phase vocoder in various aspects (Section 5.2.3).

5.2.1 STFT-based audio signal processing

We first implement a basic STFT-based processing scheme using the same weighting window for both analysis and synthesis and with very simple processing of the intermediate DFTs.

Weighting windows

Since weighting windows play an important part in STFT-based signal processing, it is interesting to check their features first. MATLAB proposes a handy tool for that: the `wvtool` function. Let us use it for comparing the rectangular, Hanning, and Blackman windows. Clearly, the spectral leakage of the Hanning and Blackman windows is lower than those of the rectangular and sqrt(Hanning) windows. By zooming on the main spectral lobes, one can check that this is compensated by a higher lobe width (Fig. 5.12): the spectral width of the rectangular window is half that of the Hanning window, and one-third of that of the Blackman window.

```
N=100;
wvtool(boxcar(N),hanning(N),blackman(N));
```

The Constant OverLap-Add (COLA) constraint

Let us now examine the operations involved in weighted overlap-add (WOLA). When no modification of the analysis frames is performed, one can to the least expect the output signal to be very close to the input signal. In particular, a constant input signal should result in a constant output signal. This condition is termed as constant overlap-add (COLA) constraint. It is only strictly met for specific choices of the weighting window and of the window shift, as we shall see. We now check this on a chirp signal (Fig. 5.13).

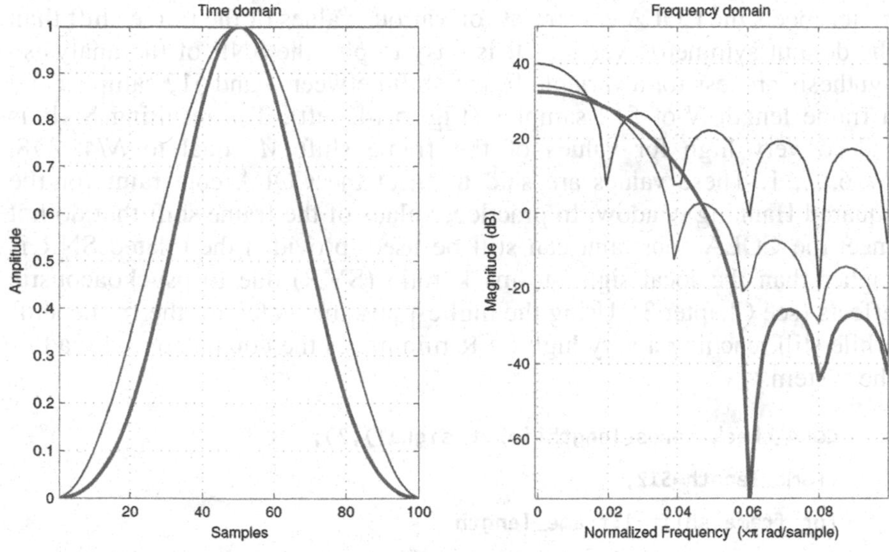

Fig. 5.12 The 100-sample rectangular, Hanning, and Blackman weighting windows (*left*) and their main spectral properties (*right*)

```
Fs=8000;
input_signal=chirp((0:Fs)/Fs,0,1,2000)';
specgram(input_signal,1024,Fs,256);
```

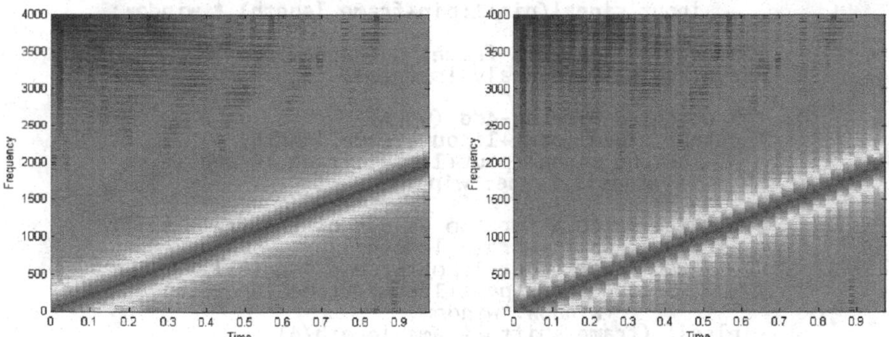

Fig. 5.13 Spectrogram of a chirp. *Left*: Original signal; *right*: after weighted overlap-add using a Hanning window and a frame shift set to half the frame length

We actually use the periodic version of the Hanning window, which better meets the COLA constraint for various values of the frame shift than the default symmetric version. It is easy to plot the SNR of the analysis–synthesis process for a variable frame shift between 1 and 512 samples and a frame length N of 512 samples (Fig. 5.14, left). The resulting SNR is indeed very high for values of the frame shift M equal to $N/4$, $N/8$, $N/16$, ..., 1. These values are said to meet the COLA constraint for the squared Hanning window. In practice, values of the frame shift that do not meet the COLA constraint can still be used, provided the related SNR is higher than the local signal-to-mask ratio (SMR) due to psychoacoustic effects (see Chapter 3). Using the highest possible value for the frame shift while still reaching a very high SNR minimizes the computational load of the system.

```
COLA_check=zeros(length(input_signal),2);

frame_length=512;

for frame_shift=1:frame_length

    % Amplitude normalization for imposing unity COLA
    window=hanning (frame_length,'periodic');
    COLA_ratio=sum(window.*window)/frame_shift;
    window=window/sqrt(COLA_ratio);

    output_signal=zeros(length(input_signal),1);
    pin=0;     % current position in the input signal
    pout=0;   % current position in the output signal

    while pin+frame_length<length(input_signal)

        % Creating analysis frames
        analysis_frame = ...
            input_signal(pin+1:pin+frame_length).* window;

        % Leaving analysis frames untouched
        synthesis_frame=analysis_frame;

        % Weighted OverLap-Add (WOLA)
        output_signal(pout+1:pout+frame_length) = ...
            output_signal(pout+1:pout+frame_length)+...
            synthesis_frame.*window;

        % Checking COLA for two values of the frame shift
        if (frame_shift==frame_length/2)
            COLA_check(pout+1:pout+frame_length,1)=...
                COLA_check(pout+1:pout+frame_length,1)...
                +window.*window;
        elseif (frame_shift==frame_length/4)
            COLA_check(pout+1:pout+frame_length,2)=...
                COLA_check(pout+1:pout+frame_length,2)...
                +window.*window;
        end;

        pin=pin+frame_shift;
        pout=pout+frame_shift;

    end;
```

```
    % Storing the output signal for frame shift = frame_length/2
    if (frame_shift==frame_length/2)
        output_half=output_signal;
    end;

    % Using the homemade snr function introduced in Chapter 2,
    % and dropping the first and last frames, which are only
    % partially overlap-added.
    snr_values(frame_shift)=snr(...
        input_signal(frame_length+1:end-frame_length),...
        output_signal(frame_length+1:end-frame_length),0);

end

plot((1:frame_length)/frame_length,snr_values);
```

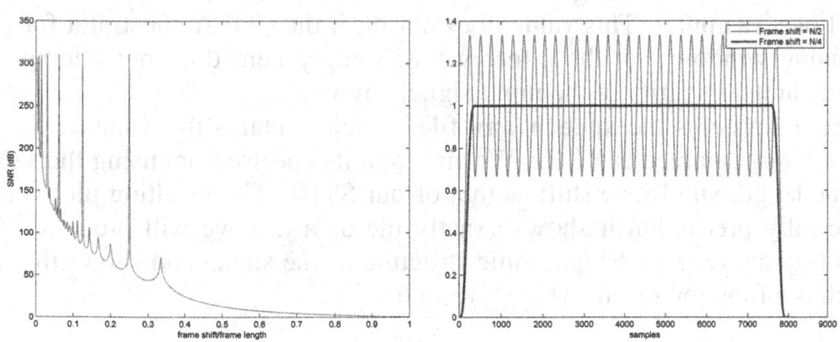

Fig. 5.14 Examining the COLA constraint for a Hanning window. *Left*: SNR for various values of the ratio of frame shift to frame length; *right*: result of the OLA with M=N/2 (non-COLA) and N/4 (COLA)

When the input signal is a unity constant, the output signal can indeed be close to unity... or not depending on the value of the frame shift. For the Hanning window, obviously, a frame shift of N/4 meets the COLA constraint, while N/2 leads to an important amplitude ripple (Fig. 5.14, right). These values also depend on the weighting window.[11]

```
plot(COLA_check(:,1)); hold on;
plot(COLA_check(:,2),'k--','linewidth',2); hold off;
```

A bad choice of the frame shift may lead to important perceptual degradation of the output audio signal, as confirmed by listening to the output produced for M=N/2 (see Fig. 5.13, right for the related spectrogram).

---

[11] See Appendix 1 in the **audio_effects.m** file for a test on the square-root Hanning window.

```
soundsc(output_half, Fs);
specgram(output_half,1024,Fs,256);
```

STFT-based signal processing

We now add an STFT/ISTFT step. The choice of the frame length *N*, of the type of weighting window, and of the frame shift depends on the type of processing applied to frequency bands. For frequency-selective processing, a large value of *N* is required, as the bandwidth of the DFT channels is inversely proportional to *N*. In this section, we will implement a simple robotization effect on a speech signal, which is not a frequency-selective modification. We therefore set *N* to 256 samples (but many other values would match our goal). We choose a Hanning window with a frame shift of 80 samples. This value does not meet the COLA constraint for the Hanning window, but the effect we will apply here does not attempt to maintain the integrity of the input signal anyway.

Let us process the **speech.wav** file, which contains the sentence *Paint the circuits* sampled at 8 kHz. We first plot its spectrogram using the same frame length and frame shift as that of our STFT. The resulting plot is not especially pretty, but it shows exactly the data that we will process. The spectrogram reveals the harmonic structure of the signal and shows that its pitch is a function of time (Fig. 5.15, left).

```
frame_length=256;
frame_shift=80;
window=hanning (frame_length,'periodic');
COLA_ratio=sum(window.*window)/frame_shift;
window=window/sqrt(COLA_ratio);

[input_signal,Fs]=wavread('speech.wav');
specgram(input_signal,frame_length,Fs,window);
```

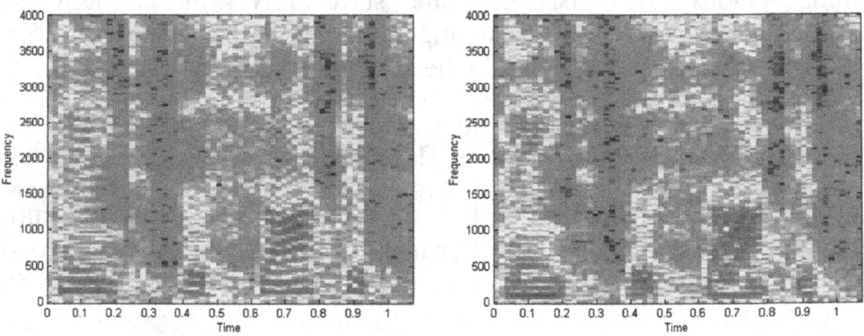

Fig. 5.15 Spectrogram of the sentence *Paint the circuits*. *Left*: Original signal, as produced by the STFT; *right*: robotized signal

When processing of the intermediate DFTs is not performed, the output signal is thus equal to the input signal. It is easy to modify the amplitudes or the phases of the STFT to produce audio effects. Let us test a simple robotization effect, for instance, by setting all phases to zero. This produces a disruptive perceptual effect at the periodicity of the frame shift (Fig. 5.16). When the shift is chosen small enough and the effect is applied to speech, it is perceived as artificial constant pitch. This simple effect appears on the spectrogram as a horizontal reshaping of the harmonics (Fig. 5.15, right). It reveals the importance of phase modifications (and more specifically of phase discontinuities) in the perception of a sound.

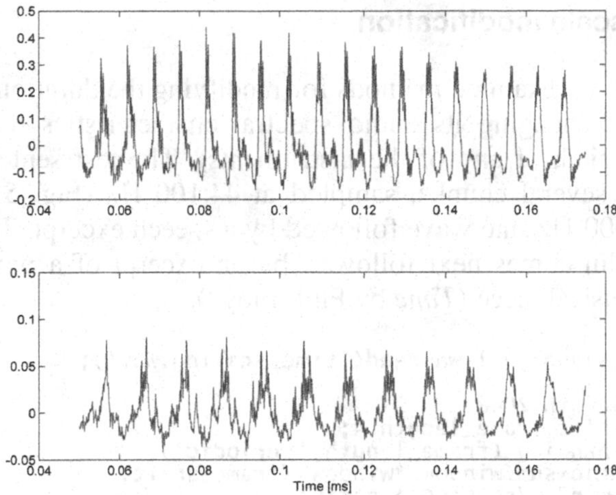

Fig. 5.16 Zoom on the first vowel of *Paint the circuits*. *Top*: Original signal; *bottom*: robotized signal with frame shift set to 10 ms (hence producing an artificial pitch of 100 Hz)

```
output_signal=zeros(length(input_signal),1);
pin=0;pout=0;

while pin+frame_length<length(input_signal)

    % STFT
    analysis_frame=input_signal(pin+1:pin+frame_length).* window;
    dft=fft(analysis_frame);

    % Setting all phases to zero
    dft=abs(dft);

    % ISTFT
    synthesis_frame=ifft(dft).*window;
     output_signal(pout+1:pout+frame_length) = ...
```

```
        output_signal(pout+1:pout+frame_length)+synthesis_frame;

    pin=pin+frame_shift;
    pout=pout+frame_shift;

end;

% Zoom on a few ms of signal, before and after robotization
clf;
range=(400:1400);
subplot(211);
plot(range/Fs,input_signal(range))
subplot(212);
plot(range/Fs,output_signal(range))

specgram(output_signal,frame_length,Fs,window);
```

5.2.2 Time-scale modification

In this section, we examine methods for modifying the duration of an input signal without changing its audio spectral characteristics (e.g., without changing the pitch of any of the instruments). The test sentence we use here contains several chunks, sampled at 44,100 Hz (Fig. 5.17, left). It starts with 1,000-Hz sine wave followed by a speech excerpt. The sound of a (single) violin comes next followed by an excerpt of a more complex polyphonic musical piece (*Time* by Pink Floyd).

```
[input_signal,Fs]=wavread('time_scaling.wav');

frame_length=2048;
frame_shift=frame_length/4;
window=hanning (frame_length,'periodic');
COLA_ratio=sum(window.*window)/frame_shift;
window=window/sqrt(COLA_ratio);

specgram(input_signal,frame_length,Fs,window);
```

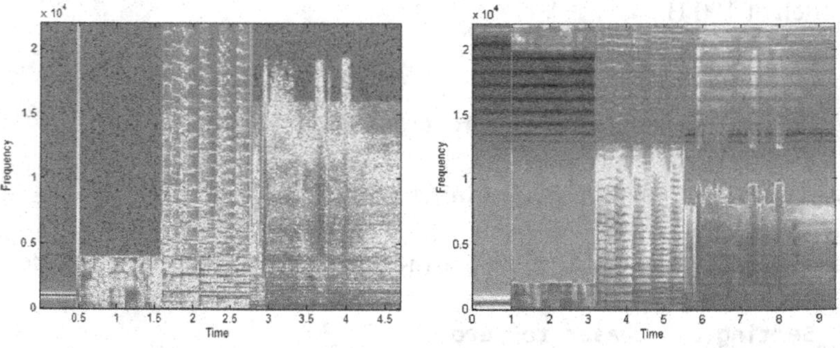

Fig. 5.17 Spectrogram of the test signal. *Left*: Original signal; *right*: after time stretching by a factor of 2

Interpolating the signal

Increasing the length of the input signal by a factor of 2, for instance, is easily obtained by interpolating the signal, using the `resample` function provided by MATLAB, and not telling the digital-to-analog converter (DAC) about the sampling frequency change. This operation, however, also changes the frequency content of the signal; all frequencies are divided by two. Similarly, speeding up the signal by a factor 2 would multiply all frequencies by two. This is sometimes referred to as the "chipmunk effect." Note incidentally the aliasing introduced by the imperfect low-pass filter used by `resample` (Fig. 5.17, right).

```
resampled_signal=resample(input_signal,2,1);

specgram(resampled_signal,frame_length,Fs,window);
```

Time-scale modification with weighted overlap-add (WOLA)

It is also possible to modify the time scale by decomposing it into overlapping frames, changing the analysis frame shift into a different synthesis frame shift, and applying weighted overlap-add (WOLA) to the resulting synthesis frames. While this technique works well for unstructured signals, its application to harmonic signals produces unpleasant audio artifacts at the frequency of the synthesis frame rate due to the loss of synchronicity between excerpts of the same partials in overlapping synthesis frames (Fig. 5.18). This somehow reminds us of the robotization effect obtained in Section 1.3.

```
frame_length=2048;
synthesis_frame_shift=frame_length/4;
window=hanning (frame_length,'periodic');
COLA_ratio=sum(window.*window)/synthesis_frame_shift;
window=window/sqrt(COLA_ratio);

time_scaling_ratio=2.85;
analysis_frame_shift=...
   round(synthesis_frame_shift/time_scaling_ratio);

pin=0;pout=0;
output_signal=zeros(time_scaling_ratio*length(input_signal),1);

while (pin+frame_length<length(input_signal)) ...
      && (pout+frame_length<length(output_signal))

   analysis_frame = input_signal(pin+1:pin+frame_length).*...
      window;
   synthesis_frame = analysis_frame.* window;
   output_signal(pout+1:pout+frame_length) = ...
       output_signal(pout+1:pout+frame_length)+synthesis_frame;

   pin=pin+analysis_frame_shift;
   pout=pout+synthesis_frame_shift;

   % Saving frames for later use
   if (pin==2*analysis_frame_shift) % 3rd frame
```

```
        frame_3=synthesis_frame;
    elseif (pin==3*analysis_frame_shift) % 4th frame
        frame_4=synthesis_frame;
    end;

end;

% Plot two overlapping synthesis frames and show the resulting
output signal
clf;
ax(1)=subplot(211);
range=(2*synthesis_frame_shift:2*synthesis_frame_shift+...
    frame_length-1);
plot(range/Fs,frame_3)
hold on;
range=(3*synthesis_frame_shift:3*synthesis_frame_shift+...
    frame_length-1);
plot(range/Fs,frame_4,'r')
ax(2)=subplot(212);
range=(2*synthesis_frame_shift:3*synthesis_frame_shift+...
    frame_length-1);
plot(range/Fs,output_signal(range))
xlabel('Time [ms]');
linkaxes(ax,'x');
set(gca,'xlim',[0.045 0.06]);

soundsc(output_signal,Fs);
```

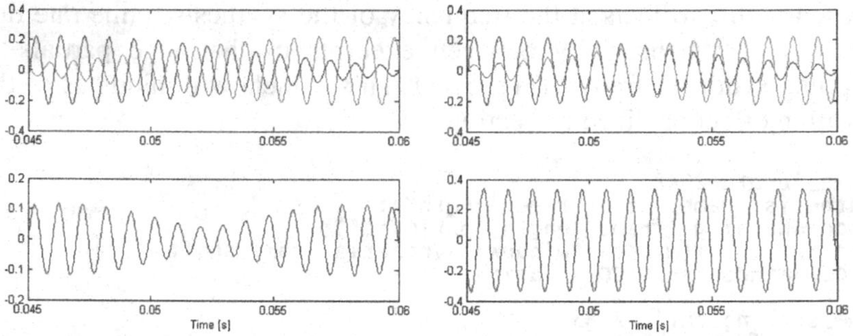

Fig. 5.18 Time-scale modification of a sinusoidal signal. *Top*: Zoom on the overlap of two successive synthesis frames; *bottom*: the resulting output signal; *left*: using WOLA alone; *right*: using the phase vocoder

Time-scale modification with the phase vocoder

Let us now modify the time scale of the input signal without affecting its frequency content by using a phase vocoder, which is a STFT-based signal processing system with specific hypotheses on the STFT. Time scaling is again achieved by modifying the analysis frame shift without changing the synthesis frame shift, but the STFT has to be modified so as to avoid creating phase mismatches between overlapped synthesis frames.

We use the periodic Hanning window, which provides a low spectral leakage while its main spectral lobe width is limited to $4*F_s/N$. The choice of the frame length N is dictated by a maximum bandwidth constraint for the DFT sub-band filters, since the phase vocoder is based on the hypothesis that each DFT bin is mainly influenced by a single partial, i.e., a sinusoidal component with constant frequency. For a Hanning window, each sinusoidal component will drive four DFT bins; the width of each bin is F_s/N. Clearly, the higher the value of N, the better the selectivity of the sub-band filters. On the other hand, a high value of N tends to break the stationarity hypothesis on the analysis frame (i.e., the frequency of partials will no longer be constant). We set N to 2,048 samples here, which imposes the width of each DFT bin to 21 Hz and the bandwidth of the DFT sub-band filters to 86 Hz. We set the frame shift to $N/4$=512 samples, which meets the COLA constraint for the Hanning window. What is more, this choice obeys the Shannon theorem for the output of each DTF bin, seen as the output of a sub-band filter. The result is much closer to what is expected (Fig. 5.19).

```
NFFT=2048;
frame_length=NFFT;
synthesis_frame_shift=frame_length/4;
window=hanning (frame_length,'periodic');
COLA_ratio=sum(window.*window)/synthesis_frame_shift;
window=window/sqrt(COLA_ratio);

time_scaling_ratio=2.65;
analysis_frame_shift=round(synthesis_frame_shift/...
   time_scaling_ratio);
% Central frequency of each DFT channel
DFT_bin_freqs=((0:NFFT/2-1)*2*pi/NFFT)';

pin=0;pout=0;

output_signal=zeros(time_scaling_ratio*length(input_signal),1);
last_analysis_phase=zeros(NFFT/2,1);
last_synthesis_phase=zeros(NFFT/2,1);

while (pin+frame_length<length(input_signal)) ...
      && (pout+frame_length<length(output_signal))

    % STFT
    analysis_frame=input_signal(pin+1:pin+frame_length).* window;
    dft=fft(fftshift(analysis_frame));
    dft=dft(1:NFFT/2);

    % PHASE MODIFICATION
    % Find phase for each bin, compute by how much it increased
    % since last frame.
    this_analysis_phase = angle(dft);
    delta_phase = this_analysis_phase - last_analysis_phase;
    phase_increment=delta_phase-analysis_frame_shift...
        *DFT_bin_freqs;

    % Estimate the frequency of the main partial for each bin
    principal_determination=mod(phase_increment+pi,2*pi)-pi;
```

```
        partials_freq=principal_determination/analysis_frame_shift+...
            DFT_bin_freqs;

        % Update the phase in each bin
        this_synthesis_phase=last_synthesis_phase+...
            synthesis_frame_shift*partials_freq;

        % Compute DFT of the synthesis frame
        dft= abs(dft).* exp(j*this_synthesis_phase);

        % Remember phases
        last_analysis_phase=this_analysis_phase;
        last_synthesis_phase=this_synthesis_phase;

        % ISTFT
        dft(NFFT/2+2:NFFT)=fliplr(dft(2:NFFT/2)');
        synthesis_frame = fftshift(real(ifft(dft))).* window;
        output_signal(pout+1:pout+frame_length) = ...
            output_signal(pout+1:pout+frame_length)+synthesis_frame;

        pin=pin+analysis_frame_shift;
        pout=pout+synthesis_frame_shift;

        % Saving the estimated frequency of partials for later use
        if (pin==2*analysis_frame_shift) % 3rd frame
            partials_freq_3=partials_freq;
      end;

end;

specgram(output_signal(1:end-5000),frame_length,Fs,window);
```

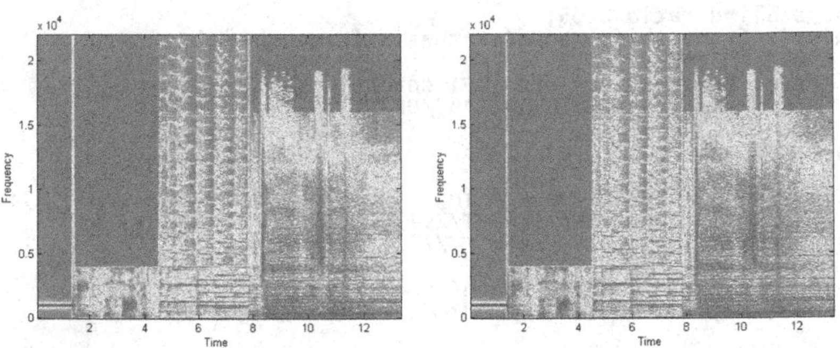

Fig. 5.19 Spectrogram of the test signal after time stretching by a factor 2.85 using a phase vocoder with horizontal phase locking (*left*) and with both horizontal and vertical phase locking (*right*)

The phasing effect we had observed in the previous method has now disappeared (Fig. 5.18, right).

Let us check how the normalized frequency of the initial sinusoid (i.e., 1,000*2*π/F_s = 0.142 rad/s) has been estimated. The first spectral line of the sinusoid mainly influences DTF bins around index 1000/(44100/2048)=46.43. As observed in Fig. 5.20 (left), bins 42–51 (which correspond to the main spectral lobe) have indeed measured the correct frequency.

```
subplot(2,1,1);
plot((0:NFFT/2-1),partials_freq_3);
set(gca,'xlim',[25 67]);

subplot(2,1,2);
third_frame=input_signal(2*analysis_frame_shift:2*analysis_fra
me_shift+frame_length-1);
dft=fft(third_frame.* window);
dft=20*log10(abs(dft(1:NFFT/2)));
plot((0:NFFT/2-1),dft);
set(gca,'xlim',[25 67]);
```

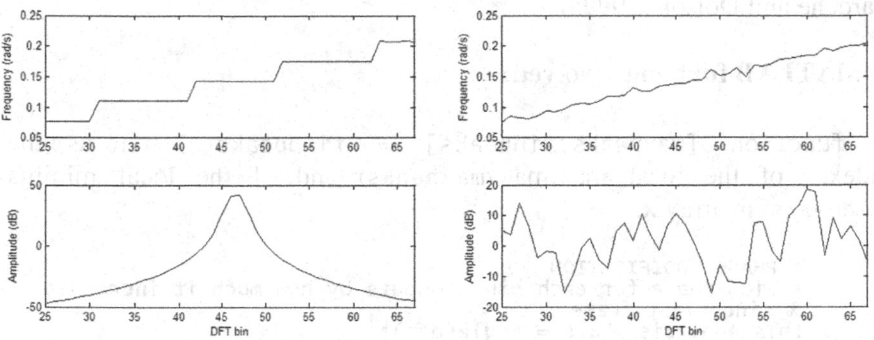

Fig. 5.20 Estimated (normalized) frequency of the partials in DFT bins 25–65 (*top*) and spectral amplitude of the same bins (*bottom*), in a phase vocoder. *Left*: In the sinusoid; *right*: in more complex polyphonic music

However, for more complex sounds (as in the last frame of our test signal), the estimated partial frequency in neighboring bins vary a lot (Fig. 5.20, right).

```
subplot(2,1,1);
plot((0:NFFT/2-1),partials_freq);
set(gca,'xlim',[25 67]);

subplot(2,1,2);
dft=fft(analysis_frame);
dft=20*log10(abs(dft(1:NFFT/2)));
plot((0:NFFT/2-1),dft);
set(gca,'xlim',[25 67]);
```

As a result, although the spectrogram of the time-scaled signal looks similar to that of the original signal (except, of course, for the time axis), it exhibits significant phasiness and transient smearing.

The phase-locked vocoder

We have seen in the previous paragraphs that in the case of a sinusoidal input signal, the estimated frequency of the main partial in several neighboring DFT bins (around the frequency of the sinusoid) is somehow constant. One of the reasons of the phasiness in the previous version of the phase vocoder comes from the fact that it does not enforce this effect. If other small-amplitude partials are added to the sinusoid, each DFT bin computes its own partial frequency so that these estimations in neighboring bins will coincide only by chance. The phase-locked vocoder changes this by locking the estimation of partial frequencies in bins surrounding spectral peaks. This is called *vertical phase locking*. In the following MATLAB lines, we give an implementation of the phase modification stage in the identity phase-locking scheme proposed in Laroche and Dolson (1999a).

MATLAB function involved:

- `function [maxpeaks,minpeaks] = findpeaks(x)` returns the indexes of the local maxima (`maxpeaks`) and of the local minima (`minpeaks`) in array `x`.

```
% PHASE MODIFICATION
% Find phase for each bin, compute by how much it increased
% since last frame.
this_analysis_phase = angle(dft);
delta_phase = this_analysis_phase - last_analysis_phase;
phase_increment=delta_phase-analysis_frame_shift*...
    DFT_bin_freqs;

% Find peaks in DFT.
peaks = findpeaks(abs(dft));

% Estimate the frequency of the main partial for each bin
principal_determination(peaks)=...
    mod(phase_increment(peaks)+pi,2*pi)-pi;
partials_freq(peaks)=principal_determination(peaks)/...
    analysis_frame_shift+DFT_bin_freqs(peaks);

% Find regions of influence around peaks
regions = round(0.5*(peaks(1:end-1)+peaks(2:end)));
regions = [1;regions;NFFT/2];

% Set the frequency of partials in regions of influence to
% that of the peak (this is not strictly needed; it is used
% for subsequent plots)
for i=1:length(peaks)
    partials_freq(regions(i):regions(i+1)) = ...
        partials_freq(peaks(i));
end

% Update the phase for each peak
this_synthesis_phase(peaks)=last_synthesis_phase(peaks)+...
    synthesis_frame_shift*partials_freq(peaks);
```

```
% force identity phase locking in regions of influence
for i=1:length(peaks)
    this_synthesis_phase(regions(i):regions(i+1)) = ...
        this_synthesis_phase(peaks(i)) +...
        this_analysis_phase(regions(i):regions(i+1))-...
        this_analysis_phase(peaks(i));
end
```

As a result of vertical phase locking, a much larger number of bins around the main spectral line of our sinusoid have been assigned the same partial frequency. More importantly, this property has been enforced in more complex sounds, as shown in the last frame of our test signal (Fig. 5.21).

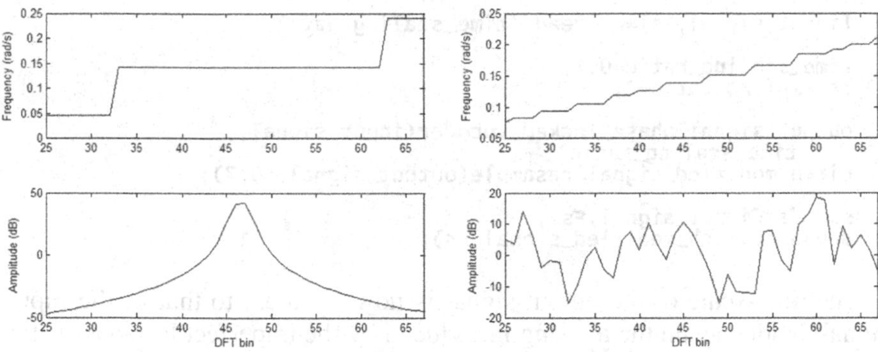

Fig. 5.21 Estimated (normalized) frequency of the partials in DFT bins 25–65 (*top*) and spectral amplitude of the same bins (*bottom*), in a phase-locked vocoder. *Left*: In the sinusoid; *right*: in more complex polyphonic music

The result is improved, although it is still far from natural. It is possible to attenuate transient smearing by using separate analysis windows for stationary and transient sounds (as done in the MPEG audio coders). We do not examine this option here.

5.2.3 Pitch modification

The simplest way to produce a modification of the frequency axis of a signal is simply to resample it, as already seen in Section 2.1. This method, however, also produces time modification. In this section, we focus on methods for modifying the pitch of an input signal without changing its duration.

Time-scale modification and resampling
In order to avoid the time-scaling effect of a simple resampling-based approach, a compensatory time-scale modification can be applied using a phase vocoder. Let us multiply the pitch by a factor 0.7, for instance. We first multiply the duration of the input signal by 0.7 and then resample it by 10/7 so as to recover the original number of samples.

MATLAB function involved:

function [output_signal] = phase_locked_vocoder(input_signal, time_scaling_ratio) is a simple implementation of the phase-locked vocoder described in Laroche and Dolson (1999a).

```
[input_signal,Fs]=wavread('time_scaling.wav');

time_scaling_ratio=0.7;
resampling_ratio=2;

output_signal=phase_locked_vocoder(input_signal,...
    time_scaling_ratio);
pitch_modified_signal=resample(output_signal,10,7);

soundsc(input_signal,Fs);
soundsc(pitch_modified_signal,Fs);
```

The time-scale of the output signal is now identical to that of the input signal. Notice again the aliasing introduced by the imperfect low-pass filter used by `resample` (Fig. 5.22, left).

```
specgram(pitch_modified_signal(1:end-5000),frame_length,...
    Fs,window);
```

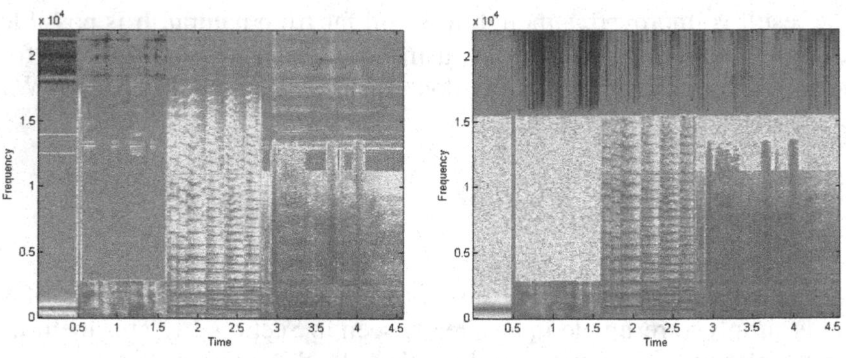

Fig. 5.22 Spectrogram of the test signal after pitch modification by a factor 0.7 using a phase vocoder followed by a resampling stage (*left*) and a fully integrated STFT-based approach (*right*)

The STFT-based approach

It is also possible, and much easier, to perform pitch modification in the frequency domain by translating partials to new frequencies, i.e., inside the phase vocoder. This technique is much more flexible, as it allows for nonlinear modification of the frequency axis.

Note that the implementation we give below is the simplest possible version; the peak frequencies are not interpolated for better precision, and the shifting is rounded to an integer number of bins. This will produce modulation artifacts on sweeping sinusoids and phasing artifacts on audio signals such as speech. Moreover, when shifting the pitch down, peaks are truncated at DC, whereas they should be reflected with a conjugation sign.

```
NFFT=2048;
frame_length=NFFT;
frame_shift=frame_length/4;
window=hanning (frame_length,'periodic');
COLA_ratio=sum(window.*window)/frame_shift;
window=window/sqrt(COLA_ratio);

pitch_scaling_ratio=0.7;

% Central frequency of each DFT channel
DFT_bin_freqs=((0:NFFT/2-1)*2*pi/NFFT)';

pin=0;pout=0;

output_signal=zeros(length(input_signal),1);
accumulated_rotation_angles=zeros(1,NFFT/2);

while pin+frame_length<length(input_signal)

  % STFT
  analysis_frame = input_signal(pin+1:pin+frame_length)...
     .* window;
  dft=fft(fftshift(analysis_frame));
  dft=dft(1:NFFT/2);

  % Find peaks in DFT.
  peaks = findpeaks(abs(dft));

  % Find regions of influence around peaks
  regions = round(0.5*(peaks(1:end-1)+peaks(2:end)));
  regions = [1;regions;NFFT/2];

  % Move each peak in frequency according to modification
  % factor.

  modified_dft = zeros(size(dft));
  for u=1:length(peaks)

     % Locate old and new bin.
     old_bin = peaks(u)-1;
     new_bin = round(pitch_scaling_ratio*old_bin);
```

```
        % Be sure to  stay within 0-NFFT/2 when
        % shifting/copying the peak bins
        if(new_bin-old_bin+regions(u) >= NFFT/2) break; end;
        if(new_bin-old_bin+regions(u+1) >= NFFT/2)
            regions(u+1) = NFFT/2 - new_bin + old_bin;
        end;
        if(new_bin-old_bin+regions(u) <= 0)
            regions(u) = 1 - new_bin + old_bin;
        end;

        % Compute the rotation angle required, which has
        % to be cumulated from frame to frame
        rotation_angles=accumulated_rotation_angles(old_bin+1)...
            + 2*pi*frame_shift*(new_bin-old_bin)/NFFT;

        % Overlap/add the bins around the peak, changing the
       % phases accordingly
       modified_dft(new_bin-old_bin+(regions(u):regions(u+1)))= ...
         modified_dft(new_bin-old_bin+(regions(u):regions(u+1)))...
         + dft(regions(u):regions(u+1)) * exp(j*rotation_angles);
       accumulated_rotation_angles((regions(u):regions(u+1))) = ...
         rotation_angles;

    end

    % ISTFT
    modified_dft(NFFT/2+2:NFFT)=fliplr(modified_dft(2:NFFT/2)');
    synthesis_frame = fftshift(real(ifft(modified_dft))).* window;
    output_signal(pout+1:pout+frame_length) = ...
       output_signal(pout+1:pout+frame_length)+synthesis_frame;

    pin=pin+frame_shift;
    pout=pout+frame_shift;

end;
```

The time-scale of the output signal is still identical to that of the input signal, and the frequency content above `Fs*time_scaling_ratio` is set to zero (Fig. 5.22, right).

```
specgram(output_signal(1:end-5000),frame_length,Fs,window);
```

5.3 Going further

For people interested in the use of transforms for audio processing, Sethares (2007) is an excellent reference.

An important research area for phase vocoders is that of transient detection, for avoiding the transient smearing effect mentioned in this chapter (see Röbel 2003, for instance).

One of the extensions of phase vocoders that have not been dealt with in this chapter is the fact that not only phases but also amplitudes should be modified when performing time-scale modifications. In the frequency domain, a chirp sinusoid has a wider spectral lobe than a constant frequency sinusoid. Exploiting this fact, however, would imply an important increase in computational and algorithmic complexity. Sinusoidal modeling techniques (McAulay and Quatieri 1986), which have been developed in parallel to the phase vocoder, are more suited to handle such effects.

For monophonic signals, such as speech, a number of time-domain techniques known as synchronized OLA (SOLA) have also been developed with great success for time-scaling and pitch modification, given their very low computational cost compared to the phase vocoder. The wavefom similarity OLA (WSOLA) and pitch-synchronous (PSOLA) techniques are presented in a unified framework in Verhelst et al. (2000). Their limitations are exposed in Laroche (1998).

The multiband resynthesis OLA (MBROLA) technique (Dutoit 1997) attempts to take the best of both worlds: it performs off-line frequency-domain modifications of the signal to make it more amenable to time-domain time-scaling and pitch modification. Recently, Laroche (2003) has also proposed a vocoder-based technique for monophonic sources, relying on a (possibly online) pitch detection and able to perform very flexible pitch and formant modifications.

5.4 Conclusion

Modifying the duration or the pitch of a sound is not as easy as it might seem at first sight. Simply changing the sampling rate only partially does the job. In this chapter, we have implemented a phase-locked vocoder and tested it for time-scale and pitch-scale modification. We have seen that processing the signal in the STFT domain provides better results, although not perfect yet. One of the reasons for the remaining artifacts lies in the interpretation of the input signal in terms of partials, which are neither easy to spot nor even always clearly defined (in transients, for instance). For pitch shifting, we have also implemented an all-spectral approach, which is simpler than using a phase coder followed by a sampling rate converter, but still exhibits artifacts.

References

Arfib D, Keiler F, Zölzer U (2002) Time-frequency processing. In: DAFX: Digital Audio Effects. U. Zölzer, Ed. Hoboken (NJ): John Wiley & Sons

Bagdasarian R (1958) The Chipmunk song (Christmas don't be late). Jet Records, UK

Dolson M (1986) The phase vocoder: A tutorial. Computer Music Journal, 10(4): 14–27

Dutoit T (1997) Time-domain algorithms. In: An Introduction to Text-To-Speech Synthesis. Dutoit T. Dordrecht: Kluwer Academic Publishers

Flanagan L, Golden RM (1966) Phase Vocoder. Bell System Technical Journal, pp 1493–1509 [online] Available: http://www.ee.columbia.edu/~dpwe/e6820/papers/FlanG66.pdf[07/06/2007]

Laroche J (1998) Time and pitch-scale modification of audio signals. In: Applications of Digital Signal Processing to Audio Signals. M. Kahrs and K. Brandenburg, Eds. Norwell, MA: Kluwer Academic Publishers

Laroche J (2003) Frequency-domain techniques for high quality. Voice Modification. In: Proceedings of DAFX-03, 322–328

Laroche J, Dolson M (1999a) Improved Phase Vocoder Time-Scale Modification of Audio. IEEE Transactions on Speech and Audio Processing, 3: 323–332

Laroche J, Dolson M (1999b) New Phase Vocoder Technique for Pitch-Shifting, Harmonizing and Other Exotic Effects. IEEE Workshop on Applications of Signal Processing to Audio and Acoustics. Mohonk, New Paltz, NY [online]Available: http://www.ee.columbia.edu/~dpwe/papers/LaroD99-pvoc.pdf [07/06/2007]

McAulay R, Quatieri T (1986) Speech analysis/Synthesis based on a sinusoidal representation. IEEE Transactions on Acoustics, Speech, and Signal Processing 34 (4): 744–754

Puckette MS (1995) Phase-locked Vocoder. Proceedings of the IEEE Conference on Applications of Signal Processing to Audio and Acoustics, Mohonk

Röbel A (2003) A new approach to transient processing in the phase vocoder. Proceedings of the 6th International. Conference on Digital Audio Effects (DAFx-03), 344–349

Sethares WA (2007) Rhythm and Transforms. London: Springer Verlag [online] Available: http://eceserv0.ece.wisc.edu/~sethares/vocoders/Transforms.pdf [20/3/08]

Smith JO (2007a) Example of overlap-add convolution. In: Spectral Audio Signal Processing, Center for Computer Research in Music and Acoustics (CCRMA), Stanford University [online] Available: http://ccrma.stanford.edu/~jos/sasp/Example_Overlap_Add_Convolution.html [14/06/07]

Smith JO (2007b) Dual of constant overlap-add. In: Spectral Audio Signal Processing, Center for Computer Research in Music and Acoustics (CCRMA), Stanford University [online] Available: http://ccrma.stanford.edu/~jos/sasp/ Example_Overlap_Add_Convolution.html [14/06/07]

Verhelst V, van Compernolle D, Wambacq P (2000). A unified view on synchronized overlap-add methods for prosodic modifications of speech. In Proc. ICSLP2000, 2: 63–66

Chapter 6

How can marine biologists track sperm whales in the oceans?

T. Dutoit (°), V. Kandia(+), Y. Stylianou (*)

(°) Faculté Polytechnique de Mons, Belgium
(+) Foundation for Research and Technology-Hellas, Heraklion, Greece
(*) University of Crete, Heraklion, Greece

Hear they are!

Whale watching has become a trendy occupation these last years. It is carried on in the waters of some 40 countries, plus Antarctica. Far beyond its touristic aspects, being able to spot the position of whales in real time has several important scientific applications, such as censusing (estimation of animal population), behavior studies, and mitigation efforts concerning fatal collisions between marine mammals and ships and exposure of marine mammals to loud sounds of anthropogenic origin (seismic surveys, oil and gas exploitation and drilling, naval and other uses of sonar).

When whales do not appear on the surface of water, one way of detecting their position is to listen to the sounds they emit through several fixed *hydrophones* (i.e., microphones specially designed for underwater use).

Hydrophone signals are then processed with a sound navigation and ranging (*sonar*) technique, which uses sound propagation under water to detect and spot objects. In this chapter, we thus plunge deep into underwater acoustics to examine a proof of concept for sperm whale tracking using *passive acoustics*,[1] and based on the estimation of the cross-correlation between hydrophone signals to spot whales on a 2D map.

---

[1] Sonars can be *active* or *passive*. Passive sonars only listen to their surrounding, while active sonars emit sounds and detect echoes from their surrounding.

T. Dutoit, F. Marqués (eds.), *Applied Signal Processing*,
DOI 10.1007/978-0-387-74535-0_6,

6.1 Background – Source localization

We start by examining sperm whale sounds (Section 6.1.1) and show how the Teager–Kaiser operator efficiently increases the signal-to-noise ratio of hydrophone signals (Section 6.1.2).

Sperm whale sounds are actually received by hydrophones at slightly different times due to the various distances between the animal and the hydrophones. Knowing the location of the hydrophones, the *time difference of arrival* (TDOA) between pairs of hydrophones can be obtained, using various techniques (Sections 6.1.3 and 6.1.4), and provided to a *multilateration* algorithm (Section 6.1.5) for determining the position of the animal.

6.1.1 Sperm whale sounds

Sperm whales are highly vocal-active animals. An adult sperm whale produces some 25,000 clicks a day. If we consider that the heart rate of such a large whale is around 15 beats per minute (i.e., 20,000 beats a day), sperm whale produces more clicks than heart beats (Madsen 2002). Their repertoire is made up almost entirely of a number of click types with different properties.

Usual (or *regular*) *clicks* (Fig. 6.1) are the most commonly heard click type during deep foraging dives and are therefore used to locate the animals using passive acoustic methods. They are impulsive broadband sounds of multipulse structure with interclick interval (ICI) between 0.5 and 1 s. Usual clicks are highly directional sounds with source levels up to 235 dB rms re 1 μPa.[2] It is believed that these properties represent adaptations for long-range echolocation.

Creak clicks (Fig. 6.2) are burst of monopulsed clicks with high repetition rate (up to 200 clicks per second). They are highly directional sounds with source level between 180 and 205 dB rms re 1 μPa. They are

---

Whales and dolphins use echolocation systems similar to active sonars to locate predators and preys; marine biologists listen to the resulting sounds and echoes (hence in a passive setup) to detect cetaceans.

[2] For underwater sound, 1 μPa is the reference pressure level. Since the measured pressure level from a particular sound source decreases with distance from the source, the convention is to use 1 m as a reference distance. Thus, the pressure level from a sound source is measured in “dB rms re 1 μPa at 1 m” (and the “at 1 m” is usually omitted).

produced during foraging dives, and it is believed that they have a function analogous to the terminal buzzes produced by bats during echolocation.[3]

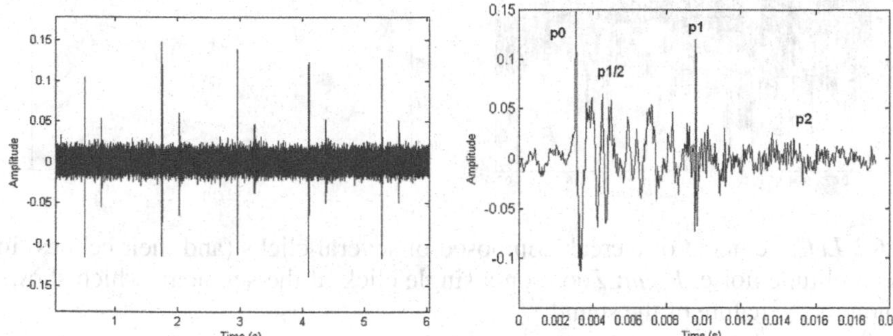

Fig. 6.1 *Left*: Sequence of usual sperm whale clicks. Each click is followed by an echo due to surface reflection.[4] *Right*: Zoom on a single click showing the multipulse structure of the signal. The notation on pulses follows that of Møhl et al. (2000) and Zimmer et al. (2005)

Several other types of sounds, such as *coda clicks*, *chirrup clicks*, *slow clicks*, *squeals*, and *trumpets,* most of which have a social communicative role, are also encountered.[5]

As shown in Figs. 6.1 and 6.2, hydrophone signals exhibit very low signal-to-noise ratios. In particular, they are often contaminated by low-frequency noise, mostly due to human activity. Shipping indeed, which accounts for more than 75% of all human sound in the sea (ICES 2005), produces low-frequency sounds (rumble of engines and propellers). Commercial shipping traffic is growing as does the tonnage (cargo capacity) of ships, adding more noise in the low-frequency band.

[3] After detecting a potential prey using low-rate echolocation calls, bats emit a characteristic series of calls at a high repetition rate (a *terminal buzz*) to localize the prey.

[4] Hydrophones are mounted at roughly 5 m off the bottom so that reflection from the bottom comes with a ~7-ms delay, i.e., inside the direct path click itself. Furthermore, the hydrophones have an upward-directed beam pattern.

[5] For more information on sperm whale sounds, please refer to Madsen (2002), Drouot (2003), and Teloni et al. (2005).

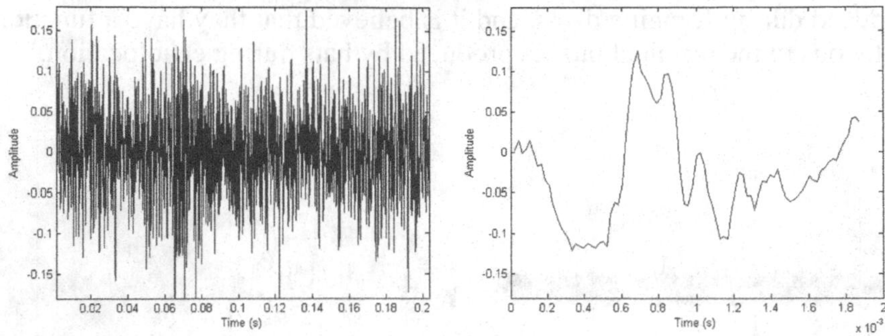

Fig. 6.2 *Left*: Segment of a creak composed of several clicks (and their echoes) in high-amplitude noise. *Right*: Zoom on a single click of the segment, which shows the monopulse nature of the sound

6.1.2 The Teager–Kaiser energy operator

The Teager–Kaiser (TK) energy operator is defined in the continuous domain as

$$TK[x(t)] = \dot{x}(t)^2 - x(t)\ddot{x}(t) \tag{6.1}$$

where $\dot{x}$ and $\ddot{x}$ denote the first and second derivatives over time, respectively. For a discrete time signal, it is shown in Kaiser (1990) that the TK energy operator is given by

$$TK[x(n)] = x^2(n) - x(n+1)x(n-1) \tag{6.2}$$

The TK operator is referred to as *energy operator* because it is related to the concept of energy in the generation of acoustic waves (Kaiser 1990).

This operator can be seen as a special case of *quadratic filters* defined by

$$y(n) = \sum_{i_1=-\infty}^{\infty} \sum_{i_2=-\infty}^{\infty} h(i_1,i_2)x(n-i_1)x(n-i_2) \tag{6.3}$$

where $h(i_1,i_2)$ is known as the second-order *Volterra kernel.*[6] As a matter of fact, the TK energy operator is obtained from Equation (6.2) when

[6] Quadratic filters are themselves a simple case of the more general *polynomial* or *Volterra filters* (Sicuranza 1992):

$$h(i_1,i_2)=\begin{cases}1 & i_1=i_2=0\\ -1/2 & (i_1,i_2)=(1,-1)\ or\ (-1,1)\\ 0 & \text{otherwise}\end{cases} \tag{6.4}$$

The analysis of such nonlinear filters is not trivial. Since their output to the sum of inputs is not the sum of their outputs to isolated inputs, the classical notions of impulse and frequency responses no longer have the usual reach. When the input is composed of a sum of sinusoids, for instance, cross-product terms appear in the output. In the case of the TK operator, it is easy to check from Equation (6.2) that the response of this filter to $\delta(n)$ is $\delta(n)$ and that its response to an isolated cosine $A\cos(n\varphi_0)$ is a constant signal $A^2\sin^2(\varphi_0)$. This response, however, cannot be interpreted as a frequency response and is not the Fourier transform of the impulse response.

In the specific case of noisy impulsive signals, the TK operator has interesting properties. Let us assume that the signal $x(n)$ recorded by a hydrophone is composed of the sum of a low-frequency interference signal $i(n)$, of the impulsive click signal $s(n)$ produced by a sperm whale, and of some background wideband noise $u(n)$:

$$x(n)=s(n)+i(n)+u(n) \tag{6.5}$$

If the interference frequency is low and the signal-to-noise ratio (SNR) between the impulsive clicks and the background noise is sufficient, it is shown in Kandia and Stylianou (2005) that the TK operator has the very interesting property of ignoring the interference component while increasing the SNR (Fig. 6.3). Moreover, it does not smear input pulses in time (Fang and Atlas 1995). As a matter of fact, if we neglect the influence of the noise $u(n)$ and identify $x(n)$ to a Dirac pulse (n) and $i(n)$ in the vicinity of this pulse to a constant value K, it is easy to show from

---

$$\begin{aligned} y(n) &= \sum_{i_1=-\infty}^{\infty} h(i_1)x(n-i_1)\\ &+\sum_{i_1=-\infty}^{\infty}\sum_{i_2=-\infty}^{\infty} h(i_1,i_2)x(n-i_1)x(n-i_2)\\ &+\dots\\ &+\sum_{i_1=-\infty}^{\infty}\dots\sum_{i_N=-\infty}^{\infty} h(i_1,\dots,i_N)x(n-i_1)\dots x(n-i_N)\end{aligned}$$

Equation (6.2) that the output $y(n)$ of the TK filter has only three nonzero samples:

$$y(n)=\begin{cases}-K & n=-1\\ 1-2K & n=0\\ -K & n=1\end{cases} \tag{6.6}$$

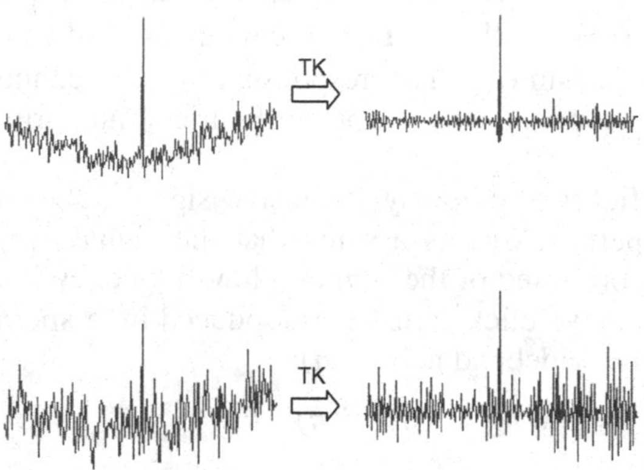

Fig. 6.3 Applying the Teager–Kaiser operator to simulated clicks corrupted by low-frequency interference and background noise with various SNR efficiently preprocesses the data for later click detection or TDOA estimation (Kandia and Stylianou 2006)

The application of the TK operator to sperm whale sounds therefore provides a good example of the efficient use of a nonlinear filter for making the data more amenable to further TDOA estimation.

6.1.3 TDOA estimation based on the generalized cross-correlation

Let us assume that signals $x_1(t)$ and $x_2(t)$ result from the propagation of signal $s(t)$ through different paths that are identified to a simple attenuation and a delay:

$$x_i(t)=a_i s(t-\tau_i)+b_i(t) \tag{6.7}$$

where $b_i(t)$ (i=1,2) are zero-mean, uncorrelated stationary random processes, which are also noncorrelated with $s(t)$.

The *cross*-power spectrum density (PSD) $S_{x_1x_2}(f)$ between x_1 and x_2 is given by

$$
\begin{aligned}
S_{x_1x_2}(f) &= \int_{-\infty}^{\infty} \phi_{x_1x_2}(t) e^{-j2\pi ft} dt \\
&= a_1 a_2 e^{-j2\pi f\tau_{12}} \int_{-\infty}^{\infty} \phi_{ss}(t) e^{-j2\pi ft} dt \\
&= a_1 a_2 e^{-j2\pi f\tau_{12}} S_{ss}(f)
\end{aligned}
\tag{6.8}
$$

in which $\phi_{x_1x_2}(t)$ is the cross-correlation between $x_1(t)$ and $x_2(t)$, $\phi_{ss}(t)$ is the autocorrelation of $s(t)$, and $S_{ss}(f)$ is therefore the PSD of $s(t)$.

Knapp and Carter (1976) have proposed to estimate the *time difference of arrival* TDOA $\tau_{12} = \tau_1 - \tau_2$ between $x_1(t)$ and $x_2(t)$ as the position of the maximum of the *generalized cross-correlation function*, defined as

$$
\psi_{x_1x_2}(\tau) = \int_{-\infty}^{+\infty} \Phi(f) S_{x_1x_2}(f) e^{j2\pi f\tau} df \tag{6.9}
$$

in which $\Phi(f)$ is a weighting function.

In particular, when $\Phi(f)$ is set to 1, $\psi_{x_1x_2}(\tau)$ is the inverse Fourier transform of the cross PSD, i.e., the standard cross-correlation function. For signals verifying Equation (6.8), this leads to

$$
\psi_{x_1x_2}(\tau) = a_1 a_2 \phi_{ss}(\tau - \tau_{12}) \tag{6.10}
$$

in which $\phi_{ss}(\tau)$ is the autocorrelation function of $s(t)$. Since the maximum of $\phi_{ss}(\tau)$ is always found at $\tau = 0$, τ_{12} can be estimated as the position of the maximum of Equation (6.10). However, if $b_1(t)$ and $b_2(t)$ exhibit some cross-correlation, Equation (6.10) becomes

$$
\psi_{x_1x_2}(\tau) = a_1 a_2 \phi_{ss}(\tau - \tau_{12}) + \phi_{b_1b_2}(\tau) \tag{6.11}
$$

whose maximum may not correspond to τ_{12}. In particular, if $b_i(t)$ are sinusoidal components with the same frequency, a sinusoidal term will appear in Equation (6.11) due to a spectral line in $S_{x_1x_2}(f)$.

When $\Phi(f)$ is set to $1/|S_{x_1x_2}(f)|$, we obtain the so-called *phase transform,* which computes the generalized cross-correlation from the phase of the cross PSD. For signals verifying Equation (6.8), the phase transform still provides a perfect estimate of τ_{12} since

$$\Phi(f)S_{x_1x_2}(f)=\frac{S_{x_1x_2}(f)}{|S_{x_1x_2}(f)|}=e^{-j2\pi f\tau_{12}} \tag{6.12}$$

which leads to

$$\psi_{x_1x_2}(\tau)=\delta(\tau-\tau_{12}) \tag{6.13}$$

Now if $b_i(t)$ are sinusoidal components with the same frequency, their contributions to $\psi_{x_1x_2}(\tau)$ will be much lower than in Equation (6.11), since the spectral line in $S_{x_1x_2}(f)$ will be canceled in Equation (6.12). This makes the phase transform an interesting estimator, provided $S_{x_1x_2}(f)$ can itself be correctly estimated.

In practice, $S_{x_1x_2}(f)$ is estimated in $[0,F_s]$ from a finite number of samples of $x_1(n)$ and $x_2(n)$, as (Fig. 6.4)

$$S_{x_1x_2}\left(k\frac{F_s}{N}\right)\cong X_1(k)X_2^*(k) \quad (k=0,\ldots,N-1) \tag{6.14}$$

where $X_i(k)$ $(i=1,2)$ is the N-point discrete Fourier transform of the sequence $[x_i(n),x_i(n-1),\ldots,x_i(n-N+1)]$. This is known to give a biased but consistent estimation of $S_{x_1x_2}(f)$.

TDOA estimation from *real* signals is not so easy. As mentioned above, hydrophone signals are polluted with interference signals and additive noise due to surrounding sources other than the one we want to spot. More annoyingly, several propagation paths may exist to each hydrophone: typically a direct path with lowest attenuation plus many secondary paths, which create echoes (one of which being due to surface reflection) and reverberations. As a result, a more realistic model for hydrophone signals x_1 and x_2 receiving a click $s(t)$ after propagation is given by

$$x_i(t)=h_i(t)*s(t)+b_i(t) \tag{6.15}$$

where * denotes convolution, $h_i(t)$ is the acoustic impulse response of the channel between the whale and the ith hydrophone (including reverberation), and $b_i(t)$ is the additive noise received by this hydrophone. Consequently, applying the phase transform to real signals does not always lead to efficient estimation of TDOAs. Several other weighting functions have been proposed for the estimation of the general cross-correlation in Equation (6.9), and proved to provide interesting results in specific cases (Knapp and Carter 1976).

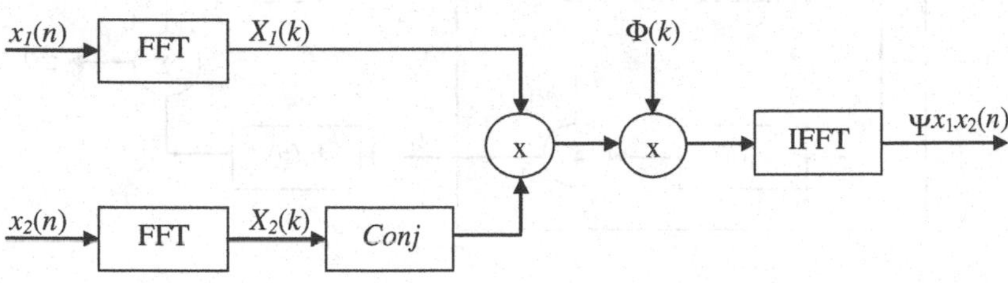

Fig. 6.4 Practical computation of the generalized cross-correlation function

6.1.4 Adaptive TDOA estimation

A radically different approach was proposed in Benesty (2000), which leads to a simple and elegant adaptive filtering implementation, in the framework of the least-means-square (LMS) algorithm. This approach is based on the estimation of the impulse responses from the source to the receivers.

If we neglect the noise components in Equation (6.15), we have (Fig. 6.5)

$$\begin{aligned} x_1(n) * h_2(n) &= s(n) * h_1(n) * h_2(n) \\ &= s(n) * h_2(n) * h_1(n) \\ &= x_2(n) * h_1(n) \end{aligned} \qquad (6.16)$$

which can be written in vector notation as

$$\mathbf{x}_1^T(n)\mathbf{h}_2 = \mathbf{x}_2^T(n)\mathbf{h}_1 \qquad \text{for all } n \qquad (6.17)$$

where

$$\mathbf{x}_i(n) = [x_i(n), x_i(n-1), \cdots, x_i(n-M+1)]^T \quad (i=1,2) \qquad (6.18)$$

and

$$\mathbf{h}_i = [h_i(0), h_i(1), \cdots, h_i(M-1)]^T \quad (i = 1,2) \tag{6.19}$$

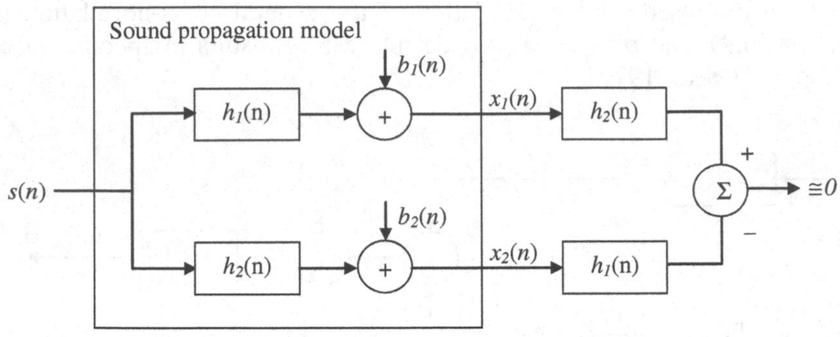

Fig. 6.5 If additive noise is neglected, passing $x_1(n)$ through h_2 gives the same result as passing $x_2(n)$ through h_1.

If we stack $\mathbf{x}_1(n)$ and $\mathbf{x}_2(n)$ as

$$\mathbf{x}(n) = \begin{bmatrix} \mathbf{x}_1(n) \\ \mathbf{x}_2(n) \end{bmatrix} \tag{6.20}$$

then the covariance matrix of $\mathbf{x}(n)$ is given by

$$\mathbf{R}_{\mathbf{xx}} = \begin{bmatrix} \mathbf{R}_{\mathbf{x}_1\mathbf{x}_1} & \mathbf{R}_{\mathbf{x}_1\mathbf{x}_2} \\ \mathbf{R}_{\mathbf{x}_2\mathbf{x}_1} & \mathbf{R}_{\mathbf{x}_2\mathbf{x}_2} \end{bmatrix} \tag{6.21}$$

with

$$\mathbf{R}_{\mathbf{x}_i\mathbf{x}_j} = E\left[\mathbf{x}_i(n)\mathbf{x}_j^T(n)\right] \quad (i, j = 1,2) \tag{6.22}$$

It then follows from Equations (6.17) and (6.21) that

$$\mathbf{Ru} = 0 \quad \text{with } \mathbf{u} = \begin{bmatrix} \mathbf{h}_2 \\ -\mathbf{h}_1 \end{bmatrix} \tag{6.23}$$

This remarkably simple result provides a simple means of estimating the impulse responses $\mathbf{h}_1$ and $\mathbf{h}_2$ from the estimation of $\mathbf{R}_{\mathbf{xx}}$.

In practice, though, accurate estimation of **u** is not trivial, as the impulse responses may be long and background noise may falsify (Equation (6.23). Benesty (2000) has therefore proposed an adaptive estimation of **u,** based on a least-mean-square (LMS) principle. The idea is to find an error function $e(n)$ such that the expectation $E[e^2(n)]$ is minimized when Equation (6.23) is verified and such that the gradient of $E[e^2(n)]$ with respect to $\mathbf{u}$ has a simple analytical form. Clearly, $e(n)=\mathbf{u}^T\mathbf{x}$ meets these requirements, as $E(e^2(n))=\mathbf{u}^T\mathbf{R}\mathbf{u}$ and $\nabla_{\mathbf{u}}(\mathbf{u}^T\mathbf{R}\mathbf{u}) = 2\mathbf{R}\mathbf{u}$. Minimizing it on $\mathbf{u}$ can then be achieved by the *gradient descent* approach (Fig. 6.6): by starting from an initial guess and updating it in the direction of the steepest descent, i.e., in the direction of the negative gradient of the cost function with respect to $\mathbf{u}$

$$\begin{aligned}\mathbf{u}^{new} &= \mathbf{u}^{old} - \frac{\mu}{2}\nabla_{\mathbf{u}}(\mathbf{u}^T\mathbf{R}\mathbf{u})\Big|_{\mathbf{u}=\mathbf{u}^{old}} \\ &= \mathbf{u}^{old} - \mu\mathbf{R}\mathbf{u}^{old}\end{aligned} \tag{6.24}$$

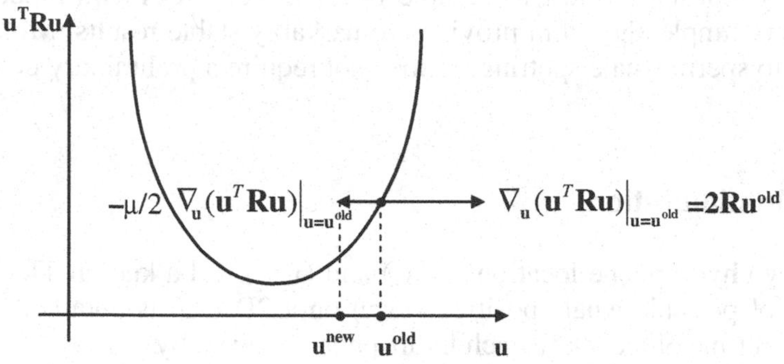

Fig. 6.6 A schematic 2D view of the gradient descent approach

The classical simplification of the LMS approach (Widrow and Stearns 1985) is then to replace $\mathbf{R}$ in Equation (6.24) by its very short term estimate $E(\mathbf{x}\mathbf{x}^T)\approx\mathbf{x}\mathbf{x}^T$ and to update $\mathbf{u}$ every sample, leading to the following update equation for $\mathbf{u}(n)$:

$$\begin{aligned}\mathbf{u}(n+1) &= \mathbf{u}(n) - \mu\mathbf{x}(n)\mathbf{x}^T(n)\mathbf{u}(n) \\ &= \mathbf{u}(n) - \mu\mathbf{x}(n)e(n)\end{aligned} \tag{6.25}$$

It is also suggested by Benesty (2000) to constrain $\mathbf{u}(n)$ to unitary norm, in order to avoid round-off error propagation and to avoid selecting the obvious $\mathbf{u}$=0 solution in Equation (6.23). This is achieved by changing Equation (6.25) into

$$\mathbf{u}(n+1) = \frac{\mathbf{u}(n) - \mu\mathbf{x}(n)e(n)}{\| \mathbf{u}(n) - \mu\mathbf{x}(n)e(n) \|} \tag{6.26}$$

Additionally, as only the TDOA is required, one only needs to estimate the direct paths in $\mathbf{h}_1$ and $\mathbf{h}_2$ (as opposed to the complete impulse responses). A simple solution is to initialize $\mathbf{h}_2$ (the first half of $\mathbf{u}$) to a single Dirac pulse. A "mirror" effect follows from Equation (6.23) in the estimate of $\mathbf{h}_1$ (the second half of $\mathbf{u}$): a negative dominant peak appears, which is an estimation of the direct path of $\mathbf{h}_1$. The expected TDOA can then be computed as time difference between these peaks within their respective impulse responses. Since the TDOA can a priori be positive or negative, the initial Dirac pulse is positioned at the center of $\mathbf{h}_2$, and the value of M is set to twice the maximum value of the TDOA.

Provided the adaptation step μ is correctly set (not too high to avoid instability, but not too low to be able to track the TDOA as a function of time), this simple algorithm provides remarkably stable results. Moreover, applied to sperm whale spotting, it does not require a preliminary detection of clicks.

6.1.5 Multilateration

Given two hydrophone locations (x_1, y_1) and (x_2, y_2) and a known TDOA τ_{12}, the loci of possible whale positions (x, y) on a 2D map is obtained easily. The travel time of a click to each hydrophone is given by

$$\tau_i = \frac{1}{c}\left(\sqrt{(x-x_i)^2 + (y-y_i)^2}\right) \tag{6.27}$$

where c is the propagation speed of sound in water (typically 1,510 m/s). The locus is thus defined by

$$\tau_{ij} = \tau_i - \tau_j = \frac{1}{c}\left(\sqrt{(x-x_i)^2 + (y-y_i)^2} - \sqrt{(x-x_j)^2 + (y-y_j)^2}\right) \tag{6.28}$$

This corresponds to a half-hyperbola whose foci are the hydrophones and whose semi-major (or transverse) axis is given by $c\tau_{ij}/2$ (Fig. 6.7), where c is the speed of sound in water.

Obtaining the 2D position of the whale therefore requires two TDOAs (i.e., three hydrophones, which implicitly offer a third TDOA) a 2D *multilateration* system solves a system of two nonlinear equations such as Equation (6.28) to find the position of a whale in real time.

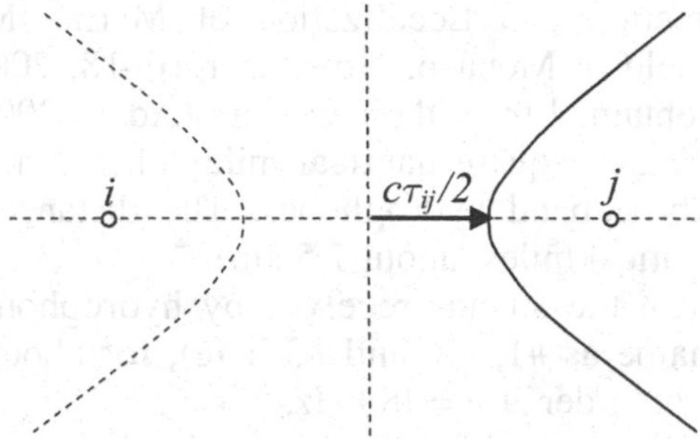

Fig. 6.7 The locus of whale positions for a given value of the TDOA $_{ij}$ between hydrophones i and j

In practice, though, the half-hyperbola defined by the third TDOA may not have the same intersection as the first two ones, since TDOA estimates may not be very precise. It is then useful to increase the number of hydrophones. The location problem must then be restated as an optimization problem and solved by a least-squares approach or with a Kalman filter (Huang et al. 2008). To keep things simple, we do not address this problem here.

6.2 MATLAB proof of concept: ASP_audio_effects.m

In this section, we analyze sperm whale clicks received by several hydrophones (Section 6.2.1), and use them to find the position of the cetacean. We start by processing the clicks with the (nonlinear) Teager–Kaiser operator (Section 6.2.2) so as to increase their signal-to-noise ratio. We then test cross-correlation estimation algorithms (Section 6.2.3) to compute the time difference of arrival (TDOA) between pairs of hydrophones and compare them to an efficient adaptive estimation algorithm based on least-mean-square minimization (Section 6.2.4). We conclude by a simple multilateration algorithm, which uses two TDOAs to estimate the position of the sperm whale (Section 6.2.5).

6.2.1 Sperm whale sounds

Throughout Section 6.2, we will work on hydrophone signals collected at the Atlantic Undersea Test and Evaluation Center (AUTEC), Andros Island, Bahamas, and made available by the Naval Undersea War-fare Center (NUWC). These signals were provided to the 2nd International

Workshop on Detection and Localization of Marine Mammals using Passive Acoustics held in Monaco, November 16–18, 2005 (Adam et al. 2006), and can be obtained from their website (Adam 2005). The Andros Island area has over 500 square nautical miles of ocean, simultaneously monitored via 68 broadband hydrophones. The distance between most hydrophones is 5 nautical miles (about 7.5 km).

We will work with the sounds received by hydrophones I, G, and H (which we will rename as #1, #2, and #3 here), for about 30 s, recorded with a digital audio recorder at $F_s = 48$ kHz.

Plotting the signals received by all three hydrophones (Fig. 6.8) shows variable attenuations and delays between clicks, as well as variable noise levels. Each direct path click is followed by an echo due to surface reflection. Obviously, clicks arrive at hydrophone #2 much earlier than at the other hydrophones. The amplitude of the signal on hydrophone #2 is also higher, and the noise level is lower. Moreover, the echo in signal #2 is much more delayed than in signals #1 and #3. All this suggests that the sperm whale is in the vicinity of hydrophone #2.

```
[hydrophone1,Fs]=wavread('hydrophone1.wav');
[hydrophone2,Fs]=wavread('hydrophone2.wav');
[hydrophone3,Fs]=wavread('hydrophone3.wav');

time=(0:length(hydrophone1)-1)/Fs;
ax(1)=subplot(3,1,1); plot(time,hydrophone1)
hold on;
ax(2)=subplot(3,1,2); plot(time,hydrophone2)
ax(3)=subplot(3,1,3); plot(time,hydrophone3)
linkaxes(ax,'x');
```

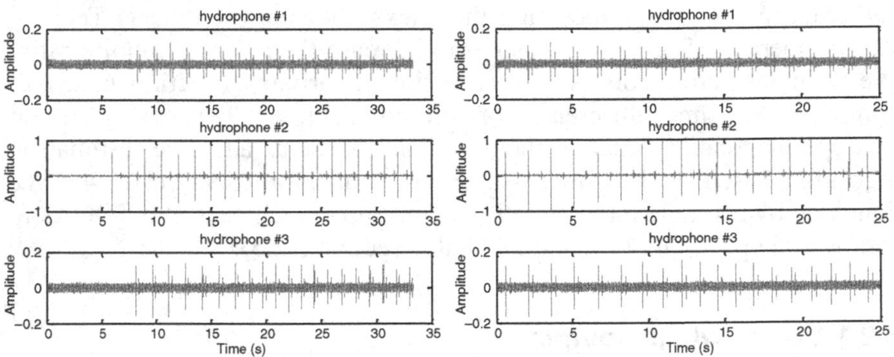

Fig. 6.8 *Left*: Clicks received by hydrophones #1, #2, and #3. Echoes are clearly visible between clicks (note that the echo received at hydrophone #2 is more delayed than for the other hydrophones). *Right*: Same signals, time-shifted so as to approximately align corresponding clicks

Since the delay between clicks is larger than the interclick interval, we will first apply a fixed time shift between signals so as to approximately align corresponding clicks. We shift the first signal by 7.72 ms to the left, the second signal by 5.4 s to the left, third signal by 7.58 s to the left, and keep 25 s of each signal. We will later account for these artificial shifts.

```
n_samples=25*Fs;
shift_1=fix(7.72*Fs);
shift_2=fix(5.40*Fs);
shift_3=fix(7.58*Fs);
[hydrophone1,Fs]=wavread('hydrophone1.wav',[1+shift_1 ...
   shift_1+n_samples]);
[hydrophone2,Fs]=wavread('hydrophone2.wav',[1+shift_2 ...
   shift_2+n_samples]);
[hydrophone3,Fs]=wavread('hydrophone3.wav',[1+shift_3 ...
   shift_3+n_samples]);
```

Zooming on a single click (Fig. 6.9) shows that whale clicks are not as simple as Dirac samples, which makes the visual estimation of TDOAs not very precise and their automatic estimation not trivial.

Let us compute the PSD of the background noise in hydrophone #3 and compute its standard deviation, for later use. It appears (Fig. 6.10) that the noise is pink,[7] with 40 dB more power around f = 0 Hz than around f=24,000 Hz.

```
real_noise=hydrophone3(1.26e4:1.46e4);
real_noise_std=std(real_noise)

pwelch(real_noise,[],[],[],Fs);
```

```
real_noise_std =    0.0073
```

Note that the data used here does not reveal LF interferences, which may, however, occur in measurements and make the estimation of TDOA still more complex.

[7] Strictly speaking, a *pink noise* is defined as a signal $b(t)$ whose power spectral density $S_{bb}(f)$ is proportional to the reciprocal of the frequency $1/f$. We use the term *pink* very loosely here, to indicate that the PSD of the noise decreases with frequency.

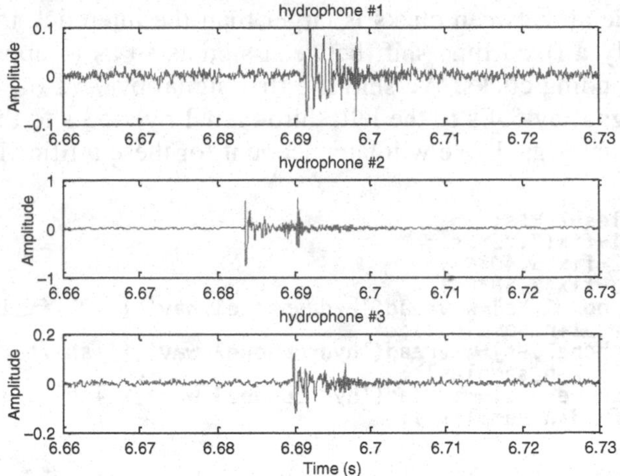

Fig. 6.9 Zoom on a single click

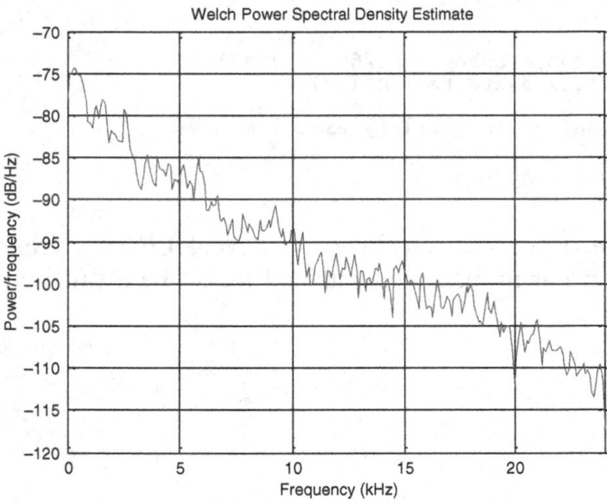

Fig. 6.10 Power spectral density of the background pink noise measured by hydrophone #3

6.2.2 Teager–Kaiser filtering

In order to better understand the Teager–Kaiser nonlinear filter, we first apply it to synthetic signals composed of clicks (assimilated to Dirac impulses), background noise (white and pink noise), and an interference component (modeled as a sinusoid). We then show that even though the effect of a general nonlinear filter on a sum of signals is not the sum of its outputs to isolated input signals, the effect of the TK operator on synthetic clicks with background noise and sinusoidal interference can still somehow be analyzed in terms of its effect on isolated components.
We finally apply the TK operator on real sperm whale sounds and compare it to a simple linear high-pass filter.

Dirac pulse input

We start with the response of the TK operator to a Dirac impulse 0.1δ (n−4,000), with the same order of magnitude as that of the clicks found in the previous section. The output is the input impulse, squared (Fig. 6.11, left).

```
click = [zeros(1,3999) 0.1 zeros(1,4000)];

% Applying TK filter
L = length(click);
click_response = click(2:L-1).^2-click(1:L-2).*click(3:L);
click_response = [click_response(1) click_response ...
    click_response(L-2)]; % lenght(output)=length(input)

subplot(2,1,1); plot(click);
subplot(2,1,2); plot(click_response);
```

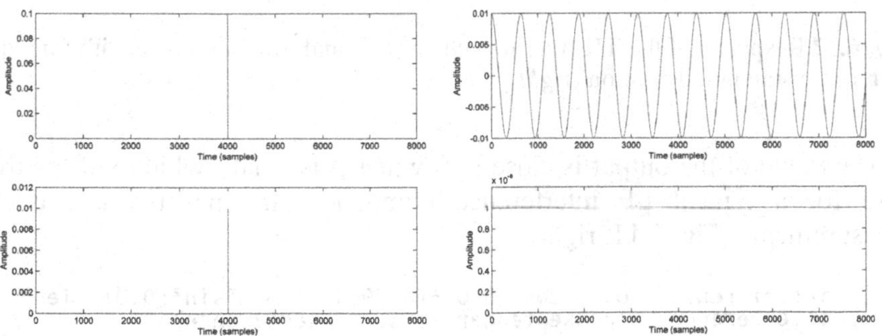

Fig. 6.11 Response of the TK filter to a Dirac pulse 0.1δ(n−4,000) (*left*) and to a cosine 0.01cos (0.01n) (*right*)

Sinusoidal input

Sinusoidal interference signals of the form $A\cos(n\phi_0)$ produce a constant output $A^2\sin^2(\phi_0)$. To check this, we generate 8,000 samples of a chirp $\cos(n\phi_0(n))$ with $\phi_0(n)=\phi_{max}n/8{,}000$ and $\phi_{max}=0.1$ (which corresponds to frequencies from 0 to $F_s\phi_{max}/2\pi = 763$ Hz). The result is plotted in Fig. 6.12 (left).

MATLAB function involved:

- `y = teager_kaiser(x)` applies the Teager–Kaiser energy operator to input signal `x`.

```
Fs=48000;
phi_max=0.1;
test=chirp(0:1/Fs:7999/Fs,0,7999/Fs,Fs*phi_max/2/pi);
test_response=teager_kaiser(test);

subplot(2,1,1); plot(test);
subplot(2,1,2); plot(test_response);
```

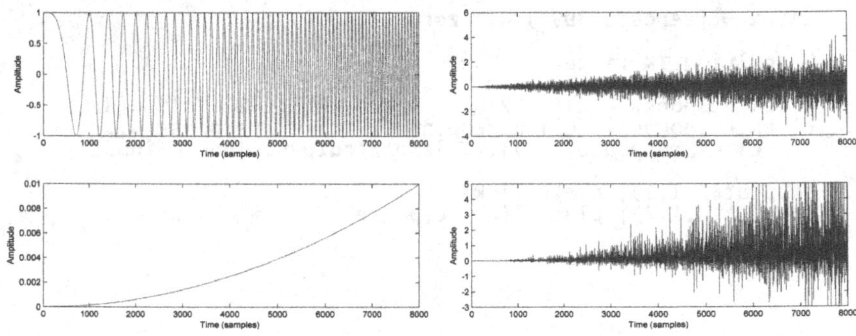

Fig. 6.12 Response of the TK filter to a chirp (*left*) and to white noise with linearly increasing standard deviation (*right*)

The value of the output is close to 0 when ϕ_0 is small, which will be the case for a typical LF interference component in sperm whale click measurements (Fig. 6.11, right).

```
interference = 0.01*cos (0.01*(0:7999)); % A²sin²(0.01)~1e-8
interference_response=teager_kaiser(interference);

subplot(2,1,1); plot(interference);
subplot(2,1,2); plot(interference_response);
```

White noise input

TK filtering of Gaussian white noise $N(0,\sigma)$ produces noise with mean of the order of σ^2 and standard deviation proportional to σ^2. We check this on white noise with linearly increasing standard deviation (see Fig. 6.12, right).

```
test = (1/8000:1/8000:1).*randn(1, 8000);
test_response = teager_kaiser(test);

subplot(2,1,1); plot(test);
subplot(2,1,2); plot(test_response);
```

The output noise is still white, but it is not Gaussian (Fig. 6.13).

```
noise = randn(1, 8000);
noise_response = teager_kaiser(noise);

subplot(2,1,1); pwelch(noise,[],[],[],Fs);
subplot(2,1,2); pwelch(noise_response,[],[],[],Fs);

subplot(2,1,1); hist(noise,30);
subplot(2,1,2); hist(noise_response,60);
```

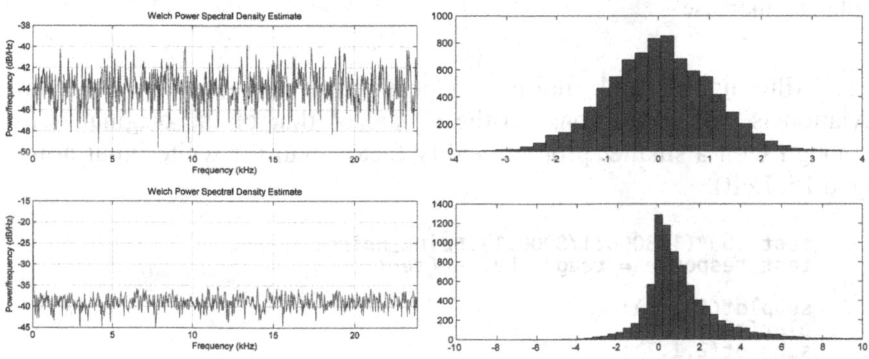

Fig. 6.13 *Left*: PSD of white noise input and its filtering through the TK filter. *Right*: Histograms of the corresponding samples

Pink noise input

Let us now test the TK filter on pink noise similar to the one we found on real hydrophone signals in 0. We first create a filter with linear frequency response from −30 dB at $f = 0$ to −65 dB at $f = F_s/2$. Applying this filter to white noise $N(0,1)$ produces a realistic pink noise component (Fig. 6.14).

```
w=0:0.1:1;
a_dB=-35*w-30;
```

```
a=10.^(a_dB/20);
[B,A] = fir2(20,w,a);
freqz(B,A,512,48000);

pink_noise=filter(B,A,noise);
pwelch(pink_noise,[],[],[],Fs);
pink_noise_std=std(pink_noise)
```

pink_noise_std =   0.0108

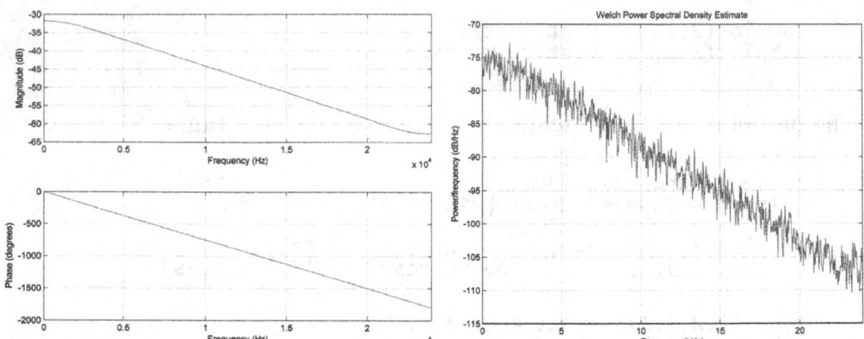

Fig. 6.14 *Left*: Frequency response of a noise shaping filter. *Right*: PSD of the resulting pink noise

TK filtering of pink noise produces pink noise, whose standard deviation is still proportional to the square of that of the original noise (although with a smaller proportionality factor than for white input noise; Fig. 6.15, Left).

```
test=100*(1/8000:1/8000:1).*pink_noise;
test_response = teager_kaiser(test);

subplot(2,1,1);
plot(test);
subplot(2,1,2);
plot(test_response);
```

The output noise is whiter; the PSD of the input noise has been decreased by about 50 dB in low frequencies and by about 30 dB in high frequencies (Fig. 6.15, Right).

```
pink_noise_response = teager_kaiser(pink_noise);

subplot(2,1,1);
pwelch(pink_noise,[],[],[],Fs);
subplot(2,1,2);
pwelch(pink_noise_response,[],[],[],Fs);
```

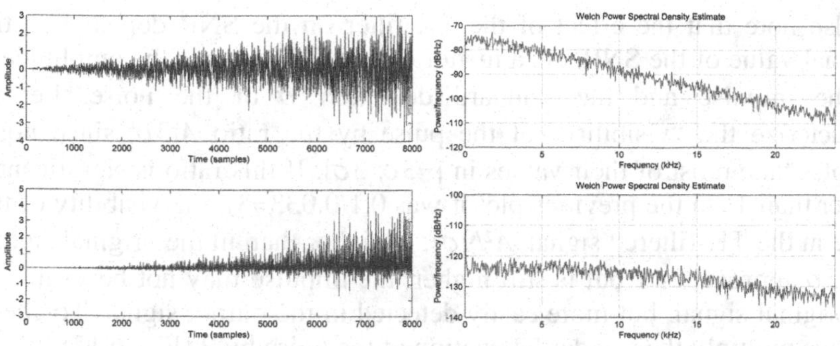

Fig. 6.15 *Left*: TK filtering of pink noise with linearly increasing standard deviation. *Right*: PSDs of input pink noise (with constant variance) and of its filtering through the TK filter

Complex input

We now apply the TK operator to the complete synthetic signal obtained by summing all three components: click, interference, and pink noise (Fig. 6.16, left). Thanks to the TK operator, the impulse is highlighted in the signal. The LF signal is removed, and the pink noise is very much attenuated relative to the impulse. The SNR is therefore highly increased.

```
input=click+interference+pink_noise;
output_tk=teager_kaiser(input);

subplot(2,1,1); plot(input);
subplot(2,1,2); plot(output_tk);
set(gca,'xlim',[3980 4020]);
```

Note that the TK operator did not smear the impulsive part of the input waveform (Fig. 6.16, right).

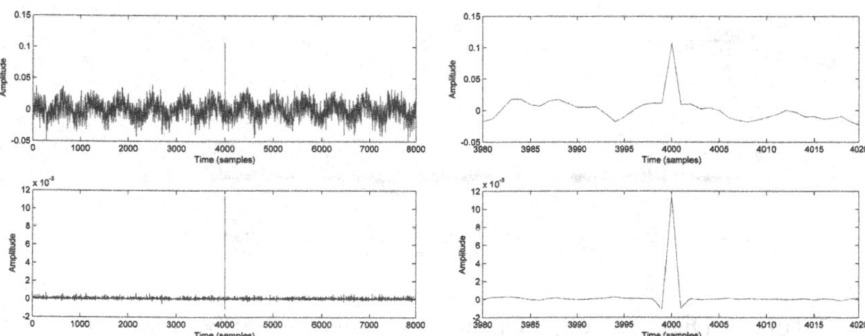

Fig. 6.16 *Top left*: Synthetic hydrophone signal composed of the sum of a Dirac pulse, sinusoidal interference, and background noise; *bottom left*: same signal filtered by the TK operator; *right*: zoom around the Dirac pulse

Also note that the effect of the TK filter on the SNR depends on the original value of the SNR. As a matter of fact, TK squares the amplitude A of the impulse and the standard deviation σ of the noise. Let us characterize the "visibility" of the pulse by the ratio $A/3\sigma$, since noise samples take most of their values in $[-3\sigma, 3\sigma]$. If this ratio is significantly higher than 1 (in the previous plot it was 0.1/0.033=3), the visibility of the pulse in the TK-filtered signal, $A^2/9\sigma^2$, is higher than in the original signal. If $A/3\sigma$ approaches 1 but is still higher, the impulse may not be visible in the original signal, but more easily detected in the output signal. To check this, we multiply the standard deviation of the noise by 2 (Fig. 6.17).

```
input=click+2*pink_noise+interference;
output_tk=teager_kaiser(input);
```

In summary, the TK operator has a high-pass filtering effect on background pink noise (and it squares its standard deviation), a notch effect on possible low-frequency interference (while a DC component appears), and almost no effect on Dirac pulses (apart from squaring their amplitude): it acts as a signal-dependent filter and amplifier.[8]

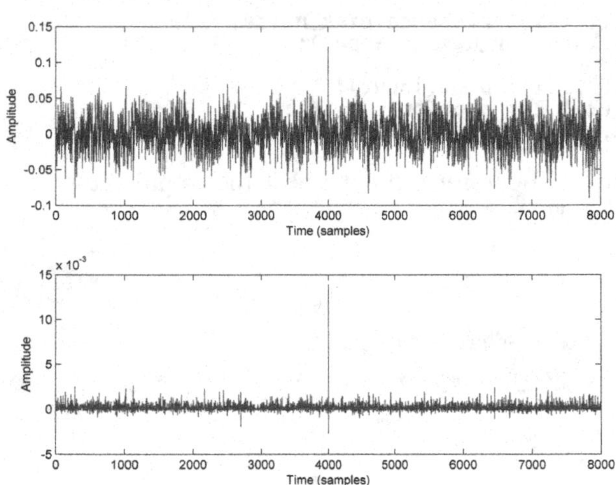

Fig. 6.17 Top: A Dirac pulse hidden in background noise; bottom: Same signal after TK filtering

[8] In practice, it also produces cross terms when several components are summed at the input. We did not examine them here, as they are not essential for the kind of signal we deal with.

Hydrophone signal

Finally we apply the TK operator to a real click taken from a hydrophone signal, as examined in Section 6.1 and compare its effect to that of a linear filter.

We first create a linear-phase, high-pass symmetric FIR filter whose frequency response has approximately the same effect on the pink background noise as the TK filter: a frequency-linear attenuation from −50 dB close to f=0 to −30 dB close to f=F_s/2 (Fig. 6.18, Left). The length of its impulse response is 31 (its order is 30). It therefore has a delay of 15 samples.

```
clf;
w=0:0.1:1;
a_dB=20*w-50;
a=10.^(a_dB/20);
N=30;
[B,A] = fir2(N,w,a);
freqz(B,A,512,Fs);
```

Obviously, the TK operator does a very good job at highlighting the click. Although the linear filter has a similar effect on background noise, it also decreases the power of the click, thereby decreasing the overall SNR (Fig. 6.18, Right).

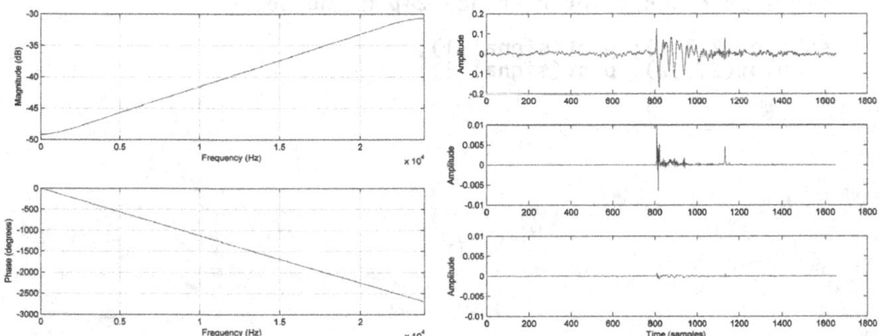

Fig. 6.18 *Left*: Frequency response of a linear filter having the same effect on background noise as the TK filter. *Right*: A real click (*top*), filtered by the TK operator (*middle*) and by the linear filter (*bottom*).

```
input=wavread('hydrophone3.wav',[464750 466650]);
output_tk=teager_kaiser(input);
% Taking the filter dealy into account
output_lin=[filter(B,A,input(16:end)) ; zeros(15,1)];

subplot(3,1,1); plot(input);
subplot(3,1,2); plot(output_tk);
subplot(3,1,3); plot(output_lin);
```

6.2.3 TDOA estimation using generalized cross-correlation

We will now show how generalized cross-correlation (GCC) can be used for estimating TDOAs. We start again by applying it to simple synthetic signals and end with real hydrophone signals.

Synthetic input

We first create a simple synthetic signal, as in Section 6.2.2, but with more severe interference and noise, and a second signal, with other noise samples, a phase-shifted interference signal, and an attenuated click, delayed by 60 samples (Fig. 6.19, Left).

```
click = [zeros(1,499) 0.1 zeros(1,500)];
interference = 0.03*cos (0.01*(0:999));
noise = 2*randn(1, 1000);
w=0:0.1:1;
a_dB=-35*w-30;
a=10.^(a_dB/20);
[B,A] = fir2(20,w,a);
pink_noise=filter(B,A,noise);

signal_1=click+interference+pink_noise;

click_2=circshift(click'*0.5,60)';
interference_2 = 0.02*cos (pi/4+0.01*(0:999));
noise_2 = randn(1, 1000); % new noise samples
pink_noise_2=filter(B,A,noise_2);

signal_2=click_2+interference_2+pink_noise_2;

subplot(2,1,1); plot(signal_1);
subplot(2,1,2); plot(signal_2);
```

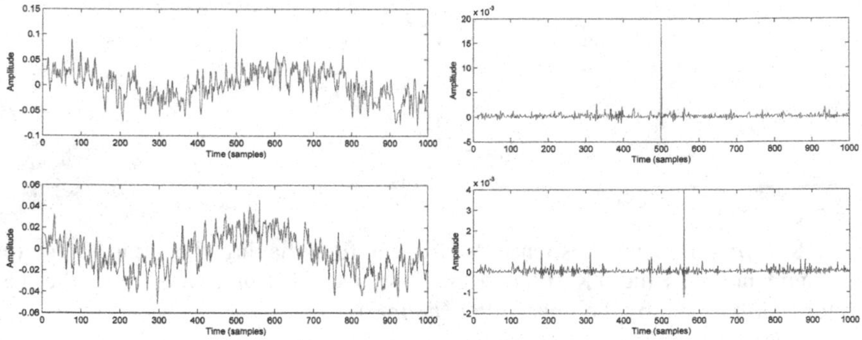

Fig. 6.19 *Left*: Two synthetic hydrophone signals with a delay of 60 samples between their respective Dirac pulses. *Right*: Same signal filtered by the TK operator

Estimating the time difference of arrival (TDOA) between the filtered signals with the standard cross-correlation does not provide the expected value (–60 samples), as the total cross-correlation function is dominated by the correlation between the sinusoidal interference components. Note that the (biased) cross-correlation is obtained via FFT-IFFT (Fig. 6.20) after making sure that the number of FFT samples is high enough for linear convolution to be identical to circular convolution (as obtained by multiplying FFTs).

```
% Computing the standard biased cross-correlation via FFT-IFFT
M = min(length(signal_1),length(signal_2));
NFFT = 2*M-1; % so that linear convolution = circular convolution
x1 = signal_1 - mean(signal_1);
x2 = signal_2 - mean(signal_2);
X1 = fft(x1,NFFT);
X2 = fft(x2,NFFT);
S_x1x2 = X1.*conj(X2);
phi_x1x2 = ifft(S_x1x2);
% re-arranging the IFFT
standard_xcorr = [phi_x1x2(NFFT-M+2:NFFT) phi_x1x2(1:M)];
[val,ind]=max(standard_xcorr);
TDOA_standard_xcorr = ind-M  % in samples
```

```
TDOA_standard_xcorr =    41
```

```
subplot(2,1,1);
plot((0:1/M:(M-1)/M), 20*log10(abs(S_x1x2(1:M))) );
subplot(2,1,2);
plot((0:1/M:(M-1)/M), unwrap(angle(S_x1x2(1:M))) );
```

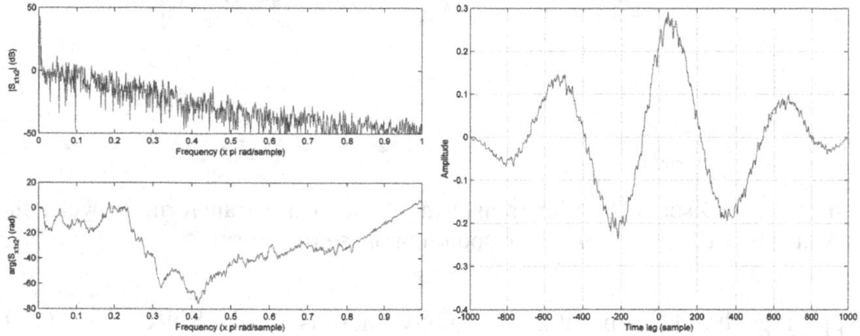

Fig. 6.20 *Left*: Cross PSD between the signals in Fig. 6.19 (right). *Right*: Corresponding cross-correlation function

```
plot((-M+1:M-1),standard_xcorr); grid;
```

Note that the linear-phase component in the HF part of Fig. 6.20, (left), where the pink noise is weak, reveals a delay. But since the amplitude of the noise and of the LF interference strongly dominates that of the HF part of the cross-PSD, this delay cannot be observed in the cross-correlation function.

Using the phase-transform version of the generalized cross-correlation removes the LF interference and gives the same importance to all frequency bands in the phase spectrum. It produces a more prominent maximum, which leads to a correct estimate of the TDOA (Fig. 6.21).

```
phi_x1x2 = ifft(S_x1x2 ./ max(abs(S_x1x2),eps));
% re-arranging the IFFT
phase_transform = [phi_x1x2(NFFT-M+2:NFFT) phi_x1x2(1:M)];

[val,ind]=max(phase_transform);
TDOA_phase_transform = ind-M

subplot(2,1,1);
plot((0:1/M:(M-1)/M), 20*log10(abs(PT(1:M))));
subplot(2,1,2);
plot((0:1/M:(M-1)/M), unwrap(angle(PT(1:M))));

plot((-M+1:M-1),phase_transform); grid;
```

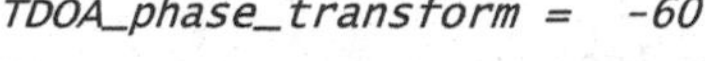

```
TDOA_phase_transform =  -60
```

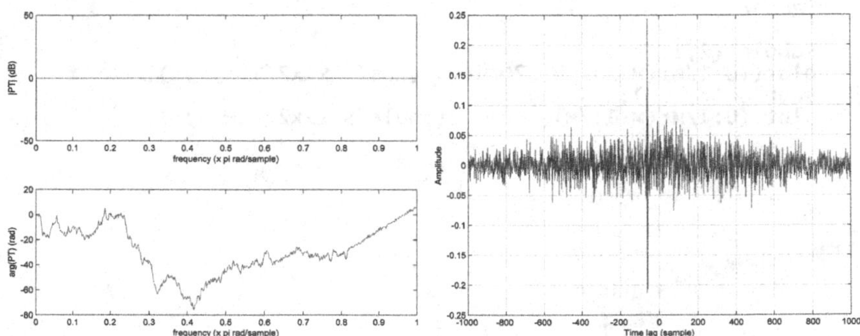

Fig. 6.21 *Left*: Discrete Fourier transform of the phase transform between the signals in Fig. 6.19 (left); *right*: corresponding phase transform

Applying the TK operator obviously increases the SNR (Fig. 6.19, right).

```
signal_1_tk=teager_kaiser(signal_1);
signal_2_tk=teager_kaiser(signal_2);

subplot(2,1,1);
plot(signal_1_tk);
subplot(2,1,2);
plot(signal_2_tk);
```

As a result of the suppression of the sinusoidal component, GCC produces an accurate result (Fig. 6.22, Left).

MATLAB function involved:

- `gcc(z1, z2, flag)` computes the generalized cross-correlation (GCC) between signals z1 and z2, from FFT/IFFT, as specified in Knapp and Carter (1976). `[flag]` makes it possible to choose the type of cross-correlation: standard cross-correlation if `flag='cc'`; phase transform if `flag='phat'`.

```
standard_xcorr_tk = gcc(signal_1_tk, signal_2_tk, 'cc');
[val,ind]=max(standard_xcorr_tk);
TDOA_standard_xcorr_tk = ind-M  % in samples

plot((-M+1:M-1),standard_xcorr_tk); grid;
```

```
TDOA_standard_xcorr_tk =   -60
```

Using the phase transform on TK-filtered data still produces an even more prominent maximum (Fig. 6.22, right).

```
phase_transform_tk = gcc(teager_kaiser(signal_1),
teager_kaiser(signal_2),  'phat');
[val,ind]=max(phase_transform_tk);
TDOA_phase_transform_tk = ind-M

plot((-M+1:M-1),phase_transform_tk); grid;
```

```
TDOA_phase_transform_tk =   -60
```

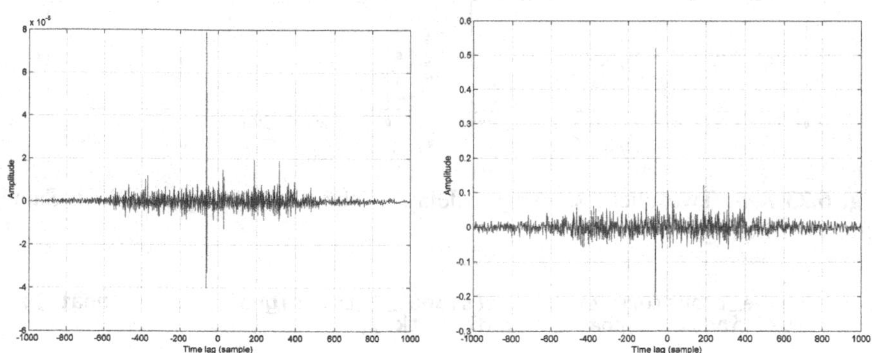

Fig. 6.22 *Left*: Cross-correlation computed on TK-filtered signals; *right*: corresponding phase transform

Hydrophone signals

We now apply GCC to real hydrophone signals. Clicks are separated by about 400 samples (Fig. 6.23, left).

```
Fs=48000;
shift_2=fix(5.40*Fs);
shift_3=fix(7.58*Fs);

[signal_1,Fs]=wavread('hydrophone2.wav',[460000+shift_2 ...
    460000+shift_2+2000]);
[signal_2,Fs]=wavread('hydrophone3.wav',[460000+shift_3 ...
    460000+shift_3+2000]);
```

Again, applying the TK operator has a positive effect on the SNR (Fig. 6.23, right).

```
signal_1_tk=teager_kaiser(signal_1);
signal_2_tk=teager_kaiser(signal_2);

subplot(2,1,1); plot(signal_1_tk);
subplot(2,1,2); plot(signal_2_tk);
```

Applying the phase transform to these signals delivers a correct estimate of the TDOA: −404 samples (negative, since the click hits hydrophone #1 before hydrophone #2).

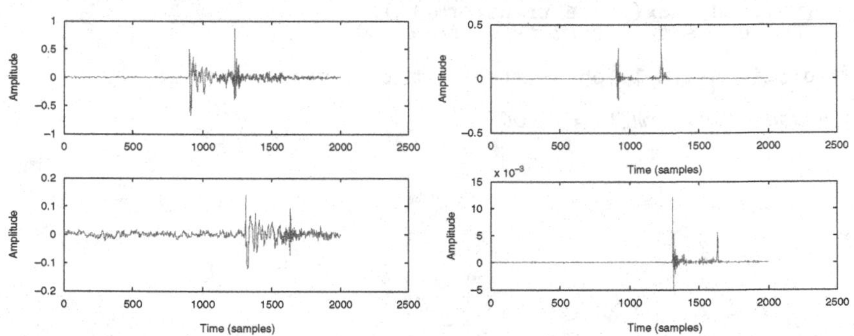

Fig. 6.23 *Left*: Two clicks showing a delay of about 400 samples. *Right*: Same after TK filtering

```
phase_transform_tk  = gcc(signal_1_tk, signal_2_tk,  'phat');
[val,ind]=max(phase_transform_tk );

M = min(length(signal_1),length(signal_2));
TDOA_phase_transform_tk= ind-M
```

TDOA_phase_transform_tk = -404

6.2.4 TDOA estimation using least-mean squares

In this section, we will apply the least-mean-square (LMS)-adaptive approach to TDOA estimation, again first on synthetic signals (with imposed TDOA) and then on real hydrophone signals.

Synthetic input

We start with the same synthetic signals as in Fig. 6.19, apply TK preprocessing, and run the Benesty's adaptive filtering algorithm for TDOA estimation within [−600,600] (samples) and step mu=0.01 (for full-scale signals in [−1,+1]).

```
load synthetic_signals signal_1 signal_2
signal_1_tk=teager_kaiser(signal_1);
signal_2_tk=teager_kaiser(signal_2);

% Normalizing max signal amplitudes to +1
x1=signal_1_tk/max(signal_1_tk);
x2=signal_2_tk/max(signal_2_tk);

% LMS initialization
M = 600; % max value of the estimated TDOA
x1c = zeros(M,1);
x2c = zeros(M,1);
u = zeros(2*M,1);
u(M/2) = 1;
N = length(x1);
e = zeros(1,N);
tdoa = zeros(1,N);
peak = zeros(1,N);
mu = 0.01; % LMS step

% LMS loop
for n=1:N

    x1c = [x1(n);x1c(1:length(x1c)-1)];
    x2c = [x2(n);x2c(1:length(x2c)-1)];
    x = [x1c;x2c];

    e(n) = u'*x;
    u = u-mu*e(n)*x;
    u(M/2) = 1; %forcing g2 to an impulse response at M/2
    u = u/norm(u); %forcing ||u|| to 1

    [peak(n),ind] = min(u(M+1:end));
    peak(n)=-peak(n); % (positive) impulse in g1
    TDOA(n) = ind-M/2;

end

% Estimated TDOA(n), with values of the maximum peak in g1
subplot(2,1,1);
plot(TDOA);
xlabel('Time (samples)'); ylabel('TDOA (samples)');
subplot(2,1,2);
plot(peak);
xlabel('Time (samples)'); ylabel('peak');
```

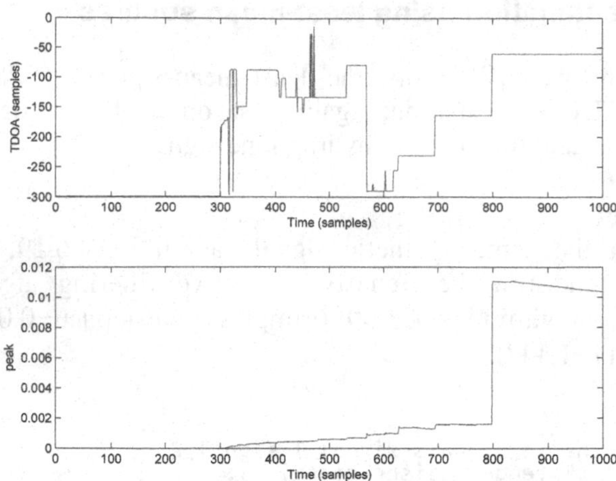

Fig. 6.24 *Top*: TDOA estimated by the LMS approach on the signals in Fig. 6.19 on an adaptive, sample-by-sample basis; *bottom*: value of the peak found in **h1** (the higher the peak, the more reliable the TDOA estimate)

It appears that the TDOA estimate is wrong at the beginning of the frame, as the adaptive filter must have started processing a click to start converging (Fig. 6.24). The value of the peak in **h1** (the impulse response between the source and the first signal) is very low for these wrong TDOA values. The best TDOA estimate is the one that produced the most prominent peak in **h1**. This estimate is correct: −60 samples. After a click has been processed, the peak in **h1** starts decreasing again, but the value of TDOA remains correct.

```
[val,ind]=max(peak);
Best_estimate_TDOA=TDOA(ind)
```

```
Best_estimate_TDOA =  -60
```

Hydrophone signals

Applying the same algorithm to hydrophone signals #1 and #2 is straightforward.

```
Fs=48000;
n_samples=25*Fs;
shift_1=fix(7.72*Fs);
shift_2=fix(5.40*Fs);
[hydrophone1,Fs]=wavread('hydrophone1.wav',[1+shift_1 ...
   shift_1+n_samples]);
[hydrophone2,Fs]=wavread('hydrophone2.wav',[1+shift_2 ...
   shift_2+n_samples]);
```

```
% Subsampling by 6, for decreasing computational cost
signal_1=resample(hydrophone1,1,6);
signal_2=resample(hydrophone2,1,6);

% Applying Teager-Kaiser filter
signal_1_tk=teager_kaiser(signal_1);
signal_2_tk=teager_kaiser(signal_2);

subplot(2,1,1); plot(signal_1_tk);
subplot(2,1,2); plot(signal_2_tk);
```

The estimated TDOAs change by a few tens of milliseconds in our 25-s recording (Fig. 6.25).[9] This shows that the sperm whale moved during the recording. Plotting our TODAs with the ones obtained by visual inspection of the signals proves that the LMS algorithm provided accurate results (the error is less than a millisecond).

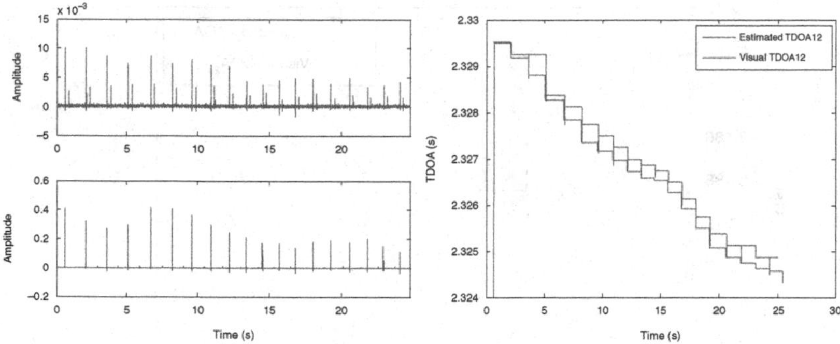

Fig. 6.25 *Left*: Hydrophone signals #1 and #2 after TK filtering. *Right*: TDOA estimated from these signals, as a function of time, compared to the estimate obtained by visual inspection of the signals

MATLAB function involved:

- `[tdoa, peak] = TDOA_LMS(x1, x2, max_tdoa, mu)` is an implementation of Benesty (2000) for source localization. It returns the time difference of arrival, in samples, between signals `x1` and `x2`, as a function of time, together with the value of the peak found in the estimate of the main propagation path for signal `x1`. This value can be used as a confidence value for the TDOA. `max_tdoa` is the maximum TDOA value returned and `mu` is the step weight.

[9] Again, the first TDOA estimates are not significant.

```
TDOA_12 = TDOA_LMS(signal_1_tk,signal_2_tk,600,0.01);
TDOA_12 = TDOA_12*6; % taking downsampling into account
% Accounting for preliminary time-shifts
TDOA_12 = (TDOA_12 + shift_1-shift_2)/Fs;

load clicks; % clicks_1, clicks_2, and clicks_3 give the
             % positions of clicks in the respective
             % hydrophone signals, as estimated by visual
             % inspection.
TDOA_12_visual=(clicks_1-clicks_2)/Fs;

plot((0:length(TDOA_12)-1)*6/Fs,TDOA_12);
hold on;
stairs((clicks_2-shift_2)/Fs,TDOA_12_visual,'--r');
```

We repeat the operation for hydrophone signals #2 and #3 (Fig. 6.26).

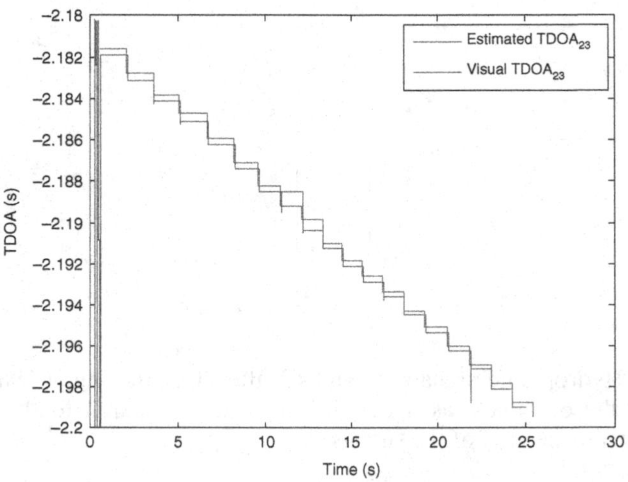

Fig. 6.26 TDOA estimated from hydrophone signals #2 and #3, as a function of time, compared to the estimate obtained by visual inspection of the signals

6.2.5 Multilateration

In this final section, we use the TDOAs estimated in the previous section from real hydrophone signals for tracking a sperm whale on a 2D map. We develop a rudimentary multilateration system on the basis of two TDOAs only.

The position of the hydrophones, taking some common point as a reference, is given in Table 6.1.

Table 6.1 Hydrophone relative locations and depth

Hydrophone	*X (m)*	*Y (m)*	*Z (m)*
#1	14318.86	−16189.18	−1553.58
#2	10658.04	−14953.63	−1530.55
#3	12788.99	−11897.12	−1556.14

We first show the position of the hydrophones on the map (Fig. 6.27) and then plot the hyperbolas corresponding to each pair of TDOA, assuming that the speed of sound in water is 1,510 m/s. Note that we drop the first TDOA estimates, which the LMS algorithm used for converging, and use only 10 estimates for our plot.

```
TDOA_12=TDOA_12(6000:length(TDOA_12)/10:end);
TDOA_23=TDOA_23(6000:length(TDOA_23)/10:end);

sound_speed=1510; % m/s
for i=1:length(TDOA_12)
    PlotHyp(hydrophone_pos(1,1), hydrophone_pos(1,2), ...
            hydrophone_pos(2,1), hydrophone_pos(2,2), ...
            -TDOA_12(i)*sound_speed/2,'b');
    PlotHyp(hydrophone_pos(2,1), hydrophone_pos(2,2), ...
            hydrophone_pos(3,1), hydrophone_pos(3,2), ...
            -TDOA_23(i)*sound_speed/2,'r');
end;
```

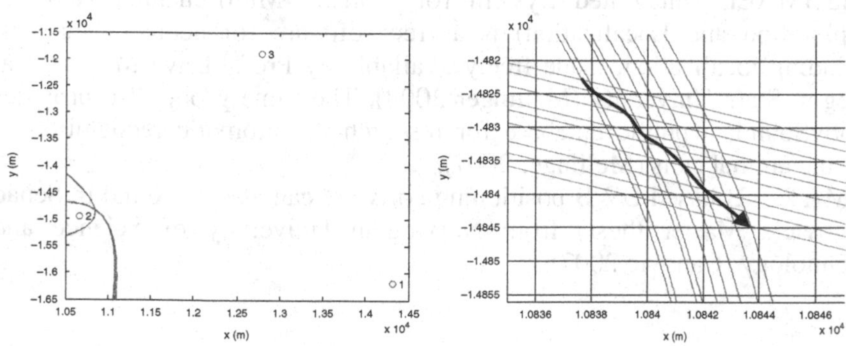

Fig. 6.27 Multilateration hyperbolas on the 2D hydrophone map. *Left*: General view; *right*: zoom on the intersection area and observed movement

Since the maximum speed of a whale is close to 30 km/h (i.e., 8 m/s), it obviously could not move much around hydrophone #2 in the 25 s of data we have used for this proof of concept (Fig. 6.27, left).

Zooming around hydrophone #2 (Fig. 6.27, right) reveals its trajectory; it moved from (10837.5,−14822) to (10844, −14847), i.e., by about 26 m in 25 s, i.e., at a speed of about 1 m/s.

```
clf;
axis([10835, 10847, -14856, -14816]);
hold on;
for i=1:length(TDOA_12)-1
    PlotHyp(hydrophone_pos(1,1), hydrophone_pos(1,2), ...
            hydrophone_pos(2,1), hydrophone_pos(2,2), ...
            -TDOA_12(i)*sound_speed/2,'b');
    PlotHyp(hydrophone_pos(2,1), hydrophone_pos(2,2), ...
            hydrophone_pos(3,1), hydrophone_pos(3,2), ...
            -TDOA_23(i)*sound_speed/2,'r');
end;
distance = sqrt((10844-10837.5)^2+(14847-14822)^2)
```

```
distance = 25.8312
```

6.3 Going further

Readers interested in methods for estimating the position of a source from multiple TDOA estimations should read Huang et al. (2008).

MATLAB code for localization, including more elaborate multilateration algorithms, is available from the Beam Reach Marine Science and Sustainability School (Beam Reach 2007).

ISHMAEL (Integrated System for Holistic Multi-channel Acoustic Exploration and Localization) is a free software for acoustic analysis, including localization, made freely available by Prof. Dave Mellinger at Oregon State University (Mellinger 2008). The same group also provides MobySound, a public database for research in automatic recognition of marine animal calls (Mellinger 2007).

MATLAB-based UWB positioning software can also be found in Senad Canovic's Master thesis from Norwegian University of Science and Technology (Canovic 2007).

6.4 Conclusion

In this chapter, we have provided signal processing solutions for cleaning impulsive signals mixed in noise and low-frequency interferences, and for automatically estimating TDOAs based on either generalized cross-correlation or adaptive filtering.

Although we have examined these techniques in the specific context of whale spotting, they can easily be generalized to many positioning problems, such as the localization of speakers in a room, that of GSM mobile phones, global positioning system (GPS) receivers, or aircrafts or vehicles using either sonar or radar.

Last but not least, this chapter also highlights one of the many “natural” abilities of human beings: that of using both ears to locate sound sources.

References

Adam O (2005) Website of the 2nd International Workshop on Detection and Localization of Marine Mammals using Passive Acoustics [online] Available: http://www.circe-asso.org/workshop/ [26/4/2008]

Adam O, Motsch JF, Desharnais F, DiMarzio N, Gillespie D, Gisiner RC (2006) Overview of the 2005 workshop on detection and localization of marine mammals using passive acoustics, Applied Acoustics 67:1061–1070

Beam Reach (2007) Acoustic Localization [online] Available: http://beamreach.org/soft/AcousticLocation/AcousticLocation-041001/ [23/6/2008]

Benesty J (2000) Adaptive eigenvalue decomposition algorithm for passive acoustic source localization. Journal of the Acoustical Society of America 107(1): 384–391

Canovic S (2007) Application of UWB Technology for Positioning, a Feasibility Study. Master thesis, Norwegian University of Science and Technology [online] Available: http://www.diva-portal.org/ntnu/undergraduate/abstract.xsql?dbid=2091 [23/06/08]

Drouot V (2003) Ecology of sperm whale (Physeter macrocephalus) in the Mediterranean Sea. Ph.D. Thesis, Univ. of Whales, Bangor, UK

Fang J, Atlas LE (1995) Quadratic detectors for energy estimation. IEEE Transactions on Signal Processing 43–11:2582–2594

Huang Y, Benesty J, Chen J (2008) Time delay estimation and source localization. In: Springer Handbook of Speech Processing, Benesty J, Sondhi MM, Huang Y (Eds.). Springer: Berlin, 1043–1063

Kaiser JF (1990) On a simple algorithm to calculate the ‘‘Energy’’ of a signal. Proceedings of the IEEE ICASSP, 381–384, Albuquerque, NM, USA

Kandia V, Stylianou Y (2005) Detection of creak clicks of sperm whales in low SNR conditions. CD Proceedings of the IEEE Oceans, Brest, France

Kandia V, Stylianou Y (2006) Detection of sperm whale clicks based on the Teager–Kaiser energy operator, Applied Acoustics 67:1144–1163

Knapp CH, Carter GC (1976) The generalized correlation method for estimation of time delay. In: IEEE Transactions on Acoustic, Speech and Signal Processing 24:320–327

ICES (International Council for the Exploration of the Sea–Advisory Committee on Ecosystems, ICES CM 2005/ACE:01) (2005) Report of the Ad-hoc Group on the Impact of Sonar on Cetaceans and Fish (AGISC).

Madsen PT (2002) Sperm whale sound production – in the acoustic realm of the biggest nose on record. In Madsen P.T., PhD. Dissertation, Sperm Whale Sound Production, Department of Zoophysiology, University of Aarhus, Denmark

Mellinger D (2007) MobySound [online] Available: http://hmsc.oregonstate.edu/projects/MobySound/ [23/6/2008]

Mellinger D (2008) Ishmael Integrated System for Holistic Multi-channel Acoustic Exploration and Localization [online] Available: http://www.pmel.noaa.gov/vents/acoustics/whales/ishmael [23/6/2008]

Morrissey RP, Ward J, DiMarzio N, Jarvis S, Moretti DJ (2006) Passive acoustic detection and localization of sperm whales (Physeter macrocephalus) in the tongue of the ocean. Applied Acoustics 67:1091–1105

Møhl B, Wahlberg M, Madsen PT, Heerfordt A, Lund A (2000) Sperm whale clicks: Directionality and source level revisited. Journal of the Acoustical Society of America 107(1):638–648.

Sicuranza G (1992) Quadratic filters for signal processing. Proceedings of the IEEE, 80(8):1263–1285

Teloni V (2005) Patterns of sound production in diving sperm whales in the northwestern Mediterranean, *Marine Mammal Science* 21(3):446–457

Widrow B, Stearns SD (1985) Adaptive Signal Processing. Prentice-Hall,Inc., Upper Saddle River, NJ

Zimmer WMX, Madsen PT, Teloni V, Johnson MP, Tyack PL (2005) Off-axis effects on the multi-pulse structure of sperm whale usual clicks with implications for the sound production. Journal of the Acoustical Society of America 118:3337–3345

Chapter 7

How could music contain hidden information?

C. Baras ([+]), N. Moreau (*), T. Dutoit (°)

([+]) GIPSA-lab (Grenoble Image Parole Signal et Automatique), Grenoble, France
(*) Ecole Nationale Supérieure des Télécommunications, Paris, France
(°) Faculté Polytechnique de Mons, Belgium

*W*hy not *A*ttempt *T*o hid*E* the very p*R*esence
of a *M*ess*A*ge in an audio t*R*ac*K* ?

Audio watermarking started in the 1990s as a modern and very technical version of playing cat and mouse.[1] The music industry, dominated by the "big four" record groups, also known as the "Majors" (Sony BMG, EMI, Universal, and Warner), quickly realized that the availability of digital media for music recordings and the possibility to transfer it fast and degradation-free (thanks to the availability of high data-rate networks and of efficient data compression standards) would not only offer many benefits in terms of market expansion, but also expose their business to a great danger: that of piracy of intellectual property rights. Being able to insert proprietary marks in the media without affecting audio quality (i.e., in a "transparent way") was then recognized as a first step toward solving

[1] The real origin of watermarking can be found in the old art of steganography (literally, "hidden writing", in ancient Greek). Although the goal of cryptography is to render a message unintelligible, steganography attempts to hide the very presence of a message by embedding it into another information (Petitcolas et al. 1999)

T. Dutoit, F. Marqués (eds.), *Applied Signal Processing*,
DOI 10.1007/978-0-387-74535-0_7,

this issue. Additionally, ensuring the robustness of the proprietary mark to not only to usual media modifications (such as cropping, filtering, gain modification, or compression) but also to more severe piracy attacks quickly became a hot research topic worldwide.

Things changed radically in 2007 (i.e., before audio watermarking could be given a serious commercial trial). As a matter of fact, the digital rights management (DRM) employed by the Majors to restrict the use of digital media (which did not make use of watermarking techniques) became less of a central issue after the last Majors stopped selling DRM-protected CDs in January 2007 and after S. Jobs' open letter calling the music industry to simply get rid of DRM in February of that same year (Jobs 2007).

Audio watermarking techniques are now targeted to more specific applications, related to the notions of "augmented content" or "hidden channel." One such application is audio fingerprinting,[2] in which a content-specific watermark is inserted in the media and can thus be retrieved for the automatic generation of the playlist of radio stations and music television channels for royalty tracking. Audio watermarking has also been proposed for the next generation of digital TV audience analysis systems; each TV channel would insert its own watermark in the digital audio and video streams it delivers, and a number of test TV viewers would receive a watermark detection system connected to a central computer.[3]

In the next section, we will first introduce the principles of spread spectrum and audio watermarking (Section 7.1). This will be followed by a MATLAB-based proof of concept, showing how to insert a text message in an audio signal (Section 7.2) and by a list of pointers to more advanced literature and software (Section 7.3).

---

[2] The term *fingerprinting* is also used in the literature for specifying the technique that associates a short numeric sequence to an audio signal (i.e., some sort of *hashing;* Craver et al. 2001). We do not use it here with this meaning.

[3] With analog TV, audience analysis is much simpler; since each TV channel uses a specific carrier frequency, measuring this frequency through a sensor inserted in TV boxes gives the channel being watched at any time. This is no longer possible with digital TV, since several digital TV channels often share the same carrier frequency, thanks to digital modulation techniques.

7.1 Background – Audio watermarking seen as a digital communication problem

Watermarking in an audio signal can be seen as a means of transmitting a sequence of bits ("1"s and "0"s) through a very particular noisy communication channel, whose "noise" contribution is the audio signal itself. With this view in mind, creating an audio watermarking system amounts to designing an emitter and a receiver adapted to the specificities of the channel; the emitter has to insert the watermark in the audio signal in such a way that the watermark cannot be heard, and the receiver must extract this hidden information from the watermarked audio signal[4] (Fig. 7.1).

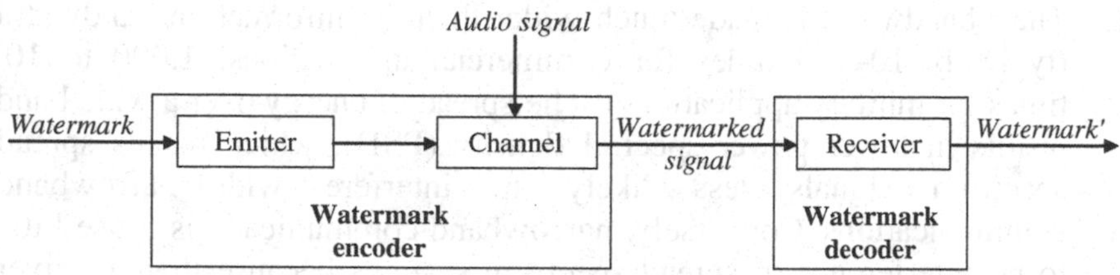

Fig. 7.1 Audio watermarking seen as a communication problem

The problem therefore looks very similar to that of sending information bits through the Internet with an ADSL connection; one has to maximize the bit rate while minimizing the bit error rate, even in the presence of communication noise and distortion.[5] In other words, watermarking directly benefits from 50 years of digital communication theory, after the pioneering work of Claude Shannon.

However, watermarking techniques are very specific in terms of bit rate, error rate, and signal-to-noise ratio. SNR is typically made very low so as to make the watermark inaudible. This results in very low bit rates (several hundred bits per second at best) and high error rates (typically 10^{-3}), as opposed to the megabits per second of ADSL modems and the associated low error rates (10^{-6}). In that respect, watermarking techniques are closer

[4] This is often referred to as *blind* watermarking, in which the receiver has no a priori information on the audio signal.

[5] Watermark distortion is typically produced by filtering and encoding operations applied to the audio signal while being transmitted, such as MP3 encoding.

to those used for communication between interplanetary satellites and earth. In particular, spread spectrum techniques are used in both cases.

Spread spectrum signals will be reviewed in Section 7.1.1. In the next sections, we will consider the emitter/receiver design issues in a blind (Section 7.1.2) and informed watermarking system (Section 7.1.3) respectively.

7.1.1 Spread spectrum signals

Spread spectrum communications (Peterson et al. 1995) were originally used for military applications,[6] given their robustness to narrowband interference, their secrecy, and their security. As a matter of fact, spread spectrum signals are characterized by two unique features:

1. Their bandwidth is made much wider than the information bandwidth (typically 10–100 times for commercial applications; 1,000 to 10^6 times for military applications). The spread of energy over a wide band results in lower power spectral density (PSD), which makes spread spectrum signals less likely to interfere with narrowband communications. Conversely, narrowband communications cause little to no interference to spread spectrum systems because their receiver integrates over a very wide bandwidth to recover a spread spectrum signal. As we shall see below, another advantage of such a reduced PSD is that spread spectrum signal PSD can be pushed on purpose below communication noise, thereby making these signals unnoticeable.
2. Some *spreading sequence* (also called *spreading code* or *pseudo-noise*) is used to create the wide-band spread spectrum signal from the information bits to send. This code must be shared by the emitter and the receiver, which makes spread spectrum signals hard to intercept.

The price to pay for these features is a reduced spectral efficiency, defined as the ratio between the bit rate and the bandwidth; the number of bits per second and per hertz provided by spread spectrum techniques is typically less than 1/5, while "standard" techniques offer a spectral efficiency close to 1 (Massey 1994).

In the 1990s, spread spectrum techniques have found a major application in code division multiple access (CDMA), now used in satellite positioning systems (GPS, hopefully in Galileo), and for wireless digital phone communications systems (UMTS).

[6] Spread spectrum communications dates back to World War II.

Direct sequence spread spectrum (DSSS)

Common spread spectrum systems are of the *time hopping* (also called *direct sequence*) or *frequency hopping* type. In the latter case, as its name implies, the carrier of a frequency hopping system hops from frequency to frequency over a wide band, in an order defined by the spreading code. In contrast, direct sequence spread spectrum (DSSS) signals are obtained by multiplying each bit-sized period (of duration T_b) of the emitted signal (the watermark signal in our case) by a (unique) spreading sequence composed of a random sequence of ±1 rectangular pulses (of duration T_c) (Fig. 7.2, left).

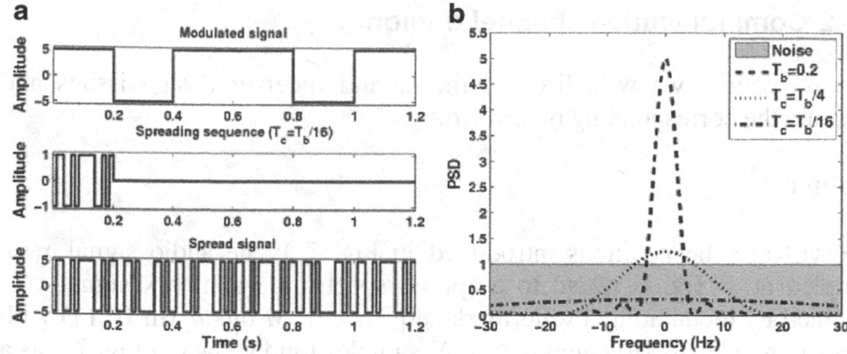

Fig. 7.2 *Left*: Emitted signal, spreading sequence, and spread signal waveforms for T_b=0.2s and T_c=T_b/16. *Right*: PSD of the input signal compared to that of the spread signal for T_c=T_b/4 and T_c=T_b/16

It is easy to compute the PSD of such a spread spectrum signal if we assume that the emitted signal is a classical NRZ random binary code composed of rectangular pulses of duration T_b and of amplitude $\pm 1/T_b$. As a matter of fact, the PSD of such a signal is given by

$$S_{NRZ}(f) = \frac{1}{T_b}\text{sinc}^2(fT_b) \tag{7.1}$$

and since the spread signal is itself an NRZ signal with pulse duration T_c and pulse amplitude $\pm 1/T_{b,}$ its PSD is given by the following equation

$$S_{spread}(f) = \frac{T_c}{T_b^{\ 2}}\text{sinc}^2(fT_c) \tag{7.2}$$

Both PSDs are plotted in Fig. 7.2 (right) for two values of T_c. Clearly, when T_c decreases, the bandwidth of the spread signal increases, and its maximum PSD level decreases accordingly.

As mentioned above, the resulting spread spectrum signal offers an increased robustness to narrowband interference and distortion; while a low-pass filter with cutoff frequency of 5 Hz would almost completely delete the input NRZ signal, its effect on the spread signal would be mild. Spread spectrum techniques also make it possible to hide information in communication noise, as suggested in Fig. 7.2 (right) with a background noise of unitary variance.

7.1.2 Communication channel design

In this section, we will discuss emitter and receiver design issues and estimate the corresponding bit error rate.

Emitter

We will use the notations introduced in Fig. 7.3; the audio signal $x(n)$, sampled at F_s Hz, is added to a spread spectrum watermark signal $v(n)$ obtained by modulating a watermark sequence of M bits b_m (in {0,1}) with a spreading signal $c(n)$ composed of N_b samples (and shown in Fig. 7.3 as a vector $\mathbf{c}=[c(0),c(1),\ldots,c(N_b-1)]^T$).

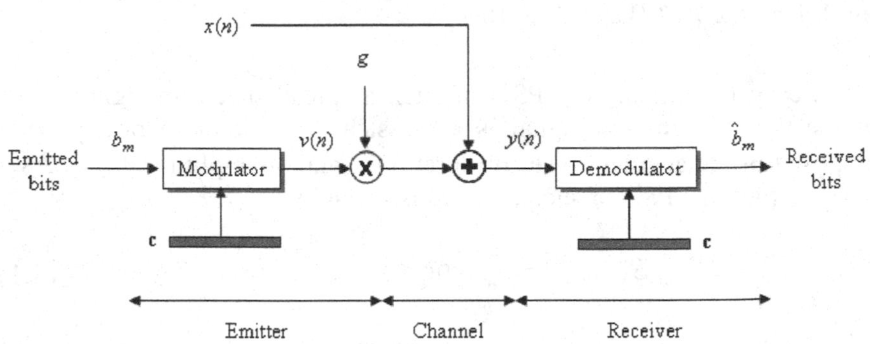

Fig. 7.3 A more detailed view of a watermarking system seen as a communication system with additive channel noise

Vector **c** is synthesized by a pseudo-random sequence generator as a Walsh–Hadamard sequence or a Gold sequence with values in {−1,+1}.

Following the time-hopping principle exposed in the previous Section, the spread spectrum signal *v(n)* is given by

$$v(n) = \sum_{m=0}^{M-1} a_m c(n - mN_b) \tag{7.3}$$

where a_m is a symbol in {−1,+1} given by $a_m=2b_m-1$. One can also see this signal as a concatenation of vectors:

$$\mathbf{v_m} = a_m \mathbf{c} = \begin{cases} +\mathbf{c} & \text{if } a_m = 1 \\ -\mathbf{c} & \text{if } a_m = -1 \end{cases} \tag{7.4}$$

It is finally amplified by some gain *g,* which makes it possible to control the signal-to-noise ratio (SNR) between the watermarked signal *v(n)* and the audio signal *x(n).*

Since 1 bit is emitted every N_b samples, the bit rate is simply given by $R=F_s/N_b$ (bits/s), where F_s is the sampling frequency (typically 44.1 kHz).

Receiver

In the case of the additive white Gaussian noise (AWGN) channel shown in Fig. 7.3, the signal *y*(*n*) available at the receiver is a noisy version of the emitted signal *v(n)*, with channel noise PDF given by $N(0,\sigma_X^2)$.

We will further assume that the received samples *y(n)* are perfectly synchronized with the emitted samples *x(n)*, i.e., the received samples can be organized in vectors $\underline{y}_m=[y(mN_b), y(mN_b+1), \ldots, y(mN_b+N_b-1)]^T$ related to vectors $\underline{v}_m$ by

$$\mathbf{y_m} = g\mathbf{v_m} + \mathbf{x_m} = \begin{cases} +g\mathbf{c} + \mathbf{x_m} & \text{if } a_m = 1 \\ -g\mathbf{c} + \mathbf{x_m} & \text{if } a_m = -1 \end{cases} \tag{7.5}$$

where $\mathbf{x_m}=[x(mN_b), x(mN_b+1), \ldots, x(mN_b+N_b-1)]^T$ is a vector of channel noise samples and *m* is the frame index.

Since the modulation operation is assimilated to raw concatenation,[7] and thereby produces no symbol interference, symbol detection can be performed on a symbol-by-symbol basis (no Viterbi algorithm is needed). The optimal detector in this case is the *maximum a posteriori* (MAP) detector (Proakis 2001),[8] which maximizes the probability of each received bit $\tilde{b}_m$, given the corresponding received vector $\mathbf{y_m}$:

[7] This is the simplest possible case of a memoryless linear modulation scheme.
[8] The same as the one used in Chapter 4 for vowel classification.

$$\tilde{b}_m = \arg\max_{b \in \{0,1\}} P(b \mid \mathbf{y_m}) \tag{7.6}$$[9]

In the next paragraphs, we will show that this is equivalent to estimating $\tilde{b}_m$ from the sign of the normalized scalar product α_m between the received vector $\mathbf{y_m}$ and the spreading code $\mathbf{c}$:

$$\alpha_m = \frac{<\mathbf{y_m}, \mathbf{c}>}{\|\mathbf{c}\|^2} = \frac{<\mathbf{y_m}, \mathbf{c}>}{N_b} \tag{7.7}$$

As a matter of fact, following the same development as in Chapter 4, Section 4.1, Equation (7.6) is equivalent to

$$\begin{aligned}\tilde{b}_m &= \arg\max_{b \in \{0,1\}} \frac{P(\mathbf{y_m} \mid b)P(b)}{P(\mathbf{y_m})} \\ &= \arg\max_{b \in \{0,1\}} P(\mathbf{y_m} \mid b)P(b)\end{aligned} \tag{7.8}$$

in which $P(\mathbf{y_m})$ is the a priori probability of $\mathbf{y_m}$, which does not influence the estimation of $\tilde{b}_m$, and $P(b)$ is the a priori probability of each possible value of b. Furthermore, if the symbols have the same probability of occurrence, Equation (7.8) reduces to

$$\tilde{b}_m = \arg\max_{b \in \{0,1\}} P(\mathbf{y_m} \mid b) \tag{7.9}$$

known as the *maximum likelihood* decision.

In the AWGN channel case, since $\mathbf{x_m}$ is an N_b-dimensional multivariate Gaussian random variable with zero mean and covariance matrix equal to $\sigma_x^2\mathrm{I}$ (where I is the N_b-dimensional unity matrix), Equation (7.5) shows that $\mathbf{y_m}$ is also a multivariate Gaussian random variable, with mean equal to $a_m g\mathbf{c}$ and the same covariance matrix as $\mathbf{x_m}$. Hence

$$P(\mathbf{y_m} \mid b) = \frac{1}{\sqrt{(2\pi\sigma_X^2)^{N_b}}} \exp\left(-\frac{\|\mathbf{y_m} - (2b-1)g\mathbf{c}\|^2}{2\sigma_X^2}\right) \tag{7.10}$$

[9] In the sequel, we use the same shortcut notation for probabilities as in Chapter 4; $P(a|b)$ denotes the probability $P(A=a|B=b)$ that a discrete random variable A is equal to a, given the fact that random variable B is equal to b will simply be written, or the probability *density* $p_{A|B=b}(a)$ when A is a continuous random variable.

It follows that

$$\tilde{b}_m = \underset{b\in\{0,1\}}{\arg\min} \| \mathbf{y_m} - (2b-1)g\mathbf{c} \|^2$$
$$= \underset{b\in\{0,1\}}{\arg\min} \left(\| \mathbf{y_m} \|^2 - 2 < \mathbf{y_m}, (2b-1)g\mathbf{c} > + (2b-1)^2 \| \mathbf{c} \|^2 \right) \quad (7.11)$$
$$= \underset{b\in\{0,1\}}{\arg\max} < \mathbf{y_m}, (2b-1)g\mathbf{c} >$$

since $(2b-1)^2=1$, whatever be the value of b in $\{0,1\}$. In other words, as mentioned, $\tilde{b}_m$ is obtained from the sign of the scalar product $<\mathbf{y_m}, \mathbf{c}>$, i.e., from the sign of α_m:

$$\tilde{b}_m = \begin{cases} 0 & \text{if} < \mathbf{y_m}, \mathbf{c} > \leq 0 \\ 1 & \text{if} < \mathbf{y_m}, \mathbf{c} > > 0 \end{cases} \quad (7.12)$$

Fig. 7.4 shows a synthetic view of Equation (7.9) in the case b_m is "1." In the left plot, $\tilde{b}_m$="1," while the right plot will wrongly result in $\tilde{b}_m$="0".

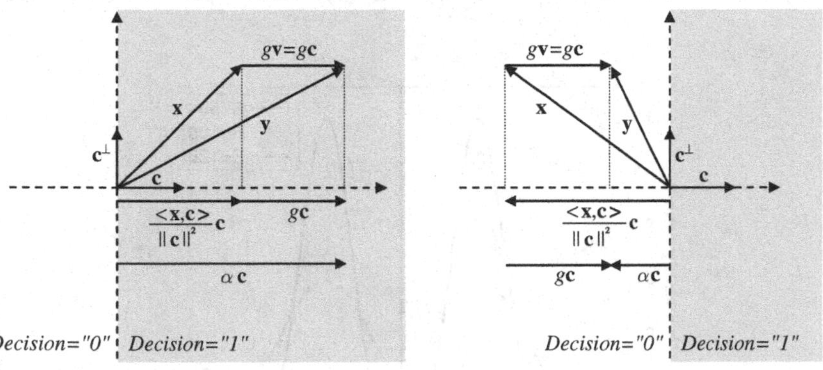

Fig. 7.4 Watermarking of audio vector **x** with symbol "1." Watermarked vector **y** is obtained by adding $g\mathbf{v}$ (i.e., $g\mathbf{c}$, given the symbol is "1") to **x**. Symbol "1" will be detected if $\alpha \geq 0$. *Left*: Since $<\mathbf{x},\mathbf{c}>$ is positive, any (positive) value of g will result in $\alpha \geq 0$. *Right*: Since $<\mathbf{x},\mathbf{c}>$ is negative, only a value of $g \geq <\mathbf{x},\mathbf{c}>/\|\mathbf{c}\|^2$ will result in $\alpha \geq 0$ (this is not the case in the figure)

Error rate

The PDF of α_m for g set to 1 and σ_x^2 set to 20 dB is given in Fig. 7.5 for various vector lengths N_b, i.e., for various bit rates R.

Projecting Equation (7.5) on the spread sequence **c**, the scalar product α_m appears as a one-dimensional Gaussian random variable with mean $(2b_m-1)g$ and variance $\sigma_x^2/\|\mathbf{c}\|^2$. The latter is that of the normalized scalar product between the N_b-dimensional multivariate Gaussian random variable $\mathbf{x}_m$ and **c**. Hence

$$P(\alpha_m \mid b_m) = \frac{1}{\sqrt{2\pi}\sigma_X/\|\mathbf{c}\|} \exp\left(-\frac{\alpha_m - (2b_m - 1)g}{2\sigma_X^2/\|\mathbf{c}\|^2} \right) \tag{7.13}$$

As expected, the probability of α_m to be positive while b_m=0 and the related probability of α_m to be negative while b_m=1 are nonzero, and these values increase with R (i.e., when N_b decreases). The probability of error is given by the following equation:

$$P_e = P(b=1)P(\alpha \le 0 \mid b=1) + P(b=0)P(\alpha > 0 \mid b=0) \tag{7.14}$$

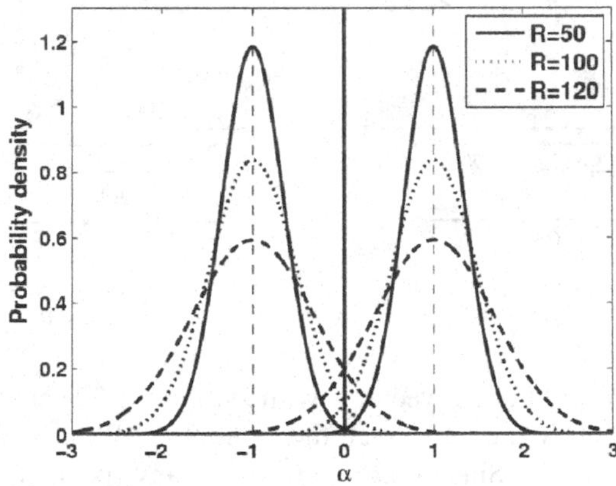

Fig. 7.5 Probability density functions for α when b=0 and 1 for various values of the bit rate R (and with F_s=44,100, g=1, σ_x^2 =20 dB)

When symbols have the same a priori probability, this probability of error can be estimated in Fig. 7.5 as the area under the intersection of $P(\alpha_m|b_m=0)$ and $P(\alpha_m|b_m=1)$. It is also possible to compute P_e analytically (Proakis 2001), which leads to

$$P_e = Q\left(\sqrt{\frac{g^2\|\underline{c}\|^2}{\sigma_X^2}}\right) = Q\left(\sqrt{\frac{F_s}{R}\frac{g^2\sigma_C^2}{\sigma_X^2}}\right) \quad \text{with} \quad Q(u) = \int_u^\infty \frac{1}{\sqrt{2\pi}}\exp\left(-\frac{t^2}{2}\right)dt \tag{7.15}$$

in which we have used $\sigma_C^2 = \|\mathbf{c}\|^2/N_b$, an estimate of the power of the spreading sequence.

The probability of error is thus finally a function of the watermarking SNR (i.e., the ratio between the power of the emitted signal, $g^2\sigma_C^2$, and that of the audio signal, σ_X^2) and of the bit rate R. Figure 7.6 shows a plot of $Q\left(\sqrt{u}\right)$ and the resulting plot of $P_e(R)$ for various values of the SNR. It appears that $P_e=10^{-3}$ can be reached with a bit rate of 50 bits/s and a SNR of –20 dB. As we will see below, this SNR approximately corresponds to a watermark just low enough not to be heard (or, consequently, to the kind of SNR introduced by MPEG compression; see Chapter 3).

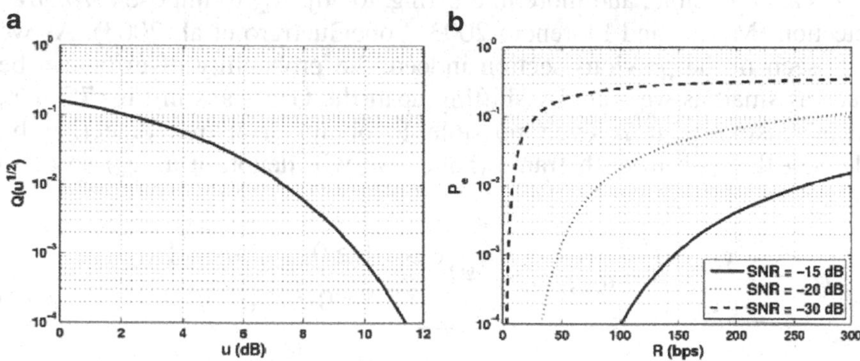

Fig. 7.6 *Left*: $Q\left(\sqrt{u}\right)$ with u in dB; *right*: $P_e(R)$ for various values of the SNR and F_s=44,100

7.1.3 Informed watermarking

Although it is generally not possible to adapt the encoder to the features of the channel noise,[10] such a strategy is perfectly possible in watermarking systems, as the communication noise itself is available at the encoder. In other words, watermarking is a good example of a communication channel with side information (Shannon 1958, Costa 1983). This opens the doors to

[10] In some communications systems (like in ADSL modems), the communication SNR is estimated at the receiver and sent back to the emitter through a communication back channel.

the second generation of watermarking systems: that of *informed watermarking* (Cox et al. 1999, Chen and Wornell 2001, Malvar and Florencio 2003) and its triple objectives; maximal bit rate, minimal distortion, and maximal robustness. In the next paragraphs, we will consider two steps in this direction: gain adaptation and perceptual filtering.

Gain adjustment for error-free detection

The most obvious way to turn the encoder into an informed encoder is to adapt the gain g as a function of time so as to compensate for the sudden power variations in the audio signal and keep the SNR constant. This can be achieved by estimating σ_X^2 on audio frames (typically 20 ms long, in which music is assumed to be stationary) and imposing g to be proportional to σ_X.

It is also possible, and more interesting, to adjust g to impose *error-free* detection (Malvar and Florencio 2003, LoboGuerrero et al. 2003). As we have seen in the previous section indeed, the probability of error can be made as small as we want by shifting apart the Gaussians in Fig. 7.5, i.e., by increasing g. It is even possible to obtain error-free detection by adapting the gain to each frame (hence we will denote it as g_m) so as to yield

$$\alpha_m = \frac{<\mathbf{y_m},\mathbf{c}>}{\|\mathbf{c}\|^2} = a_m g_m + \underbrace{\frac{<\mathbf{x_m},\mathbf{c}>}{\|\mathbf{c}\|^2}}_{\beta_m} = \begin{cases} \geq 0 & if\ a_m = 1 \\ < 0 & if\ a_m = -1 \end{cases} \tag{7.16}$$

As seen in Fig. 7.4, this condition is automatically verified for any (positive) value of g_m if β_m has the same sign as a_m. In the opposite case, it is verified by imposing $g_m = -\beta_m / a_m$.

In practice though, the transmission channel itself will add noise to the watermarked vector $\mathbf{y_m}$, leading to received vectors equal to $\mathbf{y_m}+\mathbf{p_m}$. Condition (7.16) becomes

$$\begin{aligned} \alpha_m &= <\mathbf{y_m}+\mathbf{p_m},\mathbf{c}> \\ &= a_m g_m + \underbrace{\frac{<\mathbf{x_m},\mathbf{c}>}{\|\mathbf{c}\|^2}}_{\beta_m} + \underbrace{\frac{<\mathbf{p_m},\mathbf{c}>}{\|\mathbf{c}\|^2}}_{\gamma_m} = \begin{cases} \geq 0 & if\ a_m = 1 \\ < 0 & if\ a_m = -1 \end{cases} \end{aligned} \tag{7.17}$$

Some security margin Δ_g is then required to counterbalance the projection of $\mathbf{p_m}$ on $\mathbf{c}$ (Fig. 7.7):

$$\begin{cases} g_m + \beta_m \geq \Delta_g & if\ a_m = 1 \\ -g_m + \beta_m \leq -\Delta_g & if\ a_m = -1 \end{cases} \tag{7.18}$$

The value of Δ_g is set according to the estimated variance of the transmission channel noise.

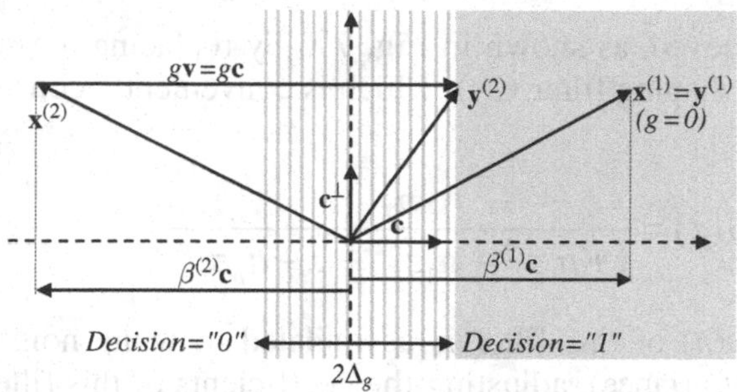

Fig. 7.7 Informed watermarking of audio vector **x** with symbol "1." Watermarked vector **y** is obtained by adding $g\mathbf{v}$ (i.e., $g\mathbf{c}$, given the symbol is "1") to **x**, with g chosen such that **y** falls in the gray region, where decision will be "1." Two cases are shown: for $\mathbf{x}^{(1)}$, g can be set to zero; for $\mathbf{x}^{(2)}$, g must have a high value. The gray region has been shifted from the white region by a $2\Delta_g$-wide gap to account for possible transmission channel noise

The efficiency of this watermarking strategy stems from the fact that the gain g is directly proportional to the correlation between the audio signal **x** and the spread sequence **c**. However, it does not ensure the perceptual transparency of the watermark: in case **x** is very much in the opposite direction of **c** while the emitted symbol is "1" (as it is the case of $\mathbf{x}^2$ in Fig. 7.7), g must be set to a high value. In such a case, chances are that the watermark may significantly distort the audio signal.

Perceptual shaping filter

A solution to the perceptual transparency requirement is to use properties of the human auditory system to ensure inaudible watermarking. As we have shown in Chapter 3, psychoacoustic modeling (PAM) has established the existence of a masking threshold, assimilated to a PSD $\Phi(f)$ (see Chapter 3, Sections 3.1.3 and 3.1.4) below which the PSD of the watermark $S_W(f)$ must lie for making it inaudible. This threshold is signal-dependent and must therefore be updated frequently, typically every 20 ms.

Since the efficiency of a watermarking system is a function of the SNR ratio (Fig. 7.6), the best performance is reached when

$$S_W(f) = \Phi(f) \tag{7.19}$$

This can be achieved, as shown in Fig. 7.8, by replacing the scalar gain g by a perceptual shaping filter $G(f)$, which is conveniently constrained to be an all-pole filter:

$$G(z) = \frac{b_0}{1 + a_1 z^{-1} + a_2 z^{-2} + \ldots + a_p z^{-p}} \tag{7.20}$$

If the input $v(n)$ of this filter is assimilated to white noise (with zero mean and unity variance), adjusting the coefficients of this filter so that its output $w(n)$ will have a given PSD $S_W(f)$ is a typical filter synthesis problem, which has already been examined in Chapter 1. It is based on solving a set of p linear equations, the so-called *Yule-Walker* Equations (1.5), whose coefficients are the first p values of the autocorrelation function $\phi_w(k)$ of $w(n)$. These values are related to the PSD of $w(n)$ by the inverse discrete-time Fourier transform formula

$$\phi_w(k) = \int_{-1/2}^{1/2} \Phi(f) e^{j2\pi fk} df \tag{7.21}$$

which can be approximated by an inverse discrete Fourier transform using N samples of $\Phi(f)$:

$$\phi_w(k) \simeq \frac{1}{N} \sum_{n=0}^{N-1} \Phi\left(\frac{n}{N}\right) e^{j\frac{2\pi}{N}nk} \tag{7.22}$$

The receiver has to be modified accordingly (Fig. 7.8) by inserting an inverse, zero-forcing equalizer $1/G(f)$ before the demodulator (Larbi et al. 2004). However, since the audio signal $x(n)$ used to derive $G(f)$ is not available at the receiver, an estimate $1/\tilde{G}(f)$ is computed by psychoacoustic modeling of the watermarked signal $y(n)$. The accuracy of this estimation depends on the robustness of the PAM to the watermark *and* to possible transmission channel noise, such as MPEG compression.

This specific constraint on the PAM used in watermarking distinguishes it from the PAMs used in audio compression (such as in MPEG).

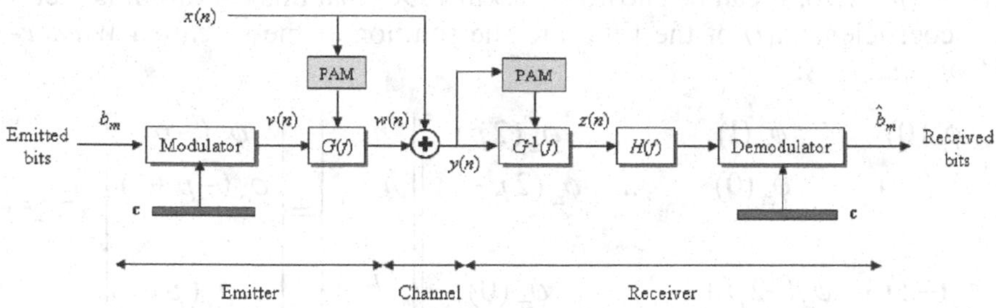

Fig. 7.8 Informed watermarking using a psychoacoustic model (PAM), a perceptual shaping filter *G(f)*, and a Wiener equalizer *H(f)*

Once perceptual shaping has been inverted, detection is achieved as previously, except that it is based on estimating the sign of $<\mathbf{z_m},\mathbf{c}>$, where *z(n)* is the output of the inverse filter. Assuming $1/\tilde{G}(f)$ is close enough to $1/G(f)$, $\mathbf{z_m}$ is given by

$$\mathbf{z_m} = \mathbf{v_m} + \mathbf{r_m} \tag{7.23}$$

where *r*(*n*) is the audio signal *x*(*n*) filtered by $1/\tilde{G}(f)$. Equation (7.23) simply replaces Equation (7.5).

Wiener filtering

As we will see in Section 7.2.3, the SNR at the output of the zero-forcing equalizer is very low; the power of the spread spectrum signal *v(n)* is small compared to that of the equalized audio signal *r(n)*. Since the PSDs of *v(n)* and *r(n)* are different, it is possible to filter *z(n)* so as to increase the SNR, by enhancing the spectral components of *z(n)* dominated by *v(n)* and attenuating those dominated by *r(n)*. This can even be done optimally by a symmetric FIR *Wiener filter H(z)* at the output of the zero-forcing equalizer (Fig. 7.8):

$$H(z) = \sum_{i=-p}^{p} h(i) z^{-i} \tag{7.24}$$

The output of the FIR Wiener filter *H(z)* is made maximally similar to *v(n)* in the minimum mean-square-error (MMSE) sense (Fig. 7.9); its coefficients are computed so as to minimize the power of the error signal $e(n)=z(n)-v(n)$. It can be shown (Haykin 1996) that this condition is met if the coefficients *h(i)* of the filter are the solution of the so-called *Wiener–Hopf equations*:

$$\begin{bmatrix} \phi_{zz}(0) & \phi_{zz}(1) & \cdots & \phi_{zz}(2p) \\ \phi_{zz}(-1) & \phi_{zz}(0) & \cdots & \phi_{zz}(2p-1) \\ \cdots & \cdots & \cdots & \cdots \\ \phi_{zz}(-2p) & \phi_{zz}(-2p+1) & \cdots & \phi_{zz}(0) \end{bmatrix} \begin{bmatrix} h_{-p} \\ h_{-p+1} \\ \cdots \\ h_{p} \end{bmatrix} = \begin{bmatrix} \phi_{vv}(-p) \\ \phi_{vv}(-p+1) \\ \cdots \\ \phi_{vv}(p) \end{bmatrix} \tag{7.25}$$

where $\phi_{zz}(k)$ and $\phi_{vv}(k)$ are the autocorrelation functions of *z(n)* and *v(n)*, respectively.

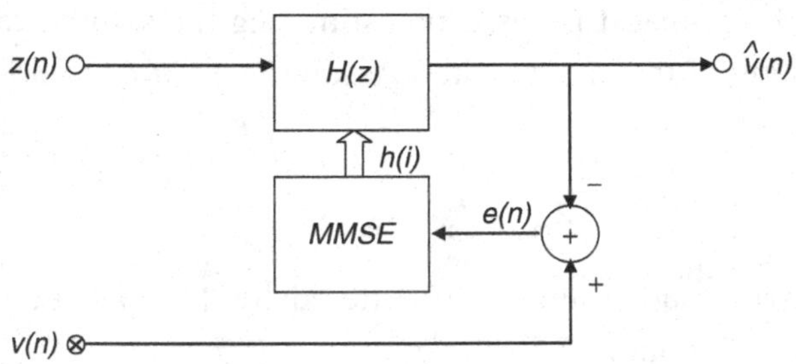

Fig. 7.9 Conceptual view of a Wiener filter[11]

In practice, since *r(n)* is not stationary, the Wiener filter is regularly adapted.

7.2 MATLAB proof of concept: ASP_watermarking.m

In this section, we develop watermarking systems whose design is based on classical communication systems using spread spectrum modulation. We first examine the implementation of a watermarking system and give a brief overview of its performance in the simple and theoretical

[11] In practice, *v(n)* is not available. Nevertheless, only its autocorrelation is required to solve the Wiener–Hopf equations.

configuration where the audio signal is a white Gaussian noise (Section 7.2.1). Then we show how this system can be adapted to account for the audio signal specificities, while still satisfying the major properties and requirements of watermarking applications (namely inaudibility, correct detection, and robustness of the watermark); we extend the initial system by focusing on the correct detection of the watermark (Section 7.2.2), on the inaudibility constraint (Section 7.2.3), and on the robustness of the system to MP3 compression (Section 7.2.4).

7.2.1 Audio watermarking seen as a digital communication problem

The following digital communication system is the direct implementation of the theoretical results detailed in Section 7.1. The watermark message is embedded in the audio signal at the binary rate of `R=100` bps. To establish the analogy between watermarking system and communication channel, the audio signal is viewed as the channel noise. In this section, it will therefore be modeled as a white Gaussian noise with zero mean and variance `sigma2_x`. `sigma2_x` is computed so that the signal-to-noise ratio (SNR), i.e., the ratio between the watermark and the audio signal powers, is equal to −20 dB.

Emitter/Embedder

Let us first generate a watermark message and its corresponding bit sequence.

```
message = 'Audio watermarking: this message will be embedded
in an audio signal, using spread spectrum modulation, and
later retrieved from the modulated signal.'

bits = dec2bin(message); % converting ascii to 7-bit string
bits = bits(:)'; % reshaping to a single string of 1's and 0's
bits = double(bits) - 48; % converting to a 0's and 1's
symbols = bits*2-1;  % a (1x1050) array of -1's and +1's
N_bits = length(bits);
```

```
message =

Audio watermarking: this message will be embedded in an audio
signal, using spread spectrum modulation, and later retrieved from
the modulated signal.
```

We then generate the audio signal, the spread spectrum sequence, and the watermarked signal. The latter is obtained by first deriving a symbol sequence (−1's and +1's) from the input bit sequence (0's and +1's) and then by modulating the spread spectrum signal by the symbol sequence

(Fig. 7.10, left). This is achieved by concatenating spread spectrum waveforms weighted by the symbol values.

```
% Random sequences generators initialization
rand('seed', 12345);
randn('seed', 12345);

% Sampling frequency |Fs|, binary rate |R|, samples per bit
% |N_samples| and |SNR|
Fs = 44100;
R = 100;
N_samples = round(Fs/R);
SNR_dB = -20;

% Spread waveform: a random sequence of -1's and +1's
spread_waveform = 2*round( rand(N_samples, 1) ) - 1;

% Audio signal: a Gaussian white noise with fixed variance
sigma2_x = 10^(-SNR_dB/10);
audio_signal = sqrt(sigma2_x)*randn(N_bits*N_samples, 1);

% Modulated signal and watermark signal
modulated_signal = zeros(N_bits*N_samples, 1);
for m = 0:N_bits-1
    modulated_signal(m*N_samples+1:m*N_samples+N_samples)= ...
        symbols(m+1)*spread_waveform;
end
watermark_signal = modulated_signal; % no gain applied

% Plotting the baseband signal (i.e., the emitted signal
% corresponding to the input bit sequence) and the watermark
% signal

emitted_signal = ones(N_samples, 1)*symbols;
emitted_signal = emitted_signal(:);

axe_ech = (4*N_samples:7*N_samples+1);
subplot(2,1,1);
plot(axe_ech/Fs, emitted_signal(axe_ech));
subplot(2,1,2);
plot(axe_ech/Fs, watermark_signal(axe_ech));
```

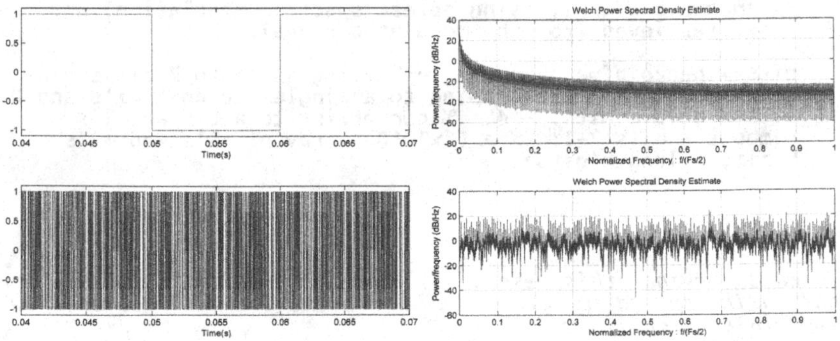

Fig. 7.10 *Left*: Three symbols in {−1,+1} and the resulting spread spectrum waveforms, or watermark signal; *right*: the corresponding PSDs

As expected, the power spectral density (PSD) of the spread spectrum sequence (i.e., the watermark signal) is much flatter than that of the baseband signal (i.e., the emitted signal) corresponding to the input bit sequence (Fig. 7.10, right), while their powers are identical: 0 dB.

```
power_baseband_dB=10*log10(var(emitted_signal))
power_spread_spectrum_dB=10*log10(var(watermark_signal))

subplot(2,1,1)
pwelch(emitted_signal,[],[],[],2);12
subplot(2,1,2)
pwelch(watermark_signal,[],[],[],2);
```

```
power_baseband_dB =  -0.0152
power_spread_spectrum_dB =   5.5639e-006
```

The embedding process finally consists in adding the result to the audio signal, yielding the audio watermarked signal. Plotting the first 100 samples of the watermark, audio, and watermarked signals shows that the watermarked signal is only slightly different from the audio signal, given the SNR we have imposed (Fig. 7.11).

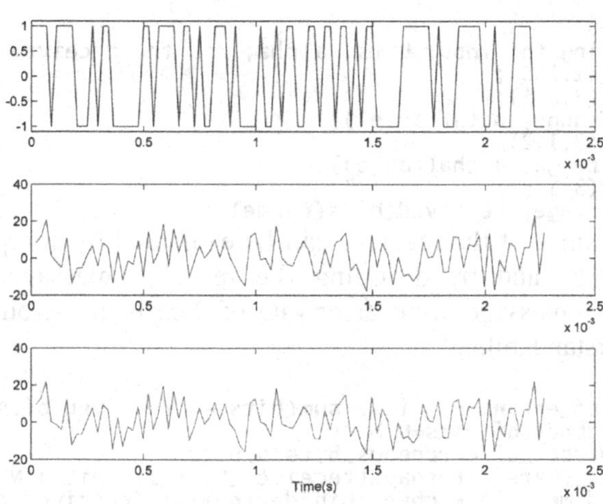

Fig. 7.11 Zoom on 100 samples. *Top*: Watermark signal; *center*: audio signal; *bottom*: watermarked signal

[12] Note that we claim F_s=2 so as to make **pwelch** return $|FFT^2/N|$, in which the variance of a white noise is directly readable. See the appendix in **ASP_audio_cd.m** for more details.

```
watermarked_signal = watermark_signal + audio_signal;

subplot(3,1,1); plot((1:100)/Fs,watermark_signal(1:100),'k-');
subplot(3,1,2);plot((1:100)/Fs,audio_signal(1:100));
subplot(3,1,3);plot((1:100)/Fs,watermarked_signal(1:100),'r');
```

Receiver

The watermark receiver is a correlation demodulator. It computes the normalized scalar product `alpha` between frames of the received signal and the spread spectrum waveform and decides on the received bits, based on the sign of `alpha`. One can see on the plot (for bits 81 to 130) that the watermark sometimes imposes a wrong sign to `alpha`, leading to erroneous bit detection (Fig. 7.12).

```
alpha = zeros(N_bits, 1);
received_bits = zeros(1,N_bits);
for m = 0:N_bits-1
    alpha(m+1) =
(watermarked_signal(m*N_samples+1:m*N_samples+N_samples)' ...
        * spread_waveform)/N_samples;
    if alpha(m+1) <= 0
        received_bits(m+1) = 0;
    else
        received_bits(m+1) = 1;
    end
end

% Plotting the input bits, alpha, and the received bits.
range=(81:130);
subplot(3,1,1);
stairs(range, bits(range));
subplot(3,1,2);
stairs(range, alpha(range));
subplot(3,1,3);
stairs(range, received_bits(range));
```

The performance of the receiver can be estimated by computing the bit error rate (BER) and by decoding the received message. As can be observed on the message, a bit error rate of 0.02 is disastrous in terms of message understandability.

```
number_of_erroneous_bits= sum(bits ~= received_bits)
total_number_of_bits=N_bits
BER = number_of_erroneous_bits/N_bits
received_chars = reshape(received_bits(1:N_bits),N_bits/7, 7);
received_message = char( bin2dec(num2str(received_chars)))'
```

```
number_of_erroneous_bits =  21
total_number_of_bits =  1050
BER =    0.0200

received_message =Cudio ⎕ateriarking: this messaGe will be embeeded
iN An audio signal( ushnG spread spectrum modulationl and later
qetrimve$ from 4he mo`uleteD!siGnal.
```

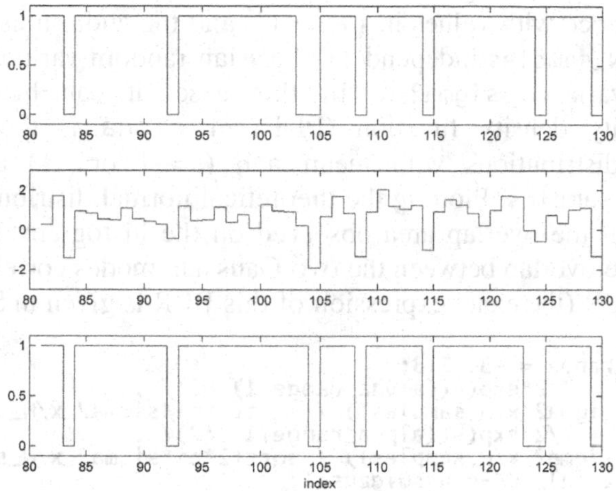

Fig. 7.12 Zoom on bits 80–130. *Top*: Input bits; *center*: **alpha**; *bottom*: received bits

System performance overview

A good overview of the system performance can be drawn from the statistical observation of alpha values through a histogram plot. As expected, a bimodal distribution is found, with nonzero overlap (Fig. 7.13). Note that the histogram gives only a rough idea of the underlying distribution, since the number of emitted bits is small.

```
hist(alpha,50);
```

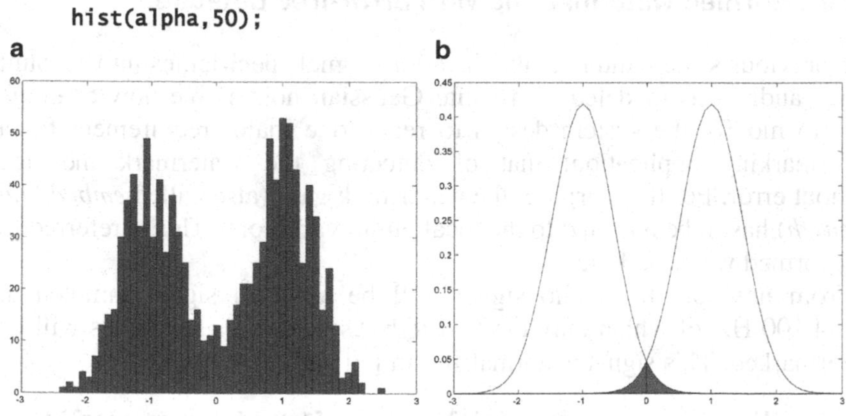

Fig. 7.13 *Left*: Histogram of alpha values. *Right*: Theoretical probability density function

In our particular configuration, the spread waveform is a realization of a random sequence with values in {+1, −1}, and the audio-noise vectors are modeled by N_samples-independent Gaussian random variables with zero mean and variance sigma2_x. In this case, it can be shown that the probability density function (PDF) of alpha is the average of two normal distributions with mean a*g (a=+1 or −1) and variance sigma2_x/N_samples. Plotting the theoretical normal distribution shows a better view of the overlap area observed on the histogram. In particular, the area of the overlap between the two Gaussian modes corresponds to the theoretical BER (the exact expression of this BER is given in Section 7.1).

```
alpha_range = -3:.1:3;
gauss0 = 1/2*exp(-((alpha_range-1).^2)/ ...
   (2*sigma2_x/N_samples)) / sqrt(2*pi*sigma2_x/N_samples);
gauss1 = 1/2*exp(-((alpha_range+1).^2)/ ...
   (2*sigma2_x/N_samples)) / sqrt(2*pi*sigma2_x/N_samples);
theoretical_PDF=gauss0+gauss1;

plot(alpha_range, gauss0); hold on
plot(alpha_range, gauss1);

ind0 = find(alpha_range<=0);
ind1 = find(alpha_range>=0);
ae1 = area(alpha_range(ind1), gauss1(ind1), ...
   'FaceColor', [0.39 0.47 0.64]);   hold on;
ae2 = area(alpha_range(ind0), gauss0(ind0), ...
   'FaceColor', [0.39 0.47 0.64]);

theoretical_BER1=2*sum(gauss1(ind1))*(alpha_range(2)-...
   alpha_range(1))
```

```
theoretical_BER1 =  0.0228
```

7.2.2 Informed watermarking with error-free detection

The previous system did not take the audio signal specificities into account (since audio was modeled as a white Gaussian noise). We now examine how to modify the system design to reach one major requirement for a watermarking application: that of detecting the watermark message without error. For this purpose, the watermark gain (also called *embedding strength*) has to be adjusted to the local audio variations. This is referred to as informed watermarking.

From now on, the audio signal will be a violin signal sampled at F_s=44.100 Hz, of which only the first N_bits*N_samples samples will be watermarked. This signal is normalized in [−1,+1].

```
audio_signal = wavread('violin.wav', [1 N_bits*N_samples]);
```

Informed emitter

To reach a zero BER, the watermark gain is adapted so that the correlation between the watermarked audio signal and the spread sequence is at least equal to some security margin `Delta_g` (if `am`=+1) and `-Delta_g` (if `am`=–1). `Delta_g` sets up a robustness margin against additive perturbation (such as the additive noise introduced by MPEG compression of the watermarked audio signal). In the following MATLAB implementation, `Delta_g` is empirically set to 0.005.

```
Delta_g = 0.005;
for m=0:N_bits-1
  beta = audio_signal(m*N_samples+1:m*N_samples+N_samples)'*...
         spread_waveform /N_samples;
  if symbols(m+1) == 1
         if  beta >= Delta_g
             gain(m+1) = 0;
         else
             gain(m+1) = Delta_g - beta;
         end
  else % if symbols(m+1) == -1
         if  beta <= -Delta_g
             gain(m+1) = 0;
         else
             gain(m+1) = Delta_g + beta;
         end
  end
  watermark_signal(m*N_samples+1:m*N_samples+N_samples,1)= ...

gain(m+1)*modulated_signal(m*N_samples+1:m*N_samples+N_samples);
end

watermarked_signal = watermark_signal + audio_signal;
```

Plotting the gain for 1 second of signal, together with the resulting watermark signal and the audio signal, shows that the encoder has to make important adjustments as a function of the audio signal (Fig. 7.14).

```
subplot(2,1,1);
plot((0:99)*N_samples/Fs, gain(1:100));
subplot(2,1,2);
plot((0:Fs-1)/Fs, watermark_signal(1:Fs));
```

The watermark, though, is small compared to the audio signal (Fig. 7.15, left).

```
plot((0:Fs-1)/Fs, audio_signal(1:Fs)); hold on;
plot((0:Fs-1)/Fs, watermark_signal(1:Fs), 'r-');
```

Plotting again a few samples of the watermark, audio, and watermarked signals shows that the watermarked signal sometimes differs significantly from the audio signal (Fig. 7.15, right). This is confirmed by a listening test: the watermark is audible.

```
range=(34739:34938);
subplot(3,1,1); plot(range/Fs, watermark_signal(range), 'k-');
subplot(3,1,2); plot(range/Fs, audio_signal(range));
subplot(3,1,3); plot(range/Fs, watermarked_signal(range),'r');
```

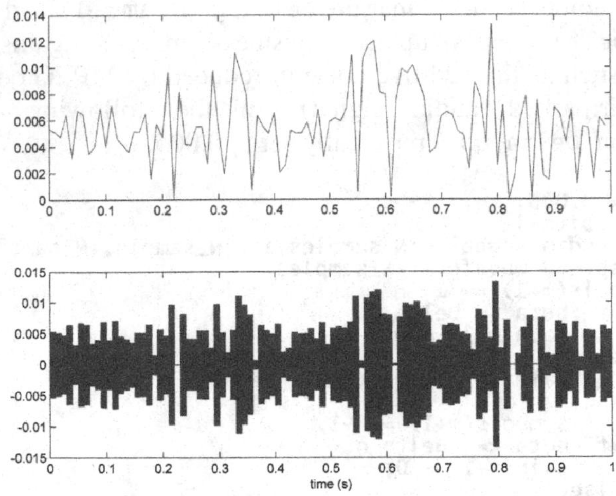

Fig. 7.14 *Top*: Watermark gain; *bottom*: watermark signal

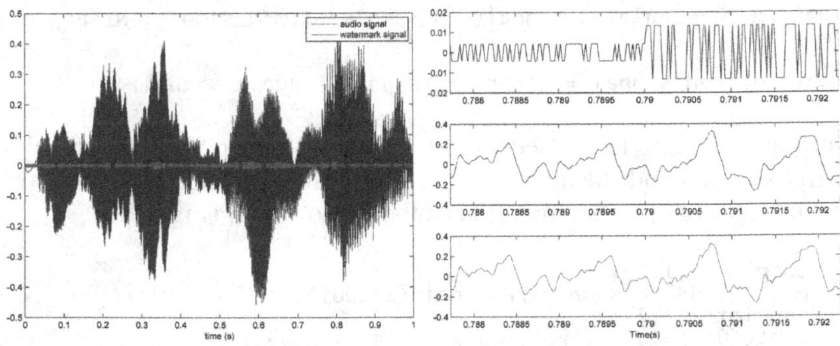

Fig. 7.15 Left: A comparison between the audio and watermark signals; *right*: zoom on the watermark signal (*top*), the audio signal (*center*), and the watermarked signal (*bottom*)

Receiver

Using the same correlation demodulator as in Section 7.1, we conclude that the transmission is now effectively error-free (Fig. 7.16), as confirmed by the resulting BER.

```
number_of_erroneous_bits= sum(bits ~= received_bits)
BER = number_of_erroneous_bits/total_number_of_bits

received_chars = reshape(received_bits(1:N_bits),N_bits/7, 7);
received_message = char( bin2dec(num2str(received_chars)))'
```

```
number_of_erroneous_bits =  0
BER =  0

received_message = Audio watermarking: this message will be embedded
in an audio signal, using spread spectrum modulation, and later
retrieved from the modulated signal.
```

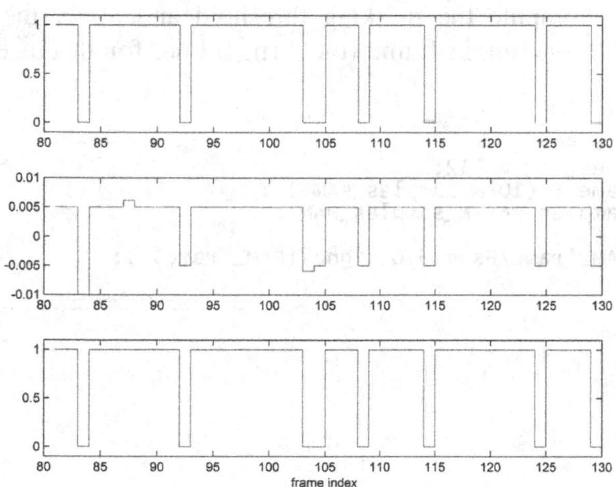

Fig. 7.16 Zoom on bits 80–130. Top: Input bits; *center*: `alpha`; *bottom*: received bits

7.2.3 Informed watermarking made inaudible

The informed watermarking system exposed in Section 7.2.2 proves that prior knowledge of the audio signal can be efficiently used in the watermarking process; error-free transmission is reached, thanks to an adaptive embedding gain. However, the resulting watermark is audible, since no perceptual condition is imposed on the embedding gain. In this Section, we examine how psychoacoustics can be put to profit to ensure the inaudibility constraint. A psychoacoustic model is used, which provides a signal-dependent masking threshold used as an upper bound for the PSD of the watermark. The watermark gain is therefore replaced by an all-pole perceptual shaping filter, and the reception process is composed of a zero-forcing equalizer, followed by a linear-phase Wiener filter.

Emitter based on perceptual shaping filtering

The perceptual shaping filter is an auto regressive (all-pole) filter with 50 coefficients `ai` and gain `b0`. It is designed so that the PSD of the watermark (obtained by filtering the `modulated_signal`) equals the `masking_threshold`. The coefficients of the filter are obtained as in Chapter 1, via the Levinson algorithm. Both the masking threshold and the shaping filter have to be updated each time the statistical properties of the `audio_signal` change (here each `N_windows=512` samples).

Let us first compute the masking threshold and apply the associated shaping filter to one audio frame (the 11th frame, for instance; Fig. 7.17 left).

```
N_coef = 50;
N_samples_PAM = 512;
PAM_frame = (10*N_samples_PAM+1 :
10*N_samples_PAM+N_samples_PAM);

plot(PAM_frame/Fs,audio_signal(PAM_frame) );
```

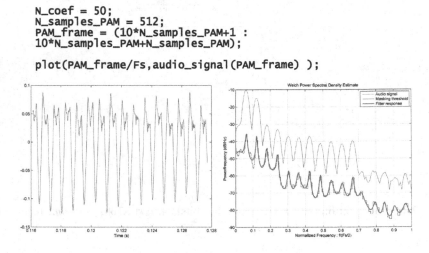

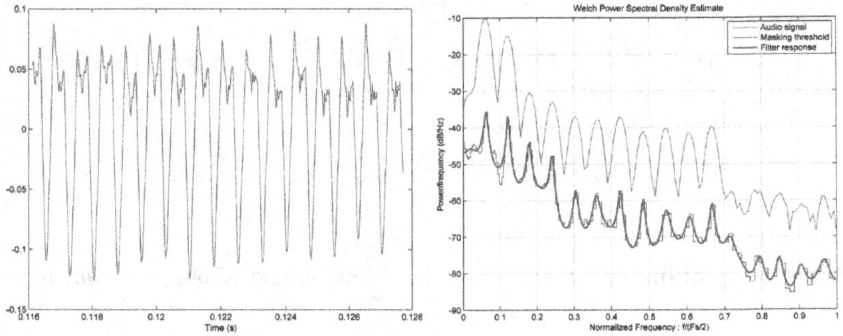

Fig. 7.17 *Left*: Audio signal of frame #11; *right*: PSD of frame #11 and the associated masking threshold and shaping filter response

Comparing the periodogram of the audio signal, the masking threshold, and the frequency response of the filter shows that the filter closely matches the masking threshold (Fig. 7.17 right). Even the absolute amplitude level of the filter is the same as that of the masking threshold. As a matter of fact, since the audio signal is normalized in [−1,+1], its nominal PSD level is 0 dB.

MATLAB functions involved:

- `masking_threshold=psychoacoustical_model(audio_signal)` returns the masking threshold deduced from a psychoacoustical analysis of the audio vector. This implementation is derived from the psychoacoustic model #1 used in MPEG-1 audio (see ISO/CEI norm 11172-3:1993 (F), pp. 122–128 or MATLAB function MPEG1_psycho_acoustic_model1.m from Chapter 3). It is based on the same principles as those used in the MPEG model, but it is further adapted here so as to make it robust to additive noise (which is a specific constraint of watermarking and is not found in MPEG).

- `[b0,ai]=shaping_filter_design(desired_frequency_response_dB,N_coef)` computes the coefficients of an auto-regressive filter

```
                                 b0
G(z) = ---------------------------------------------------
        ai(1) + ai(2)z-1      + ... + ai(N_coef)z-N_coef+1
```

  (with ai(1)=1) from the modulus of its `desired_frequency_response` (in dB) and the order `N_coef`. The coefficients are obtained as follows: if zero-mean and unity variance noise is provided at the input of the filter, the PSD of its output is given by `desired_frequency_response`. Setting the coefficients so that this PSD best matches `desired_frequency_response` is thus obtained by applying the Levinson algorithm to the autocorrelation coefficients of the output signal (computed itself from the IFFT of the `desired_frequency_response`).

```
masking_threshold = ...
   psychoacoustical_model(audio_signal(PAM_frame));
shaping_filter_response = masking_threshold;
[b0,ai]=shaping_filter_design(shaping_filter_response,N_coef);

% Plotting results. For details on how we use |pwelch|, see
% the appendix of ASP_audio_cd.m, the companion script file of
% Chapter 2.

pwelch(audio_signal(PAM_frame),[],[],[],2);
hold on;
[H,W]=freqz(b0,ai,256);
stairs(W/pi,masking_threshold,'k--');
plot(W/pi,20*log10(abs(H)),'r','linewidth',2);
hold off;
```

Applying this perceptual shaping procedure to the whole watermark signal requires to process the audio signal block per block. Note that the filtering continuity from one block to another is ensured by the `state` vector, which stores the final state of the filter at the end of one block and applies it as initial conditions for the next block.

```
state = zeros(N_coef, 1);
for m = 0:fix(N_bits*N_samples/N_samples_PAM)-1
    PAM_frame = ...
        (m*N_samples_PAM+1:m*N_samples_PAM+N_samples_PAM);

    % Shaping filter design
    masking_threshold = psychoacoustical_model(...
        audio_signal(PAM_frame) );
    shaping_filter_response = masking_threshold;
    [b0, ai]=shaping_filter_design(shaping_filter_response,...
        N_coef);

    % Filtering stage
    [watermark_signal(PAM_frame,1), state] = ...
        filter(b0, ai, modulated_signal(PAM_frame), state);
end

% Filtering the last, incomplete frame in |watermark_signal|
PAM_frame = ...
    (m*N_samples_PAM+N_samples_PAM+1:N_bits*N_samples);
watermark_signal(PAM_frame,1) = ...
    filter(b0, ai, modulated_signal(PAM_frame), state);

plot((0:Fs-1)/Fs, audio_signal(1:Fs));
hold on;
plot((0:Fs-1)/Fs, watermark_signal(1:Fs), 'r-');
```

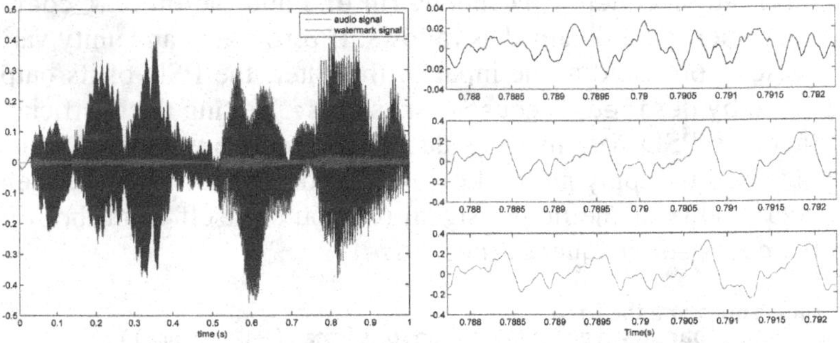

Fig. 7.18 *Left*: Audio signal and watermark signal; *right*: a few samples of the audio, watermark, and watermarked signals

The watermarked audio signal is still obtained by adding the audio signal and the watermark signal (Fig. 7.18 left). Plotting again few samples of the watermark, audio, and watermarked signals shows that the watermark signal has now been filtered (Fig. 7.18 right). As a result, the watermark is quite inaudible, as confirmed by a listening test. Its level, though, is similar to (if not higher than) that of the watermark signal in Section 7.2.2.

```
watermarked_signal = watermark_signal + audio_signal;
```

```
range=(34739:34938);
subplot(3,1,1); plot(range/Fs, watermark_signal(range), 'k');
subplot(3,1,2); plot(range/Fs, audio_signal(range));
subplot(3,1,3); plot(range/Fs, watermarked_signal(range),'r');
```

Receiver based on zero-forcing equalization and detection

The zero-forcing equalization aims at reversing the watermark perceptual shaping before extracting the embedded message. The shaping filter with frequency response shaping_filter_response is therefore recomputed from the audio watermarked_signal, since the original audio_signal is not available at the receiver. This process follows the same block processing as the watermark synthesis but involves a moving average filtering stage; filtering the watermarked_signal by the filter whose frequency response is the inverse of the shaping_filter_response yields the filtered received signal denoted by equalized_signal.

Plotting the PSD of the equalized_signal shows its flat spectral envelope, hence its name (Fig. 7.19).

```
state = zeros(N_coef, 1);
for m = 0:fix(N_bits*N_samples/N_samples_PAM)-1

    PAM_frame = ...
        m*N_samples_PAM+1:m*N_samples_PAM+N_samples_PAM;

    % Shaping filter design, based the watermarked signal
    masking_threshold =
        psychoacoustical_model(watermarked_signal(PAM_frame));
    shaping_filter_response = masking_threshold;
    [b0, ai]=shaping_filter_design(shaping_filter_response,...
        N_coef);

    % Filtering stage
    [equalized_signal(PAM_frame), state] = ...
        filter(ai./b0,1,watermarked_signal(PAM_frame),state);

    if m==10
    % Showing the frequency response of the equalizer and PSD
    % of the |equalized_signal|, for frame 10.
        pwelch(equalized_signal(PAM_frame),[],[],[],2);
        hold on;
        [H,W]=freqz(ai./b0,1,256);
        plot(W/pi,20*log10(abs(H)),'r','linewidth',2);
    end;

end
```

The equalized_signal is theoretically the sum of the original watermark (i.e., the modulated_signal, which has values in {−1,+1}) and the equalized_audio_signal, which is itself the original audio_signal filtered by the zero-forcing equalizer. In other words, from the receiver point of view, everything looks as if the watermark had been added to the equalized_audio_signal rather than to the audio_signal itself.

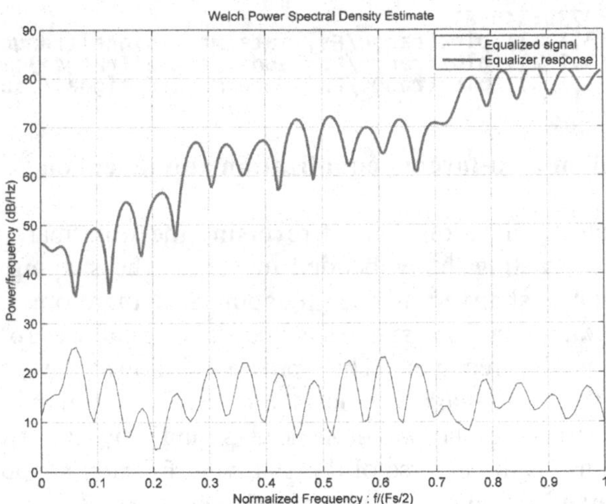

Fig. 7.19 Frequency response of the equalizer at frame #10 and PSD of the resulting equalized watermarked signal

Although the equalized_audio_signal is not available to the receiver, it is interesting to compute it and compare it to the original audio_signal (Fig. 7.20, left). Note that the level of the equalized_audio_signal is much higher than that of the original audio_signal, mostly because its HF content has been enhanced by the equalizer.

```
equalized_audio_signal = zeros(N_bits*N_samples, 1);
state = zeros(N_coef, 1);

for m = 0:fix(N_bits*N_samples/N_samples_PAM)-1

    PAM_frame = ...
       m*N_samples_PAM+1:m*N_samples_PAM+N_samples_PAM;

    % Shaping filter design, based the watermarked signal
    masking_threshold = ...
       psychoacoustical_model(watermarked_signal(PAM_frame));
    shaping_filter_response = masking_threshold;
    [b0,ai]=shaping_filter_design(shaping_filter_response, ...
       N_coef);

    % Filtering stage
    [equalized_audio_signal(PAM_frame), state] = ...
        filter(ai./b0, 1, audio_signal(PAM_frame), state);
end

% Plotting 200 samples of the |equalized_audio_signal|, and
% |modulated_signal|
range=(34739:34938);
subplot(2,1,1); plot(range/Fs, equalized_audio_signal(range));
subplot(2,1,2); plot(range/Fs, modulated_signal(range));
```

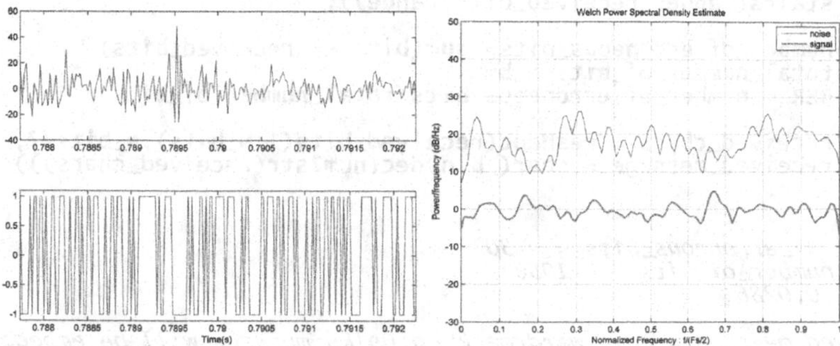

Fig. 7.20 *Left*: Zoom on the equalized audio signal and on the modulated signal; *right*: corresponding PSDs

It is also possible to estimate the SNR by visual inspection of the PSDs of the equalized audio signal and the modulated signal (Fig. 7.20, right) or to compute it from the samples of these signals.

```
range=(34398:34938);
pwelch(equalized_audio_signal(range),[],[],[],2);
[h,w]=pwelch(modulated_signal(range),[],[],[],2);
hold on;
plot(w,10*log10(h),'--r ');

snr_equalized=10*log10(var(modulated_signal(range))/...
   var(equalized_audio_signal(range)))
```

```
snr_equalized =  -19.0610
```

We can now proceed to the watermark extraction by applying the correlation demodulator to the equalized_signal and compute the BER. As expected, the obtained BER is quite high, since the SNR is low; the level of the equalized_audio_signal is high compared to that of the embedded modulated_signal (Fig. 7.20).

```
for m = 0:N_bits-1
    alpha(m+1) = (equalized_signal(m*N_samples+1: ...
        m*N_samples+N_samples)' * spread_waveform)/N_samples;
    if alpha(m+1) <= 0
        received_bits(m+1) = 0;
    else
        received_bits(m+1) = 1;
    end
end

% Plotting the input bits, alpha, and the received bits.
range=(81:130);
subplot(3,1,1);
stairs(range, bits(range));
subplot(3,1,2);
stairs(range, alpha(range));
subplot(3,1,3);
```

```
stairs(range, received_bits(range));

number_of_erroneous_bits= sum(bits ~= received_bits)
total_number_of_bits=N_bits
BER = number_of_erroneous_bits/total_number_of_bits

received_chars = reshape(received_bits(1:N_bits),N_bits/7, 7);
received_message = char( bin2dec(num2str(received_chars)))'
```

```
number_of_erroneous_bits =    30
total_number_of_bits =   1050
BER =   0.0286

received_message = udio watdrmarkyng:0plks mussage will be embedded
in e~0audik siwnql, }wi~g spread SpeCtruM(modul`tion, and
later0retrieved from the modudated qienal.
```

Wiener filtering

The Wiener filtering stage aims at enhancing the SNR between the `modulated_signal` and the `equalized_audio_signal`. This is achieved here by filtering the `equalized_signal` by a symmetric (non causal) FIR filter with `N_coef`=50 coefficients. Its coefficients `hi` are computed so that the output of the filter (when fed with the `equalized_signal`) becomes maximally similar to the `modulated_signal` in the RMSE sense. They are the solution of the so-called Wiener-Hopf equations:

`hi= equalized_signal_cov_mat`$^{-1}$ * `modulated_signal_autocor_vect`
where `equalized_signal_cov_mat` and `modulated_signal_autocor_vect` are the covariance matrix of the `equalized_signal` and the autocorrelation vector of the `modulated_signal`, respectively.

Since the `modulated_signal` is unknown from the receiver, its autocorrelation is estimated from an arbitrary modulated signal and can be computed once. To make it simple, we use our previously computed `modulated_signal` here. On the contrary, the covariance matrix of the equalized signal, and therefore the coefficients `hi`, has to be updated each time the properties of `estimated_signal` change, i.e., every `N_samples_PAM`=512 samples. Wiener filtering is then carried out by computing the covariance matrix of the `equalized_signal` and the impulse response of the Wiener filter for each PAM frame.

```
modulated_signal_autocor_vect=...
   xcorr(modulated_signal,N_coef,'biased');

state = zeros(2*N_coef, 1);

for m = 0:fix(N_bits*N_samples/N_samples_PAM)-1
    PAM_frame = ...
        (m*N_samples_PAM+1:m*N_samples_PAM+N_samples_PAM);
```

```
        % Estimating the covariance matrix of |equalized signal|,
        % as a Toeplitz matrix with first row given by the
        % autocorrelation vector of the signal.
        equalized_signal_autocor_vect= ...
            xcorr(equalized_signal(PAM_frame),2*N_coef,'biased');
        equalized_signal_cov_mat= ...
          toeplitz(equalized_signal_autocor_vect(2*N_coef+1:end));

        % Estimating the impulse response of the Wiener filter as
        % the solution of the Wiener-Hopf equations.
        hi=equalized_signal_cov_mat\modulated_signal_autocor_vect;

        % Filtering stage
        [Wiener_output_signal(PAM_frame,1), state] = ...
            filter(hi, 1, equalized_signal(PAM_frame), state);
        power = ...
            norm(Wiener_output_signal(PAM_frame))^2/N_samples_PAM;
        if (power ~= 0)
            Wiener_output_signal(PAM_frame) = ...
                Wiener_output_signal(PAM_frame)/ sqrt(power);
        end

        % Saving the Wiener filter for frame #78
        if m==78
            hi_78=hi/sqrt(power);
        end;
    end

    % Since the Wiener filter is non-causal (with |N_coeff|=50
    % coefficients for the non-causal part), the resulting
    % |Wiener_output_signal| is delayed with |N_coef| samples.
    Wiener_output_signal = ...
        [Wiener_output_signal(N_coef+1:end);  zeros(N_coef, 1)];
```

It is interesting to check how the equalized audio signal and the modulated signal have been modified by the Wiener filter (Fig. 7.21 left) and how their PSDs have evolved (Fig. 7.21 right). Obviously, the Wiener filter has enhanced the frequency bands dominated by the modulated signal in Fig. 7.20 (right), thereby increasing the SNR by more than 5 dBs.

```
range=(34398:34938);
Wiener_output_audio=filter(hi,1,equalized_audio_signal(range));
Wiener_output_modulated=filter(hi,1,modulated_signal(range));
range=(34739:34938);
subplot(2,1,1); plot(range/Fs,Wiener_output_audio(341:540));
subplot(2,1,2); plot(range/Fs,Wiener_output_modulated(341:540));

pwelch(Wiener_output_audio,[],[],[],2);
[h,w]=pwelch(Wiener_output_modulated,[],[],[],2);
hold on;
plot(w,10*log10(h),'--r ', 'linewidth',2);

snr_Wiener=10*log10(var(Wiener_output_modulated(341:530)) ...
      /var(Wiener_output_audio(341:530)))
```

snr_Wiener = -19.2023

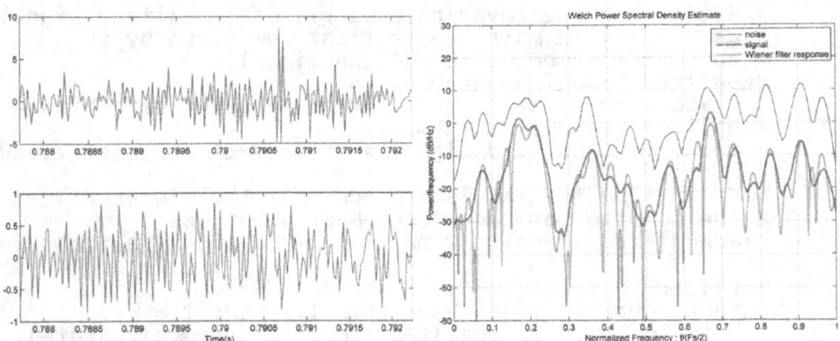

Fig. 7.21 *Left*: Zoom on the equalized audio signal and on the modulated signal passed through the Wiener filter; *right*: corresponding PSDs and frequency response of the Wiener filter

We can finally apply the correlation demodulator to the estimated modulated signal. The resulting BER is lower (Fig. 7.22), thanks to Wiener filtering; a BER of the order of 10^{-3} is reached, while the watermark is now inaudible.

```
for m = 0:N_bits-1
    alpha(m+1) = (Wiener_output_signal(m*N_samples+1: ...
          m*N_samples+N_samples)'*spread_waveform )/N_samples;
    if alpha(m+1) <= 0
          received_bits(m+1) = 0;
    else
          received_bits(m+1) = 1;
    end
end

number_of_erroneous_bits= sum(bits ~= received_bits)
total_number_of_bits=N_bits
BER = number_of_erroneous_bits/total_number_of_bits

received_chars = reshape(received_bits(1:N_bits),N_bits/7, 7);
received_message = char( bin2dec(num2str(received_chars)))'
```

```
number_of_erroneous_bits =  2
total_number_of_bits =   1050
BER =    0.0019

received_message = Audio watermarking: this message will$be embedded
in an audIo signal, using spread spectrum modulation, and later
retrieved from the modulated signal.
```

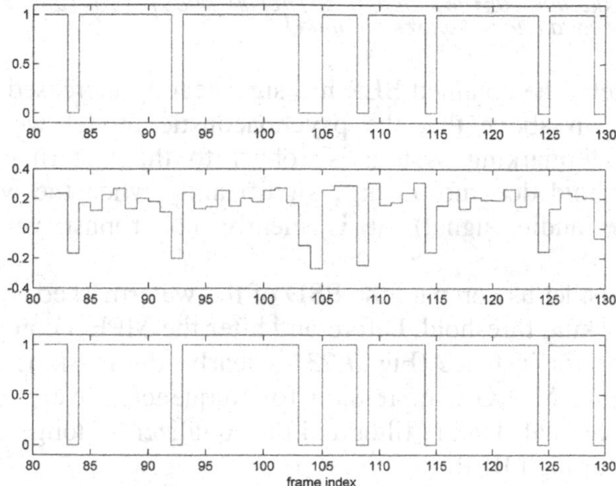

Fig. 7.22 Zoom on bits 80–130. *Top*: Input bits; *center*: `alpha`; *bottom*: received bits

Robustness to MPEG compression

We finally focus on the robustness of our system to MPEG compression. Using the MATLAB functions developed in Chapter 3, we can easily apply an mp3 coding/decoding operation to the audio watermarked signal, yielding the distorted watermarked audio signal `compressed_watermarked_signal`.

MATLAB function involved:

- `output_signal = codec_mp3( input_signal, Fs)` returns the signal `output_signal` resulting from an mp3 coding/decoding operation of the input signal `input_signal` sampled at the `Fs` frequency sampling (see chapter 3).

```
compressed_watermarked_signal=mp3_codec(watermarked_signal,Fs);
```

We then pass the compressed signal through the zero-forcing equalizer and the Wiener filter, as before, and compute the resulting BER[13] and received message.

```
BER =    0.1048
```

[13] The corresponding MATLAB code can be found in the `ASP_watermarking.m` file. We do not repeat it here.

```
received_message = )-%im$gatemapcHng"txIr }!UsagM wiLm&bD$embed`de
hN"cn yUdho skenan, Usino`sppdaD sp}btrum mOd5latio~(a^ELat
%02gtrievgd fPoMda*uOm/$uXetet4cggna}
```

Unfortunately, the obtained BER has significantly increased; although it has been shown above that the psychoacoustic model we use in our perceptual watermarking system is robust to the watermark (i.e., the masking threshold does not change significantly when the watermark is added to the audio signal), it is clearly not robust *yet* to MPEG compression.

To show this, let us compare the PSD of the watermarked signal and the associated masking threshold, before and after the MPEG compression, on the 11th frame for instance (Fig. 7.23). Clearly, the masking threshold is very sensitive to MPEG compression for frequencies over 11 kHz. As a result, the perceptual shaping filter and the equalizer no longer cancel each other, hence our high BER.

```
PAM_frame = (10*N_samples_PAM+1 : ...
    10*N_samples_PAM+N_samples_PAM);

subplot(1,2,1)
pwelch(watermarked_signal(PAM_frame),[],[],[],2);
masking_threshold = psychoacoustical_model(...
    watermarked_signal(PAM_frame) );
stairs((1:N_samples_PAM/2)*2/N_samples_PAM,masking_threshold,...
    'r','linewidth',2)
subplot(1,2,2)
pwelch(compressed_watermarked_signal(PAM_frame),[],[],[],2);
masking_threshold = psychoacoustical_model(...
    compressed_watermarked_signal(PAM_frame) );
stairs((1:N_samples_PAM/2)*2/N_samples_PAM,masking_threshold,...
    'r','linewidth',2)
```

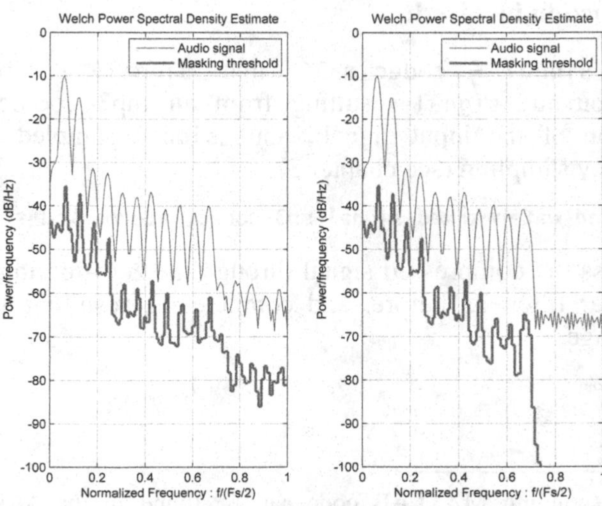

Fig. 7.23 PSD of frame #11 and the associated masking threshold. *Left*: Before MPEG compression; *right*: after MPEG compression

7.2.4 Informed watermarking robust to MPEG compression

In order to improve the robustness of the system exposed in the previous section, the watermark information should obviously be spread in the [0 Hz, 11 kHz] frequency range of the audio signal. This can be achieved with a low-pass filter with cutoff frequency set to 11 kHz (Baras et al. 2006). As we shall see below, this filter will interfere with our watermarking system in three stages.

Designing the low-pass filter

First, we design a symmetric FIR low-pass filter with `Fc=11` kHz cutoff frequency using the Parks–McClellan algorithm. Such a symmetric filter (Fig. 7.24, left) will have linear phase and therefore will not change the shape of the modulated signal more than required. We set the order of the filter to `N_coef=50`. Note that this filter is noncausal and introduces a `N_coef/2`-sample delay.

```
Fc = 11000;
low_pass_filter = firpm(N_coef, [0 Fc-1000 Fc+1000 Fs/2]*2/Fs,
    [1 1 1E-9 1E-9]);
%low_pass_filter = ...
    low_pass_filter/sqrt(sum(low_pass_filter)^2));

% Let us plot the impulse response of this filter.
plot(low_pass_filter);
```

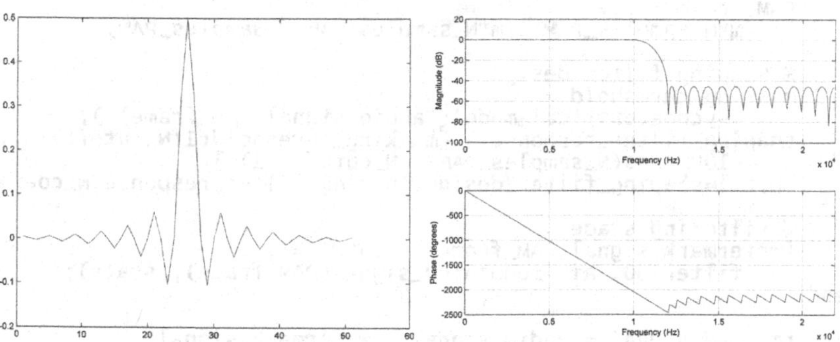

Fig. 7.24 *Left*: Impulse response of a symmetric FIR low-pass filter; *right*: its frequency response

It is easy to check that its frequency response matches our requirements (Fig. 7.24, right).

```
freqz(low_pass_filter,1,256);
```

Modifying the emitter

We first use this low-pass filter to create a spread sequence with spectral content restricted to the [0 Hz, 11 kHz] range. The modulated signal can then be designed as previously.

```
spread_waveform = ...
   filter(low_pass_filter,1,2*round(rand(N_samples+N_coef,1))-1);
% Getting rid of the transient
spread_waveform = spread_waveform(N_coef+1:end);
% Making sure the norm of the spread waveform is set to
% sqrt(N_samples), as in the previous Sections.
spread_waveform = spread_waveform/ ...
    sqrt(norm(spread_waveform).^2/N_samples);

for m = 0:N_bits-1
    modulated_signal(m*N_samples+1:m*N_samples+N_samples) = ...
        symbols(m+1)*spread_waveform;
end
```

To prevent the shaping filter from amplifying the residual frequency component of the modulated signal over 11 kHz, the `shaping_filter_response`, initially chosen to be equal to the masking threshold, is now designed to be equal to the masking threshold in the [0,11]-kHz frequency band and to zero anywhere else.

```
state = zeros(N_coef, 1);
N_cutoff = ceil(Fc/(Fs/2)*(N_samples_PAM/2));

for m = 0:fix(N_bits*N_samples/N_samples_PAM)-1
    PAM_frame = ...
       (m*N_samples_PAM+1:m*N_samples_PAM+N_samples_PAM);

    % Shaping filter design
    masking_threshold = ...
        psychoacoustical_model( audio_signal(PAM_frame) );
    shaping_filter_response = [masking_threshold(1:N_cutoff); ...
        -100*ones(N_samples_PAM/2-N_cutoff, 1) ];
   [b0,ai]=shaping_filter_design(shaping_filter_response,N_coef);

    % Filtering stage
    [watermark_signal(PAM_frame), state] = ...
        filter(b0, ai, modulated_signal(PAM_frame), state);
end

watermarked_signal = audio_signal + watermark_signal;
```

MPEG compression is then applied to the new `watermarked_signal`.

```
compressed_watermarked_signal=mp3_codec( watermarked_signal,Fs);
```

Modifying the receiver

Before starting the watermark extraction, the `compressed_watermarked_signal` is low-passed to avoid residual components over 11 kHz.

```
compressed_watermarked_signal = ...
    filter(low_pass_filter, 1, [compressed_watermarked_signal; ...
        zeros(N_coef/2, 1)]);
% Compensating tof the |N_coef/2| delay
compressed_watermarked_signal = ...
    compressed_watermarked_signal(N_coef/2+1:end);
```

Watermark extraction is finally obtained as in Section 7.2.3, except that the LP filter is again taken into account in the zero-forcing equalization.[14] The resulting BER is similar to the one we obtained with no MPEG compression; our watermarking system is thus now robust to MPEG compression.

```
BER =    0.00095

received_message = A5dio watermarking: this message will be embedded
in an audio signal, using spread spectrum modulation, and later
retrieved from the modulated signal.
```

7.3 Going further

Readers interested in a more in-depth study of spread spectrum communications will find many details, as well as MATLAB examples, in Proakis et al. (2004).

Several books on digital watermarking techniques are available. Cox et al. (2001) provides an intuitive approach, and many examples in C. Source code can also be found in Pan et al. (2004).

Also note that this chapter has mentioned only *additive watermarking* (yet only in the time domain), as opposed to the other main approach known as *substitutive watermarking* (Arnold et al. 2003). Moreover, additive watermarking can be performed in various domains, leading to various bit rate vs. distortion vs. robustness trade-offs (Cox et al. 2001): in the time domain, in the frequency domain, in the amplitude or phase domain (such as the one corresponding to the output of a modulated lapped transform, for robustness to small desynchronization; see Malvar and Florencio 2003), in the cepstral domain (often used in speech processing for its relation to the source/filter model; see Lee and Ho 2000), and in

---

[14] The corresponding MATLAB code can be easily found in the `ASP_watermarking.m` file. We do not repeat it here.

other parametric domains (such as the MPEG-compressed domain; see Siebenhaar et al. 2001). Last but not least, we have not examined synchronization problems here, which require special attention in the context of spread spectrum communications, since even a one-sample delay will have a disastrous effect on the resulting BER (Baras et al. 2006).

Security is one of the emerging challenges in watermarking (Furon et al. 2007). For more challenges, see, for instance, the BOWS challenges organized by the within European Network of Excellence ECRYPT (Bows2 2007).

7.4 Conclusion

Hiding bits of information in an audio stream is not as hard as one might think. It can be achieved by modulating the transmitted bits by a spread spectrum sequence and adding the resulting watermark signal to the audio stream. Making the watermark inaudible is a matter of dynamically shaping its spectrum so that it falls below the local masking threshold provided by a psychoacoustical model. Last but not least, ensuring that the system is robust to MPEG compression simply requires the addition of a low-pass filter in the spectrum-shaping stage.

Provided the receiver is able to invert the spectrum-shaping operation using the same psychoacoustical model as the emitter, and with the additional help of a Wiener filter to increase the SNR, a bit error rate of the order of 0.001 can be obtained.

References

Arnold M, Wolthusen S, Schmucker M (2003) Techniques and applications of digital water-marking and content protection. Artech House Publishers, Norwood, MA

Baras C, Moreau N, Dymarski P (2006) Controlling the inaudibility and maximizing the robustness in an audio annotation watermarking system. IEEE Transactions on Audio, Speech and Language Processing, 14–5:1772–1782

Bows2 (2007) Break our Watermarking system, 2nd edition [online] Available: http://bows2.gipsa-lab.inpg.fr [02/1/2008]

Chen B, Wornell G (2001) Quantization index modulation: A class of provably good methods for digital watermarking and information embedding. IEEE Transactions on Information Theory, 47:1423–1443

Costa M (1983) Writing on dirty paper. IEEE Transactions on Information Theory, 29:439–441

Cox I, Miller M, Bloom J (2001) Digital watermarking: Principles and practice Morgan Kaufmann, San Francisco, CA

Cox I, Miller M, McKellips A (1999) Watermarking as communications with side information. Proceedings of the IEEE, 87–7:1127–1141

Craver SA, Wu M, Liu B (2001) What can we reasonably expect from watermarks? Proceedings of the IEEE Workshop on Applications of Signal Processing to Audio and Acoustics, New Paltz, NY, 223–226

Furon T, Cayre F, Fontaine C (2007) Watermarking security. In: Cvejic and Seppanen (eds) Digital Audio Watermarking Techniques and Technologies: Applications and Benchmarks. Information Science Reference, Hershey, PA, USA

Haykin S (1996) Adaptive filter theory, 3rd ed. Prentice-Hall, Upper Saddle River, NJ, USA

Jobs S (2007) Thoughts on music, [online] Available: http://www.apple.com/hotnews/thoughtsonmusic/ [26/9/2007]

Larbi S, Jaïdane M, Moreau N (2004). A new Wiener filtering based detection scheme for time domain perceptual audio watermarking. In IEEE International Conference on Acoustics, Speech and Signal Processing (ICASSP), Montreal, Canada, 5:949–952

Lee SK, Ho YS (2000) Digital audio watermarking in the cepstrum domain. IEEE Transactions on Consumer Electronics, 46–3: 744–750

LoboGuerrero A, Bas P, Lienard J (2003) Iterative informed audio data hiding scheme using optimal filter. In Proc. IEEE Int. Conf. on Communication Technology, Beijing, China, 1408–1411

Malvar H, Florencio D (2003) Improved spread spectrum: A new modulation technique for robust watermarking. IEEE Transactions on Signal Processing, 51-4:898–905

Massey JL (1994) Information theory aspects of spread-spectrum communications. IEEE Third International Symposium on Spread Spectrum Techniques and Applications (ISSSTA), 1:16–21

Pan JS, Huang HC, Jain LC (2004) Intelligent watermarking techniques (Innovative Intelligence). World Scientific Publishing Company, Singapore

Peterson RL, Ziemer RA, Borth DE (1995) Introduction to spread spectrum communications. Prentice Hall, Upper Saddle River, NJ, USA

Petitcolas FAP, Anderson RJ, Kuhn MG (1999) Information hiding – a survey. Proceedings IEEE, Special Issue on Protection of Multimedia Content, 87(7):1062–1078

Proakis J (2001) Digital communications, 4th ed. McGraw-Hill, New York, USA

Proakis JG, Salehi M, Bauch G (2004) Contemporary communication systems using MATLAB and SIMULINK. Brooks/Cole-Thomson Learning, Belmont, CA, USA

Shannon C (1958) Channel with side information at the transmitter. IBM Journal of Research and Development, 2:222–293

Siebenhaar F, Neubauer C, Herre J, Kulessa R (2001) New results on combined audio compression/watermarking. In 111th Convention of Audio Engineering Society (AES), New York, USA, preprint 5442

Chapter 8

How are digital images compressed in the web?

F. Marqués(°), M. Menezes(*), J. Ruiz-Hidalgo(°)

(°) Universitat Politècnica de Catalunya, Spain
(*) Instituto Superior de Ciências do Trabalho e da Empresa, Portugal

In 2005, it was estimated that there were more than 1.6 billion images in the web.[1]

In 1992, a joint committee between the International Organization for Standardization (ISO) and the Telecommunication Standardization Sector of the International Telecommunication Union (ITU-T) known as the Joint Photographic Experts Group (JPEG) issued a standard that was approved in 1994 as ISO/IEC 10918-1 or ITU-T Recommendation T.81. This standard received the name of *Information technology – Digital compression and coding of continuous-tone still images requirements and guidelines* but it is commonly known as the *JPEG standard.*

The original goal of JPEG was to provide still image compression techniques for a large range of (i) types of images, (ii) image-reconstructed qualities, and (iii) compression ratios while allowing software implementations. The quality of the proposed solution has made the standard so successful that, nowadays, JPEG-encoded images are present in almost all areas and applications. Clear examples of this success are the constant use of JPEG-encoded images in such universal environments as the

[1] From http://hypertextbook.com/facts/2007/LorantLee.shtml

T. Dutoit, F. Marqués (eds.), *Applied Signal Processing*,
DOI 10.1007/978-0-387-74535-0_8, © Springer Science+Business Media, LLC 2009

web or the global acceptance in the digital camera industry as standard for storage.

In this chapter, we are going to present and illustrate the main concepts behind the JPEG standard. We are not going to detail specificities of the standard but we will concentrate on the common, generic tools and concepts that it uses. Moreover, since these concepts are applied in video compression as well, they will be revisited in Chapter 9.

8.1 Background–JPEG

Figure 8.1 presents the block diagram of a typical image compression and transmission system. In it, we can observe two different pairs of blocks: source encoder/decoder and channel encoder/decoder. The objective of the source encoder is to reduce the amount of data (average number of bits) necessary to represent the information in the signal. This data compression is mainly carried out by exploiting redundancy in the image information. In turn, the objective of the channel encoder is to add redundancy to the output of the source encoder to enhance the reliability on the transmission. In this chapter, we will concentrate on the problem of source coding.

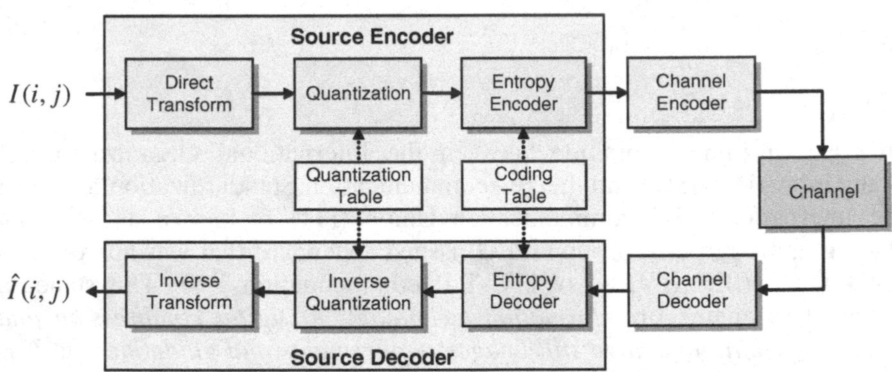

Fig. 8.1 Block diagram of an image compression and transmission system

The *direct transform* block (Noll and Jayant 1984) changes the description of the image looking for a new, less correlated set of transformed samples describing the signal. The image can be exactly recovered from the transformed samples; i.e., the transform step does not introduce losses in the description.

The *quantization* step (Gersho and Gray 1993) represents a range of values of a given transformed sample by a single value in the range.[2] The set of output values are the quantization indices. This step is responsible for the losses in the system and, therefore, determines the achieved compression.

The *entropy encoder* (Cover and Thomas 1991) produces the final bitstream. It exploits the nonuniformity of the probability distribution of quantization indices (symbols). This is carried out by assigning larger code words to less likely symbols.

In order to correctly decode the bitstream produced by the coding system, the encoder has to share some information with the decoder. The decoder should know (i) how the different code words have been stored in the bitstream (syntax definition) and (ii) which actions are associated with their different decoded values (decoder definition). This shared information defines a *standard* (Chiariglione 1999) and leads to, for example, fixing the specific transform to be used at a given stage of the decoding process or the range of possible values of a given quantization index.

Nevertheless, a standard does not fix the way in which the various parameters have to be computed, opening the door for improved implementations of standard compliant encoders (this concept is further discussed in Chapter 9 in the case of motion estimation). Moreover, a standard can introduce some additional flexibility, as some of the coding tools adopted by the standard can be adapted to the data being processed. This is the reason for explicitly representing the *quantization* and *coding tables* in Fig. 8.1. Typically, these elements can be chosen at the encoder side to improve the coding performance and are transmitted to the decoder in the bitstream.

In this section, we are going to concentrate on a very specific set of tools among all the possible ones in the context of transforming, quantizing, and entropy encoding. These tools are the basic ones for understanding the principles of the JPEG standard (and a part of the video coding standards). This way, in the transforming context, we will address the basic color transform from RGB to YCbCr (Section 8.1.1) and, as frequency transform, the discrete cosine transform (Section 8.1.2). Other transform techniques that are commonly used as well in image coding are covered in previous and subsequent chapters of this book namely linear prediction in Chapter 1, filter banks in Chapter 3, and wavelet transform in Chapter 10. Moreover, the concept of temporal prediction is addressed in the framework of video coding in Chapter 9. Regarding quantization, Chapter 2 already covers the

[2] Here, we are assuming scalar quantization. Grouping a set of transformed samples into a vector and performing vector quantization is possible as well.

concepts required in this chapter and, therefore, this topic is not further developed here. In turn, in the entropy-coding area (Section 8.1.3), we are going to present the Huffman codes, since these are the most common entropy-coding tools in image coding.[3] This section is completed with a review of a few JPEG specifications (Section 8.1.4) and a discussion on the most common quality measures (Section 8.1.5).

8.1.1 Color transform

In the context of image and video compression, a common way to represent color images is using the so-called YCbCr representation.[4] The relation between these components and the RGB ones is given by the following expressions. Given the weighting factors

$$\alpha_R \triangleq 0.299 \qquad \alpha_G \triangleq 0.587 \qquad \alpha_B \triangleq 0.114$$

the relationship between components is given by

$$\begin{aligned} Y &= \alpha_R R + \alpha_G G + \alpha_B B \\ Cb &= \frac{0.5}{1-\alpha_B}(B-Y) \\ Cr &= \frac{0.5}{1-\alpha_R}(R-Y) \end{aligned} \tag{8.1}$$

The component Y is a weighted average of the red, green, and blue components and, therefore, it is commonly referred to as the *luminance* component of the image. In turn, Cb (Cr) is a weighted difference between the blue (red) component and the luminance component. These color differences are commonly referred to as the *chrominance* components.

Mainly due to the different density of luminance and color photoreceptors (cones and rods, respectively) in the human eye, the human visual system is substantially less sensitive to the distortion in the chrominance components.

---

[3] Other entropy codes (e.g., Golomb or arithmetic codes) are currently used as well.

[4] In the case of the JPEG standard, the use of this color representation is not normative, but very common and most decoders expect the image to be represented this way.

This property is exploited in image and video encoders in different manners. For instance, chrominance components are usually subsampled by a factor of 2 in the horizontal and vertical directions.

8.1.2 Frequency transform: The discrete cosine transform

1D transforms

Finite linear transforms are commonly used in image and video compression. An efficient transform should take advantage of the statistical dependencies among the signal samples $x(n)$ (n=0, ..., N−1) (i.e., their correlation) and obtain a set of transformed samples $y(k)$ (k=0, ..., M−1), presenting, at most, local dependencies, i.e., a transform should decorrelate the original samples. Moreover, a transform should separate relevant from irrelevant information in order to identify the irrelevant samples. Last but not least, the transform should be invertible in order to recover (an approximation of) the signal from the quantized transformed samples. Note that the possibility of dropping some coefficients is included in the quantization process, i.e., the quantization can set some transformed coefficients to zero.

It is usual to study a transform by analyzing first the role of the inverse transform. The inverse transform can be seen as a linear decomposition of a signal $x(n)$ in terms of a weighted sum of elementary functions $s_k(n)$, the weights being the transformed samples $y(k)$:

$$x(n) = \sum_{k=0}^{M-1} s_k(n) y(k) \qquad \rightarrow \qquad \mathbf{x} = \mathbf{S}\mathbf{y} \tag{8.2}$$

where

$$\begin{aligned} \mathbf{x}^{\mathrm{T}} &= \left[x(0), x(1), \cdots, x(N-1)\right] \\ \mathbf{y}^{\mathrm{T}} &= \left[y(0), y(1), \cdots, y(M-1)\right] \end{aligned} \tag{8.3}$$

The vector $\mathbf{s_k}$ containing the samples of the elementary function $s_k(n)$

$$\mathbf{s}_{\mathbf{k}}^{\mathrm{T}} = \left[s_k(0), s_k(1), \cdots, s_k(N-1)\right] \tag{8.4}$$

is commonly referred to as the kth *synthesis vector*. The reason is that by means of the linear combination of the synthesis vectors the original signal is synthesized (reconstructed). In turn, the matrix **S** is known as the synthesis matrix. Note that the synthesis vectors $\mathbf{s}_k$ are the columns of the synthesis matrix **S**:

$$\mathbf{S} = \left[\mathbf{s}_0, \mathbf{s}_1, \cdots, \mathbf{s}_{M-1}\right] \tag{8.5}$$

The direct transform (and, therefore, the transformed samples) is obtained by comparing the signal $x(n)$ with a set of elementary functions $a_k(n)$:

$$y(k) = \sum_{n=0}^{N-1} a_k^*(n) x(n) \quad \rightarrow \quad \mathbf{y} = \mathbf{A}^{\mathbf{H}} \mathbf{x} \tag{8.6}$$

In that case, the vector $\mathbf{a}_k$ is formed by

$$\mathbf{a}_k^{\mathbf{T}} = \left[a_k(0), a_k(1), \cdots, a_k(N-1)\right] \tag{8.7}$$

which is commonly referred to as the k-th *analysis vector*. In an analogous manner, the matrix **A** is known as the analysis matrix:

$$\mathbf{A} = \left[\mathbf{a}_0, \mathbf{a}_1, \cdots, \mathbf{a}_{M-1}\right] \tag{8.8}$$

Note that the transform coefficients can be expressed as

$$y(k) = \mathbf{a}_k^{\mathbf{H}} \mathbf{x} \tag{8.9}$$

and this can be seen as the analysis of signal **x** with respect to vector $\mathbf{a}_k$.

In order to have a revertible transform, **S** has to be the left-inverse of $\mathbf{A}^H$, i.e., $\mathbf{S}\mathbf{A}^H=\mathbf{I}$. Nevertheless, in this chapter, we are going to restrict the study to the case of nonexpansive transforms, i.e., we are assuming that the transform maps N samples of $x(n)$ into N samples of $y(k)$. In that case, $M=N$ and $\mathbf{S}^{-1}=\mathbf{A}^H$, and, if the transform is real-valued, $\mathbf{S}^{-1}=\mathbf{A}^T$.

A common property of the image transforms is to be *orthonormal*. A transform is said to be orthonormal if all analysis vectors are orthonormal and have unit norm:

$$\mathbf{A}^{\mathbf{H}}\mathbf{A} = \mathbf{I} \rightarrow \begin{cases} \langle \mathbf{a}_i, \mathbf{a}_j \rangle = \mathbf{a}_i^{\mathbf{H}}\mathbf{a}_j = 0 \quad \forall i \neq j \\ \langle \mathbf{a}_i, \mathbf{a}_i \rangle = \|\mathbf{a}_i\|^2 = \mathbf{a}_i^{\mathbf{H}}\mathbf{a}_i = 1 \end{cases} \tag{8.10}$$

In that case, and given the previous condition of nonexpansiveness, the transform is defined by a unitary matrix **A** and it is said to be a *unitary transform*. For unitary transforms, the synthesis matrix **S** coincides with the analysis matrix **A**. As a matter of fact

$$\left.\begin{array}{lll} \mathbf{S}^{-1} = \mathbf{A}^{\mathbf{H}} & \rightarrow & \mathbf{S} = \mathbf{A}^{-\mathbf{H}} \\ \mathbf{A}^{\mathbf{H}}\mathbf{A} = \mathbf{I} & \rightarrow & \mathbf{A}^{-\mathbf{H}} = \mathbf{A} \end{array}\right\} \rightarrow \mathbf{S} = \mathbf{A} \tag{8.11}$$

Unitary transforms maintain the energy of the samples:

$$\|\mathbf{x}\|^2 = \mathbf{x}^{\mathbf{H}}\mathbf{x} = [\mathbf{A}\mathbf{y}]^{\mathbf{H}}\mathbf{A}\mathbf{y} = \mathbf{y}^{\mathbf{H}}\mathbf{A}^{\mathbf{H}}\mathbf{A}\mathbf{y} = \mathbf{y}^{\mathbf{H}}\mathbf{y} = \|\mathbf{y}\|^2 \tag{8.12}$$

This characteristic is very useful when working with the transform samples since the actions taken over a coefficient have a clear interpretation in terms of the final energy of the signal. This is a very relevant aspect when quantizing the transform coefficients in order to approximate a signal.

Nevertheless, although the total energy is preserved, it is not equally distributed among the various transformed coefficients (i.e., the components of vector **y**). It can be demonstrated (Noll and Jayant 1984) that the transform leading to the optimum energy concentration is the so-called Karhunen–Loewe transform (KLT). That is, the KLT places as much energy as possible in as few coefficients as possible. It therefore mostly "explains" the input signal in terms of a limited number of synthesis vectors (those for which **y** has high synthesis coefficients), a property which is immediately turned into profit in the quantization step. However, the analysis matrix **A** in the case of the KLT is not fixed but it depends on the statistical properties of the signal under analysis. Actually, its analysis vectors $\mathbf{a}_k$ are the eigenvectors of the signal correlation matrix $\mathbf{R}_x$:

$$\mathbf{R}_{\mathbf{x}}\mathbf{a}_{\mathbf{k}} = \lambda_k \mathbf{a}_{\mathbf{k}} \text{ with } \mathbf{R}_{\mathbf{x}} = E\{\mathbf{x}\mathbf{x}^{\mathbf{H}}\} \tag{8.13}$$

Therefore, the KLT is signal adapted and the analysis matrix that defines the transform must be estimated for each signal. Although fast algorithms have been proposed, the overall complexity of the KLT is significantly higher than the complexity of other possible transforms.

In the case of images, a good compromise between energy concentration and computational complexity is the discrete cosine transform (DCT). A common definition of the components of the analysis/synthesis DCT vectors is given by the following expression:[5]

$$a_k(n) = \sqrt{\frac{s}{N}} \cos\left(\frac{(2n+1)k\pi}{2N}\right) \qquad s = \begin{cases} 1 & k = 0 \\ 2 & k \neq 0 \end{cases} \tag{8.14}$$

The analysis/synthesis vectors consist of samples of cosine functions (Fig. 8.2). Observe that the first analysis vector ($\mathbf{a}_0$) has all samples equal and, thus, the first DCT coefficient is related to the average value of the samples in the signal. In the literature, this coefficient is referred to as the *DC coefficient*, whereas the rest of the transformed coefficients are named the *AC coefficients*.[6]

To further analyze the properties of the DCT, we are going to move first to the case of transforms in two dimensions. This way, we will be able to highlight some features of the DCT in the case of processing images.

2D transforms

To simplify the notation, we are going to assume that we have unitary transforms. The definition of a 2D transform is given by

$$y(k,l) = \sum_{m=0}^{M-1}\sum_{n=0}^{N-1} a_{k,l}^{*}(m,n)x(m,n) \tag{8.15}$$

whereas the inverse transform is given by the following equation:

$$x(m,n) = \sum_{k=0}^{M-1}\sum_{l=0}^{N-1} a_{k,l}(m,n)y(k,l) \tag{8.16}$$

[5] This is known as the Type II DCT transform

[6] These names come from the historical use of DCT for analyzing electrical circuits with direct and alternating currents.

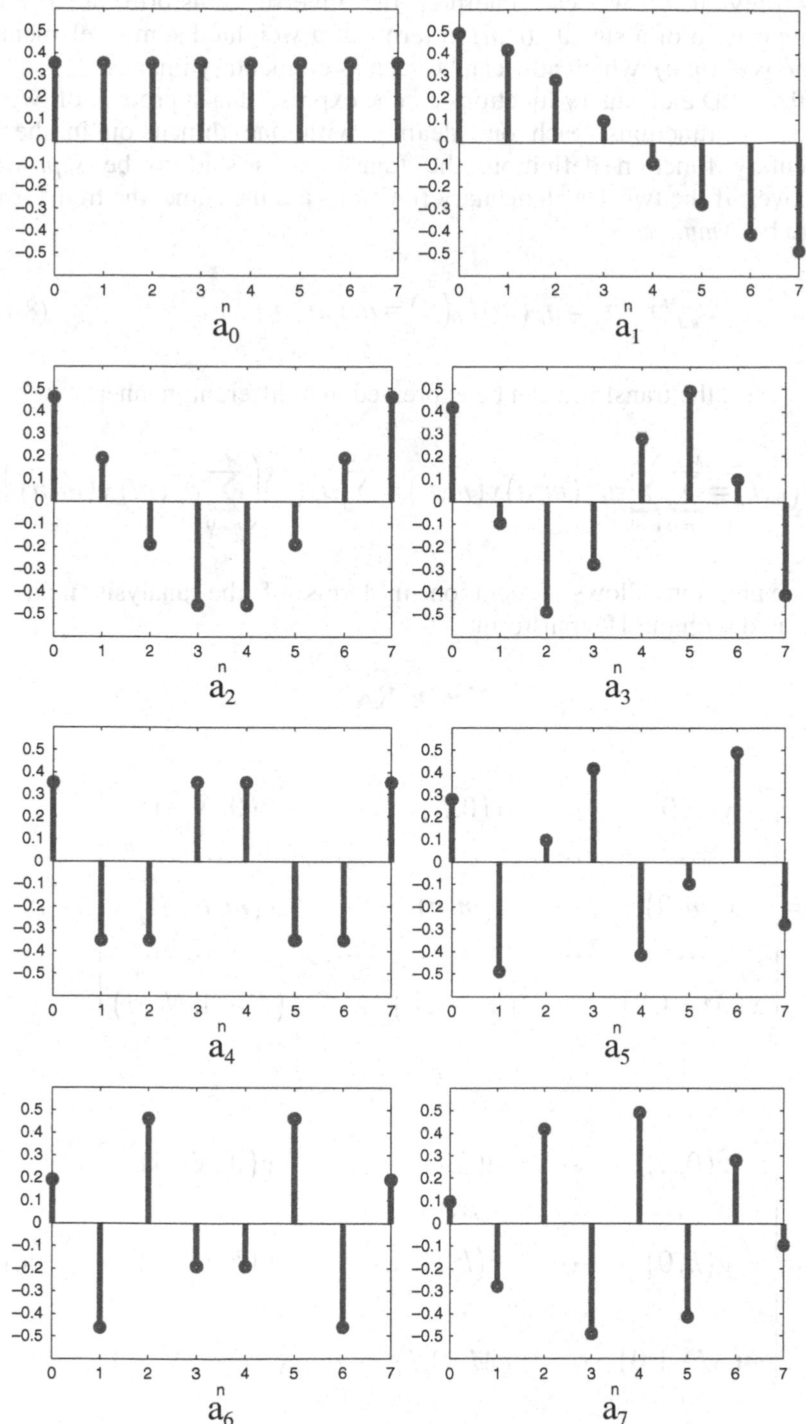

Fig. 8.2 Analysis vectors of the 1D DCT (N=8 samples)

As previously, we can interpret the inverse transform as a linear decomposition of a signal $x(m,n)$ in terms of a weighted sum of elementary functions $a_{k,l}(m,n)$, which now can be seen as elementary images.

If these 2D elementary functions can be expressed as a product of two 1D elementary functions, each one dealing with one dimension in the 2D elementary function definition, the transform is said to be *separable*. Moreover, if the two 1D elementary functions are the same, the transform is said to be *symmetric*:

$$a_{k,l}(m,n) = a_k(m)b_l(n) = a_k(m)a_l(n) \tag{8.17}$$

In that case, the transform can be expressed in a different manner:

$$y(k,l) = \sum_{m=0}^{M-1}\sum_{n=0}^{N-1} a_{k,l}^{*}(m,n)x(m,n) = \sum_{n=0}^{N-1} a_l^{*}(n)\left(\sum_{m=0}^{M-1} a_k^{*}(m)x(m,n)\right)$$

This expression allows a notation in terms of the analysis matrix **A** associated with the 1D transform:

$$\mathbf{Y} = \mathbf{A}^{*}\mathbf{X}\mathbf{A}^{\mathbf{H}}$$

where

$$\mathbf{X} = \begin{bmatrix} x(0,0) & \cdots & x(0,n) & \cdots & x(0,N-1) \\ \cdots & \cdots & \cdots & \cdots & \cdots \\ x(m,0) & \cdots & x(m,n) & \cdots & x(m,N-1) \\ \cdots & \cdots & \cdots & \cdots & \cdots \\ x(M-1,0) & \cdots & x(M-1,n) & \cdots & x(M-1,N-1) \end{bmatrix} \tag{8.18}$$

and

$$\mathbf{Y} = \begin{bmatrix} y(0,0) & \cdots & y(0,l) & \cdots & y(0,N-1) \\ \cdots & \cdots & \cdots & \cdots & \cdots \\ y(k,0) & \cdots & y(k,l) & \cdots & y(k,N-1) \\ \cdots & \cdots & \cdots & \cdots & \cdots \\ y(M-1,0) & \cdots & y(M-1,l) & \cdots & y(M-1,N-1) \end{bmatrix} \tag{8.19}$$

The separability property makes possible to compute the transform of an N×N image by means of two matrix multiplications of size N×N. If the transform is not separable, we could represent the image by means of a single vector of size $1\times N^2$ and, using Equation (8.6), compute the transform by means of one multiplication of a vector of size $1\times N^2$ with a matrix of size $N^2\times N^2$. Therefore, the complexity is reduced[7] from $O(N^4)$ to $O(N^3)$.

In the case of images, as we have previously commented, the discrete cosine transform is commonly used. The DCT presents all the properties above commented (it is separable and symmetric), while, in addition, being real-valued. The expression of the 2D DCT is therefore given by

$$y[k,l]=\sum_{n=0}^{N-1}\sqrt{\frac{s}{N}}\cos\left(\frac{(2n+1)l\pi}{2N}\right)\left(\sum_{m=0}^{M-1}\sqrt{\frac{s}{M}}\cos\left(\frac{(2m+1)k\pi}{2M}\right)x[m,n]\right)$$

with the parameter s defined as in expression (8.14).

In terms of energy concentration, the DCT reaches very high values, getting very close to the KLT performance. Actually, it can be demonstrated that for the specific case of first-order 2D Markov processes, the DCT is optimum as well (Noll and Jayant 1984). This energy concentration property is illustrated in Fig. 8.3.

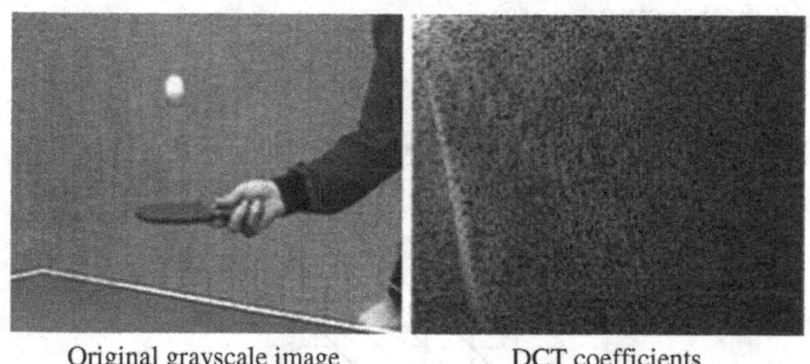

Original grayscale image DCT coefficients

Fig. 8.3 An original image and its corresponding DCT transformation. *Bright* (*dark*) values correspond to high (*low*) energy coefficients

Figure 8.3 shows an original image of 176×144 pixels and the corresponding DCT coefficients (values have been conveniently shifted and scaled for visualization). In Fig. 8.4, the percentage of energy of the image representation when using an increasing number of DCT coefficients is presented to illustrate how the DCT transform is able to compact the energy of the image pixels into fewer coefficients. In Fig. 8.4a, the DCT

[7] The KL transform is, in general, nonseparable and, therefore, cannot profit from this complexity reduction.

coefficients have been sorted by descending order of energy, whereas in Fig. 8.4b, a zigzag scan has been used to order the DCT coefficients.

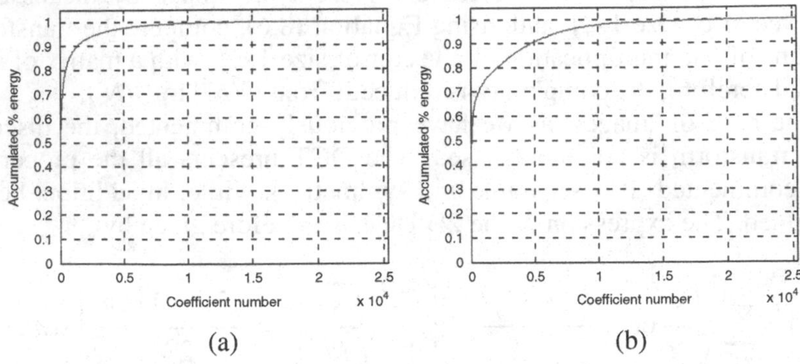

Fig. 8.4 Accumulated percentage of energy of (**a**) DCT coefficients sorted by descending order of energy and (**b**) zigzag scanned DCT coefficients

The so-called *zigzag scan* (see Fig. 8.5) is based on the observation that the transformed coefficient energy of most images tends to decrease with increasing spatial frequency. This way, the zigzag scan visits coefficients in order of roughly decreasing energy.

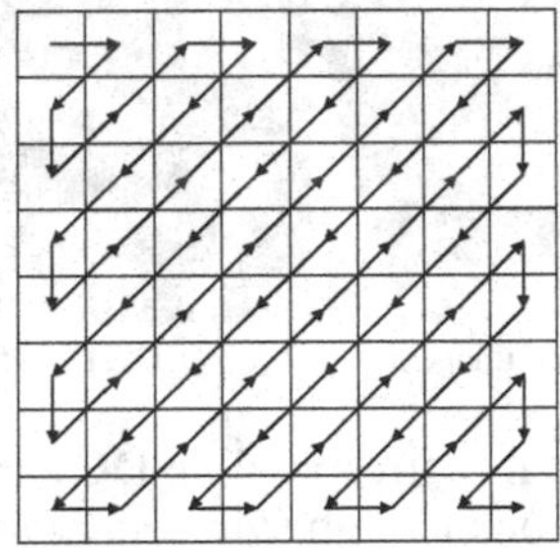

Fig. 8.5 Example of zigzag scan for the case of an image of 8×8 pixels

The following two figures illustrate the effects of reconstructing the original image after discarding (zeroing) several DCT coefficients. In Fig. 8.6, reconstructed images are shown using the N coefficients with higher energy (all other DCT coefficients are zeroed). As the DCT can compact the energy, using a few coefficients leads to good reconstructions of the original image.

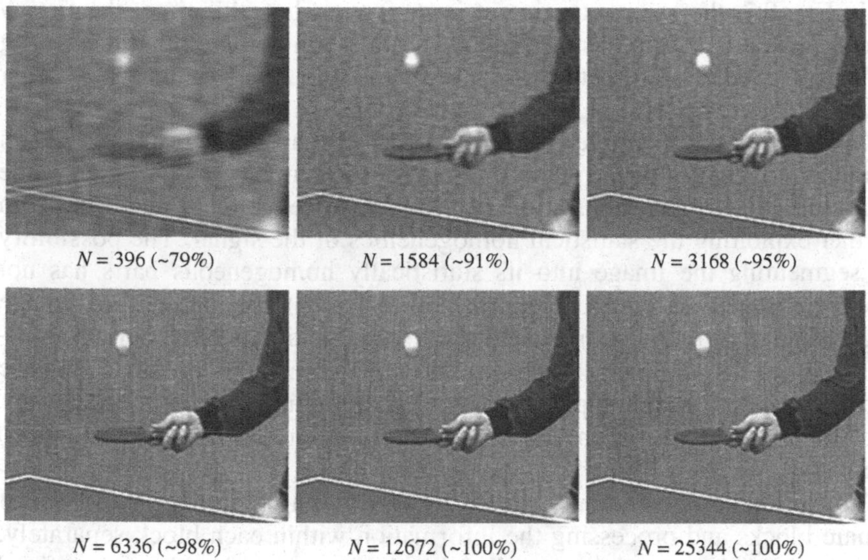

Fig. 8.6 Reconstructed images using the first N coefficients with higher energy (value in brackets represents the percentage of energy corresponding to the N selected coefficients)

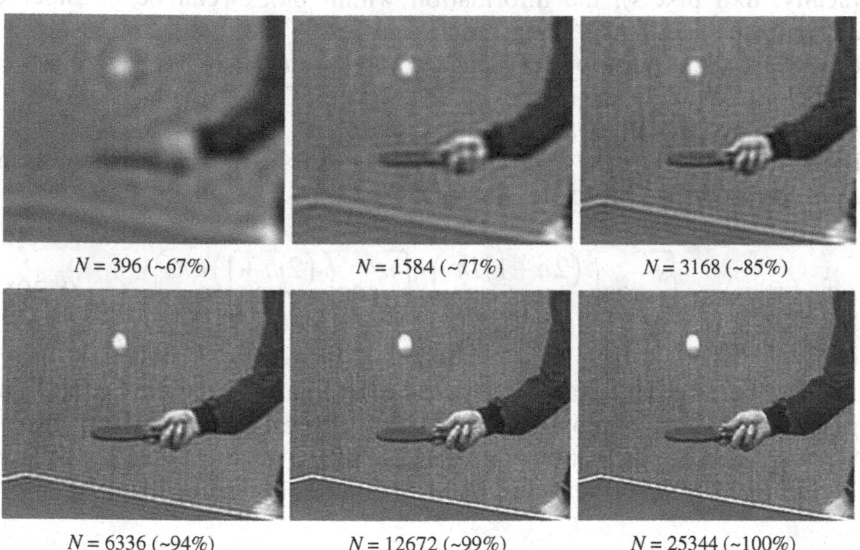

Fig. 8.7 Reconstructed images using the first zigzag scanned N coefficients (value in brackets represents the percentage of energy corresponding to the N selected coefficients)

In Fig. 8.7, the same experiment is performed but, this time, the first N coefficients are selected using a zigzag scan. It can be noted that the zigzag scan is a good approximation for the selection of coefficients with higher energy.

In the previous example, we have observed that although the DCT can compact the energy of the signal into a reduced number of coefficients, the fact that images are nonstationary signals prevents the transform from further exploiting the statistical homogeneities of the signal. The possibility of segmenting the image into its statistically homogeneous parts has not been envisaged in classical approaches since (i) the process of image segmentation is a very complicated task (segmentation is often named an ill-posed problem) and (ii) if a segmentation of the image is obtained, the use of the partition for coding purposes would require transmitting the shapes of the arbitrary regions and this boundary information is extremely difficult to compress.

This problem is commonly solved by partitioning the image into a set of square blocks and processing the information within each block separately. This leads to the use of the so-called *block transforms*. In this case, a fixed partition is used, which is known beforehand by the receiver and does not require to be transmitted. Since this partition is independent of the image under analysis, the data contained in each block may not share the same statistical properties. However, given the small size of the used blocks (typically, 8×8 pixels), the information within blocks can be considered quite homogenous. When using block transforms (i.e., when imposing an image-independent partition for image compression), it is likely to observe the block structure in the reconstructed image. This artifact is known as *block effect*, and it is illustrated in Fig. 8.22.

In the case of the DCT, when using blocks of 8×8, the analysis/synthesis vectors correspond to the expression

$$a_{k,l}(m,n) = \sqrt{\frac{s}{8}} \cos\left(\frac{(2n+1)l\pi}{16}\right) \sqrt{\frac{s}{8}} \cos\left(\frac{(2m+1)k\pi}{16}\right) \qquad (8.20)$$

which can be interpreted as a set of 64 elementary images on which the original signal is decomposed. In Fig. 8.8, we present these 64 images of 8×8 pixels each. Note that the analysis/synthesis vector values have been conveniently shifted and scaled in order to represent them as images.

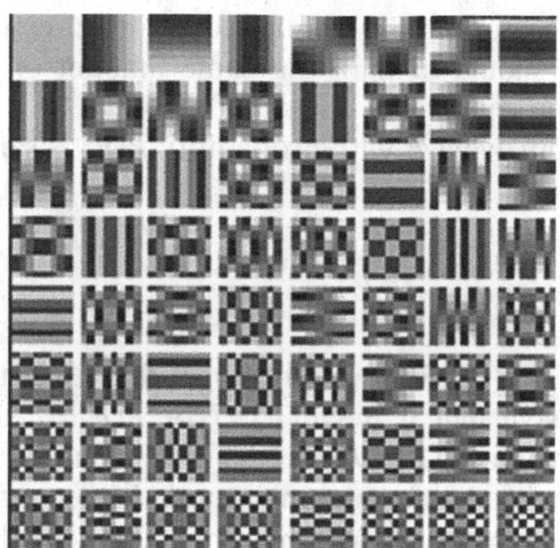

Fig. 8.8 Set of DCT analysis/synthesis vectors for the case of 8×8 pixel blocks

The study of the performance of the DCT used as a block transform for image compression is addressed in the next section.

Regarding the computational complexity of the DCT, it has to be said that, in addition to the fact of being separable and symmetric, the DCT presents a structure close to that of the FFT, which allows for highly efficient implementations. For instance, in Arai et al. (1988), a DCT implementation is presented that only requires, for 8×8 pixel blocks, 5 multiplications and 29 additions, instead of the basic 64 multiplications. Furthermore, most of the multiply operations can be performed in parallel so that the proposed algorithm is ideal for a parallel implementation.

8.1.3 Entropy coding

The entropy encoder exploits the nonuniformity of the probability distribution of quantization indices (symbols) to generate the shortest bit stream representing a given sequence of symbols. This is carried out by assigning larger code words to less likely symbols and the theoretical framework to do so is the so-called *information theory* (Cover and Thomas 1991).

The fundamental premise of information theory is that the generation of information can be modeled as a probabilistic process that can be measured so that the measure value agrees with our intuition. The information associated to a given event should thus fulfill the following conditions: (i) it should be a positive value, (ii) the probability of the event should be related to the amount of information contained in this event, and (iii) two independent events should contain the same amount of information when they appear jointly or separately.

This way, a random event E with probability $P(E)$ is said to contain

$$I(E) = \log \frac{1}{P(E)} = -\log P(E) \tag{8.21}$$

units of information. If the base of the logarithm is 2, this information is measured in bits. Note that if the $P(E)=1$ or $P(E)=0$, the event does not contain any information ($I(E)=0$), i.e., if the event is sure to happen or impossible, its associated information is null. Moreover, if $P(E)=0.5$, the information conveyed by the event is maximum and equal to 1 bit.

Let us consider now a source X of statistically independent events (symbols) $\{x_1, x_2, \ldots x_M\}$ with associated probabilities $\{P(x_1), P(x_2), \ldots, P(x_M)\}$. The average information of that source, named *entropy* and denoted by $H(X)$, is given by

$$H(X) = -\sum_{i=1}^{M} P(x_i) \log P(x_i) \tag{8.22}$$

The entropy measures the uncertainty of the source, i.e., the higher the entropy, the more uncertain the source and, therefore, the more information it conveys. It can be proven that for a source of equiprobable symbols the entropy is maximal, i.e., if all symbols have the same probability of occurrence, the uncertainty about which symbol will be produced by the source is maximal (Cover and Thomas 1991).

The goal of an entropy encoder is to assign a code word c_x from a codebook C to every symbol $\{x_1, x_2, \ldots x_M\}$, where c_x is a string of $|c_x|$ bits. The assignment has to be performed so that the average length of the produced code words R is minimal:

$$R = \sum_{m=1}^{M} P(x_m) |c_m| \tag{8.23}$$

Shannon's first theorem (the *noiseless coding theorem*) assures that the entropy of a source is the lowest bound of the average number of bits necessary to code the symbols produced by the source. Therefore, no coder can produce code words whose average length is smaller than the entropy of the source. The difference between the attained average length and the entropy of the source is called the redundancy of a code. It can be demonstrated (Cover and Thomas 1991) that for a source that generates independent symbols (a so-called *zero-memory* source) with entropy $H(X)$, it is possible to find a code whose mean length R is

$$H(X) \leq R \leq H(X) + 1 \tag{8.24}$$

Moreover, such codes satisfy the so-called *prefix condition*, i.e., no code word is a prefix of any other code word in the set. This is a very relevant property of a code, since no additional symbols are required in the stream to mark the separation between codes.

Several codes have been proposed in the literature to generate a set of code words satisfying the previous condition. Among them, Huffman codes are the most popular ones. Huffman codes provide an optimal code under the constraint that symbols are coded one at a time. Moreover, the prefix condition allows the use of a simple Look Up Table to implement the Huffman decoder (Cover and Thomas 1991).

Huffman codes require the previous knowledge of the probability density function of the source.[8] The algorithm for constructing a Huffman code has two main steps. In the first step, a series of source reductions is created by combining the two symbols with lowest probability into a single symbol. This compound symbol and its associated probability are used in a new iteration of the source reduction process, as illustrated in Fig. 8.9.

The process is repeated until a single compound symbol representing the complete source is obtained. In the second step, each reduced source is coded, starting from the smallest source (a source with only two symbols: in the example in Fig. 8.9, the *a1* symbol and the compound symbol representing the combination of the remaining original symbols: *a2* ... *a8*). The code words 0 and 1 are arbitrarily assigned to each one of the two symbols. These code words propagate as prefixes to the two symbols that have been used to create the compound symbol, and the process is iterated at each branch of the tree. In the example of Fig. 8.9, the complete Huffman code for the given source is presented. Note that the entropy of the source is $H(X)$=2.48 bits while the mean length is R=2.55 bits (and, therefore, the redundancy is only 0.07 bits).

---

[8] There exist extensions of the basic Huffman code that iteratively estimate the probability density function and can even adapt to its variations.

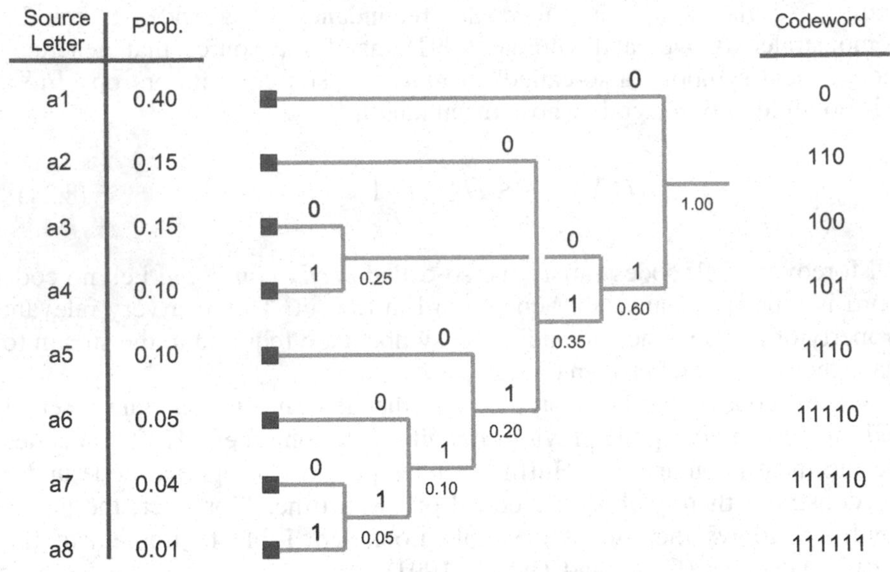

Fig. 8.9 Example of construction of a Huffman code

8.1.4 A few specificities of the JPEG standard

Figure 8.10 presents a block diagram of the main parts of the JPEG standard (Pennebaker and Mitchell 1993). Note that the basic tools used in the standard are those that have been discussed above. Nevertheless, there are a few specificities of the standard that require a more concrete explanation.

Prior to the frequency transformation, a level offset is subtracted from every image sample. If the image is represented by B bits ($B=8$ or $B=12$ in the JPEG standard), the level offset is equal to 2^B-1 to ensure that the output signal is a signed quantity in the range -2^{B-1} to $2^{B-1}-1$. This is done to ensure that all DCT coefficients are signed quantities with a similar dynamic range.

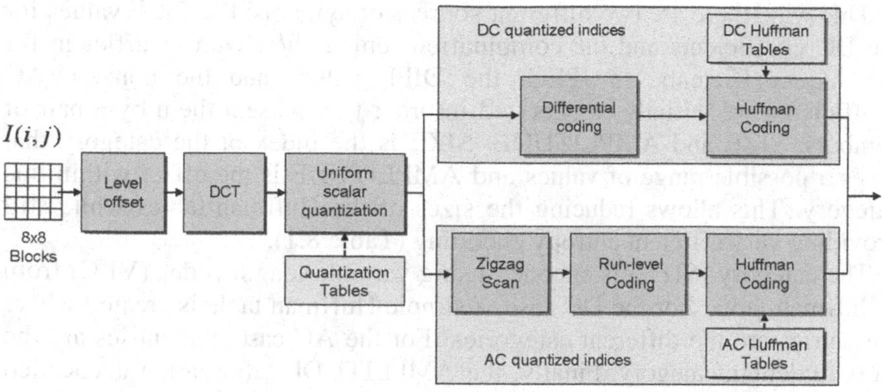

Fig. 8.10 Basic scheme of the JPEG encoder

The scalar quantization in the JPEG standard is implemented by using a quantization matrix (the so-called *Q Table*: *Q*(*k,l*)). Quantization is defined as the division of each DCT coefficient by its corresponding quantizer step size. Each component of the Q Table contains its associated quantizer step size. The final quantized coefficient is obtained by rounding the value of the division to the nearest integer:

$$\hat{y}(k,l) = round\left[\frac{y(k,l)}{Q(k,l)}\right] \tag{8.25}$$

The standard does not specify the concrete values of the Q Table but proposes the use of Lohscheller tables (Lohscheller 1984). These tables have been obtained by analyzing the relative relevancy of the different DCT coefficients for the human visual system.

Although the JPEG standard allows using arithmetic coding (Cover and Thomas 1991), the most common implementations apply Huffman Tables. As shown in Fig. 8.10, the DC coefficient is handed separately from the AC coefficients. In the case of the DC coefficient, a differential coding is initially applicd and the prediction errors (named DIFF values) are the symbols that are entropy coded. In turn, in the case of the AC coefficients, they are initially scanned following the zigzag pattern presented in Fig. 8.8. This sequence of AC coefficients is then described in a new manner; each nonzero AC coefficient is represented in combination with the "runlength" (consecutive number) of zero-valued AC coefficients, which precede it in the zigzag sequence.

This way, there are two different sources of symbols: the DIFF values for the DC coefficients and the combination runlength/nonzero coefficient for the AC coefficients. In JPEG, the DIFF values and the nonzero AC coefficients are initially categorized in order to represent them by a pair of symbols: SIZE and AMPLITUDE. SIZE is the index of the category that gives a possible range of values and AMPLITUDE is the offset within this category. This allows reducing the sizes of the Huffman tables while still providing very efficient entropy encoding (Table 8.1).

The category (SIZE) is encoded using variable-length codes (VLC) from a Huffman table. For the DC case, a simple Huffman table is created where the entries are the different categories. For the AC case, the entries are the pairs runlength/category. Finally, the AMPLITUDE information is encoded using a variable-length integer (VLI) code. VLI are variable-length codes as well, but they are not Huffman codes (and, therefore, they do not share all their properties). An important distinction is that the length of a VLC (Huffman code) is not known until it is decoded, but the length of a VLI is stored in its preceding VLC. The complete JPEG tables are not presented in this text and the reader is referred to the standard definition in order to obtain them (JPEG 1994).

Table 8.1 Categorization tables

Size	DIFF values	AC coefficients
0	0	
1	–1,1	–1,1
2	–3,–2,2,3	–3,–2,2,3
3	–7...–4,4...7	–7...–4,4...7
4	–15...–8,8...15	–15...–8,8...15
5	–31...–16,16...31	–31...–16,16...31
6	–63...–32,32...63	–63...–32,32...63
7	–127...–64,64...127	–127...–64,64...127
8	–255...–128,128...255	–255...–128,128...255
9	–511...–256,256...511	–511...–256,256...511
10	–1 023...–512,512...1 023	–1 023...–512,512...1 023
11	–2 047...–1 024,1 024...2 047	

8.1.5 Quality measures

Finally, in order to conclude this theoretical part of the chapter, let us introduce the main measures that are commonly used in image compression to assess the quality of the reconstructed images.

The most common measure of the distortion in image compression is the mean square error (MSE), defined as

$$MSE = \frac{1}{MN} \sum_{m=0}^{M-1} \sum_{n=0}^{N-1} \left\| x(m,n) - \hat{x}(m,n) \right\|^2 \quad (8.26)$$

where $\hat{x}(m,n)$ is the reconstructed image. This measure is often given in terms of the equivalent reciprocal measure, the *peak signal-to-noise ratio* (PSNR)

$$PSNR = 10\log_{10}\left(\frac{(2^B - 1)^2}{MSE} \right) dB \quad (8.27)$$

where B is the number of bits used to represent the pixel values in the image. The PSNR is expressed in decibels (dB) and good reconstructed images typically have PSNR values of 30 dB or more.

Although these measures do not take the human visual system characteristics into account, they are commonly used since they give an approximated idea of the visual quality, while being easy to compute and tractable in linear optimization problems.

8.2 MATLAB proof of concept

In this section, we are going to illustrate the main concepts behind the JPEG standard. These concepts are applied in video compression as well in the so-called hybrid approach, which is the basis for standards such as MPEG1 or MPEG2 and, therefore, these concepts will be revisited in Chapter 9. Note that in these two chapters, we will not concentrate on the specificities of any of these standards but on the common, generic tools and concepts that they use. Moreover, in order to illustrate these concepts, we will work mainly with gray level images and explain the extensions that should be introduced in the proposed systems to work with color images.

Let us start by reading a gray level image. In our case, this is the first image of a video sequence that in this chapter we will use it as an example to study the case of still (isolated) image coding and of the exploitation of the spatial redundancy. Moreover, in Chapter 9, it will be used as the initial point for the study of video sequence coding and of the exploitation of both spatial and temporal redundancies.

First we will study how the spatial redundancy can be exploited to reduce the number of bits needed to represent the image. Let us load file "table_000_g.bmp" from disk and show its pixel values (Fig. 8.11). This file contains a gray image corresponding to frame #000 of the *Table Tennis* sequence with size 176×144 pixels[9] stored in non-compressed form.

```
table = imread('seqs/table/table_000_g.bmp');
imshow(table);
```

Fig. 8.11 Table Tennis frame #000

8.2.1 Block image transformation

Image block processing

Spatial redundancy is exploited by performing a local analysis of the image, i.e., by dividing the image into nonoverlapping square blocks. Then, the information contained in each of these blocks will be processed separately. Typically, images are partitioned into blocks of 8×8 pixels (Fig. 8.12).

```
addgridtofigure(size(table),[8 8]);
```

---

[9] This image format is usually referred to as QCIF (quarter common intermediate format).

Fig. 8.12 Block processing for *Table Tennis* image

Mirror padding

Given that generic images (images of any size) may not allow an exact division in 8×8 pixel blocks, information in the right and low boundary blocks is usually padded before partitioning the image. A common way to pad this information is by mirroring it (which reduces the transitions introduced when padding with, for instance, zero values). Let us see this case with a new version of the previous *Table Tennis* image where four rows and four columns have been removed from the bottom and right sides of the original image (Fig. 8.13).

```
table_crop = table(1:end-4,1:end-4);
table_padded = padarray(table_crop,[4,4],'symmetric','post');
imshow(table_padded);
```

Fig. 8.13 Mirror padding for *Table Tennis* image

If we analyze one of the mirrored blocks, the effect of the mirroring can be easily seen (Fig. 8.14). For instance, the block in the first row and last column shows how the mirror padding is created along the vertical axis. Note as well how this padding preserves the original block contours.

```
imshow(table_padded(1:8,end-7:end));
```

Fig. 8.14 Mirror-padded block

DCT block transformation

Now, we are going to analyze how our basic coding unit (i.e., a generic 8×8 pixel block that, in the sequel, we will refer to as a *block*) is processed. Among all the different transforms that can be applied on an image block to obtain a set of less-correlated coefficients in the transformed domain, the discrete cosine transform (DCT) has been shown to present very good properties (see the discussion in the previous section).

Let us analyze one block from the *Table Tennis* image to illustrate the whole transform process. First we select a given block from the image, e.g., the block situated 12 blocks from the top and 13 from the left (Fig. 8.15):

```
posr = 12; posc = 13;
f = table*0.3;
f(8*(posr-1)+1:8*(posr-1)+8,8*(posc-1)+1:8*(posc-1)+8) = ...
    table(8*(posr-1)+1:8*(posr-1)+8,...
    8*(posc-1)+1:8*(posc-1)+8);
imshow(f);
```

Fig. 8.15 Selected block (*highlighted*)

The following image shows a magnification of the selected block. Also, the corresponding matrix of pixel values is shown (Fig. 8.16):

```
b = table(8*(posr-1)+1:8*(posr-1)+8,...
    8*(posc-1)+1:8*(posc-1)+8)
```

Fig. 8.16 Magnification of the selected block

```
b =
   73    89    90    93   179   241   227   186
  122   124    98   155   232   223   208   183
   83    69    92   198   199   183   181   153
   31    28   137   226   179   148   134   114
   15    50   184   211   180   140   110    81
   20    73   162   158   133   120   107    60
   32   102   151   135   112   107    87    47
   60   132   162   127   117   100    68    41
```

The MATLAB function `dct2` performs the DCT of a given image block. The transformed matrix (i.e., the matrix containing the DCT coefficient values) for the block under study is

```
d = dct2(double(b)-128),
```

```
d =
 -16.25 -160.49 -254.84  -12.36  -24.00   26.22   -0.51    7.40
 181.21 -184.67   69.58  109.83   46.68   -4.86  -15.38   -6.91
  11.36   13.69  117.01   62.23  -62.06  -61.07  -11.17    3.44
 -24.09  -24.65   -1.88   -8.66  -49.69   -5.43   19.46   17.67
 -19.50   -5.18  -26.64  -23.48   -7.75    5.99   23.21    5.81
 -47.70  -26.86    5.86   -8.08   -6.08   19.95    3.30  -10.85
 -19.59  -11.74   10.82    1.97   -6.76    9.26   -0.01   -2.29
  -0.12   -3.55   -4.60   -0.43   -3.76   -0.46    1.19   -0.61
```

Where, as explained in the previous section, a value of 128 has been subtracted to all pixels of block b to create a signal with zero mean and obtain a lower-energy DC coefficient.

We can observe the capability of the DCT to compact the energy of the signal with respect to the energy distribution in the original block. Fig. 8.17 shows the absolute value of the DCT coefficients (sorted in descending order):

```
v = sort(abs(d(:)),1,'descend');
plot(v);
```

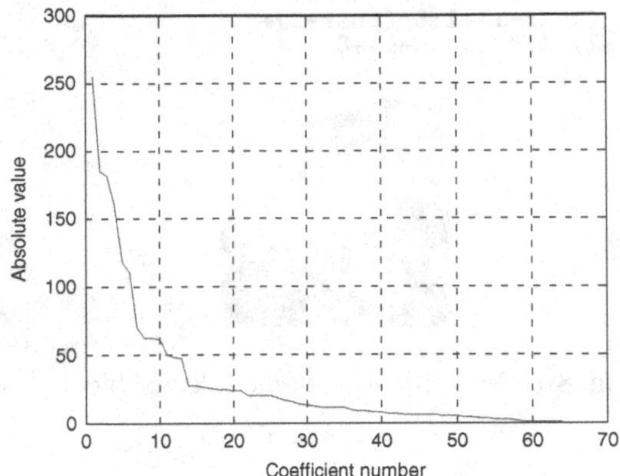

Fig. 8.17 Sorted DCT coefficients

If we plot the percentage of accumulated energy for these coefficients (see Fig. 8.18), it can be noted that, for this specific block, 95% of the energy of the signal in the transform domain is contained in the 13 coefficients with highest energy.

```
dct_energy_distribution = cumsum(v.*v)/sum(v.*v);
```

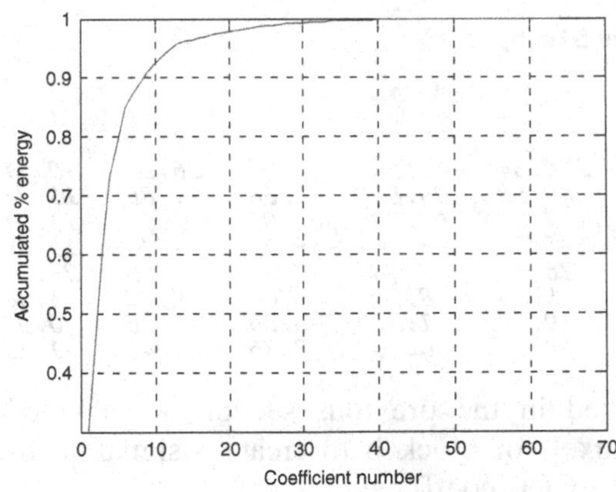

Fig. 8.18 Accumulated percentage of energy for the sorted DCT coefficients

However, in the original signal (the block b), the same percentage of energy is reached when adding up the contributions of the 47 pixels with the highest energy (as shown in Fig. 8.19).

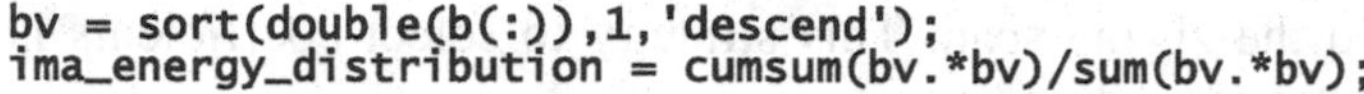

```
bv = sort(double(b(:)),1,'descend');
ima_energy_distribution = cumsum(bv.*bv)/sum(bv.*bv);
```

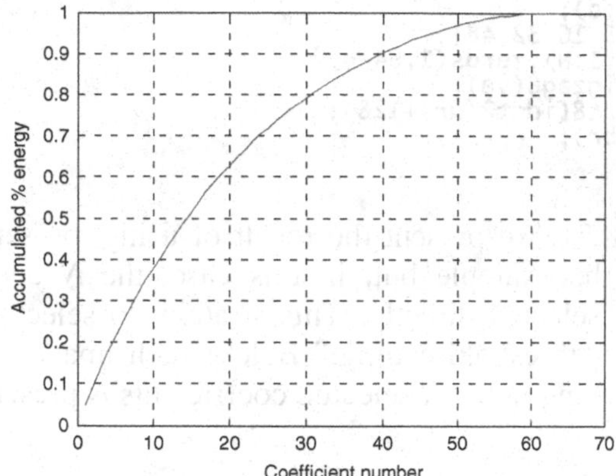

Fig. 8.19 Accumulated percentage of energy for the sorted pixels

Moreover note that, roughly speaking, coefficients in the DCT domain with higher energy are gathered around the left-top corner of matrix d (lower-frequency coefficients). This is a common behavior for all natural image block DCT coefficient matrices, which will be further exploited in the coding process when including the quantization step.

The DCT transformation is invertible and we can recover the initial pixel values of the original block b using the inverse DCT transformation (function idct2 in MATLAB):

```
br = uint8(idct2(d)+128);
b - br,
```

```
ans =
     0    0    0    0    0    0    0    0
     0    0    0    0    0    0    0    0
     0    0    0    0    0    0    0    0
     0    0    0    0    0    0    0    0
     0    0    0    0    0    0    0    0
     0    0    0    0    0    0    0    0
     0    0    0    0    0    0    0    0
     0    0    0    0    0    0    0    0
```

Elimination of DCT coefficients

Now we can see the effect of zeroing some of the DCT coefficients in the reconstruction of the original block. In Fig. 8.20 we present the result of reconstructing the image block using the first *N*=(1, 4, 8, 16, 32, 64)

coefficients in the zigzag scan. This strategy to select coefficients is often referred to as "zonal coding."

```
v = zigzag8(d);
for N=[1 4 8 16 32 48],
    ve = [v(1:N),zeros(1,64-N)];
    dr = izigzag8(ve);
    br = uint8(idct2(dr)+128);
    imshow(br);
end;
```

In turn, in Fig. 8.21, we present the result of using the same number of coefficients as in the example but, in this case, the N coefficients with highest energy are selected directly. This strategy to select coefficients is often referred to as "threshold coding." Below each image, the percentage of energy corresponding to the N selected coefficients is presented.

```
d2 = d(:);
[v,i] = sort(abs(d2),1,'descend');
v = d2(i); ii(i) = 1:64;
dct_energy_distribution = cumsum(v.*v)/sum(v.*v);
for N=[1 4 8 16 32 48],
    ve = [v(1:N);zeros(64-N,1)];
    veu = ve(ii);
    dr2 = reshape(veu,8,8);
    br = uint8(idct2(dr2)+128);
    imshow(kron(br,uint8(ones(16))));
    title(['N=' int2str(N) ' (~'...
    int2str(round(dct_energy_distribution(N)*100)) '%)']);
end;
```

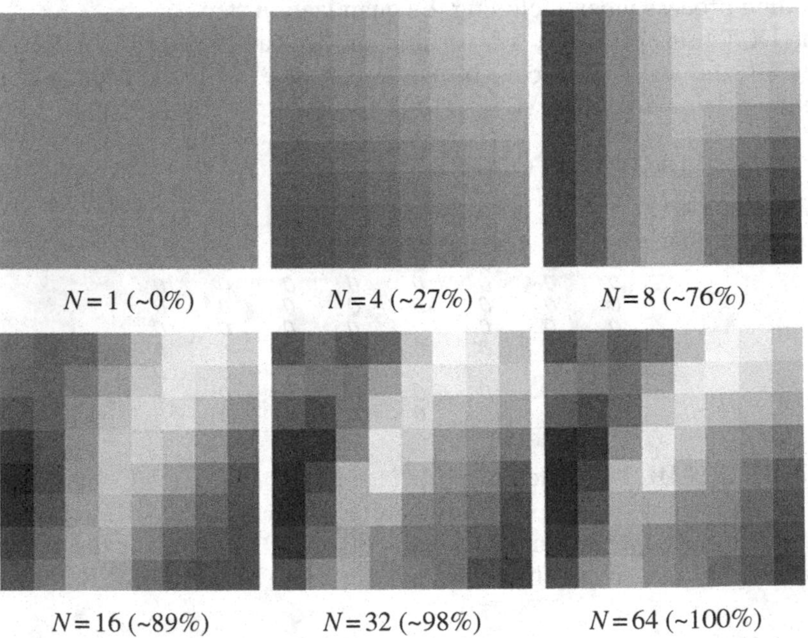

Fig. 8.20 Reconstructed blocks using the first N coefficients with a zigzag scan (in brackets the percentage of energy corresponding to the N selected coefficients)

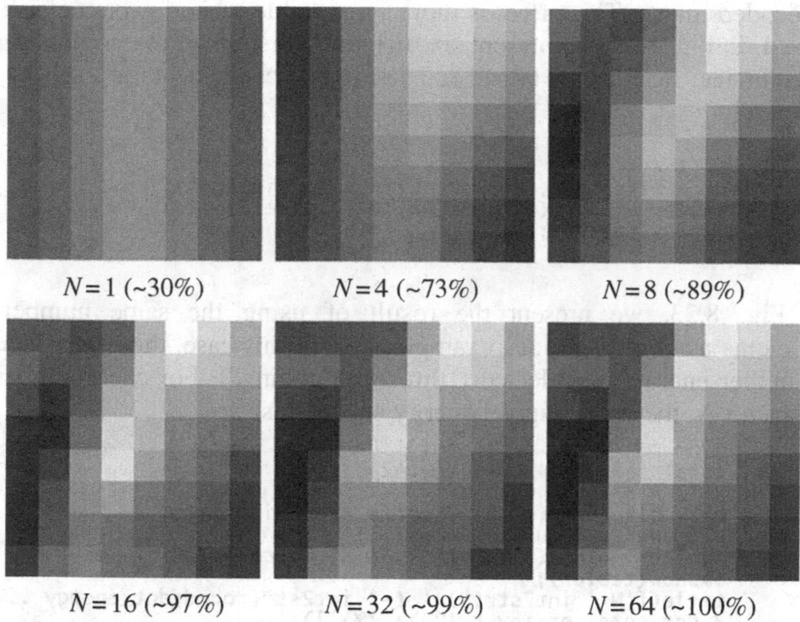

Fig. 8.21 Reconstructed blocks using the first N coefficients with higher energy (in brackets the percentage of energy corresponding to the N selected coefficients)

8.2.2 Complete image block coding

In order to draw more general conclusions, we are going to analyze some of these features in a whole image. Matrix `Ed` presents the energy distribution per DCT coefficient averaged among all the blocks in the image. As it can be seen, the previous behavior is preserved, i.e., coefficients with largest energy are those of lowest frequency (top-left matrix corner).

```
D = blkproc(double(table)-128,[8 8],@dct2);
Ed = blkmean(D.*D,[8 8]);
Ed = round(Ed/sum(sum(Ed))*100),
```

```
Ed =
  59    6    1    1    0    0    0    0
   7    2    1    0    0    0    0    0
   4    2    0    0    0    0    0    0
   4    1    0    0    0    0    0    0
   3    2    0    0    0    0    0    0
   2    1    0    0    0    0    0    0
   1    1    0    0    0    0    0    0
   0    0    0    0    0    0    0    0
```

The effect of dropping some DCT coefficients when reconstructing the complete image (the so-called *block effect*) can be observed in the following two figures. Since an image-independent block partition has been imposed and blocks have been separately coded, block boundaries are noticeable in the decoded image. This effect is more remarkable when fewer coefficients are used. In Fig. 8.22, we present six different versions of the original *Table Tennis* image: keeping N=(1, 4, 8, 16, 32, 64) coefficients at each block, respectively.

```
for N=[1 4 8 16 32 64],
    Dk = blkproc(D,[8 8],@coeffs_keep_zigzag,N);
    tabler = uint8(blkproc(Dk,[8 8],@idct2)+128);
    imshow(tabler);
                                 end;
```

In Fig. 8.23, we present the result of using the same number of coefficients as in the previous example, but, in this case, those coefficients with higher energy are selected ("threshold coding"). For each number of coefficients N, the percentage of energy selected is shown.

```
dct_total_energy = sum(sum(D.^2));
for N=[1 4 8 16 32 64],
    Dk = blkproc(D,[8 8],@coeffs_keep_higher,N);
    dct_energy = sum(sum(Dk.^2));
    tabler = uint8(blkproc(Dk,[8 8],@idct2)+128);
    imshow(tabler);
    title(['N=' int2str(N) ' (~' int2str(round(dct_energy ...
    / dct_total_energy * 100)) '%)']);
end;
```

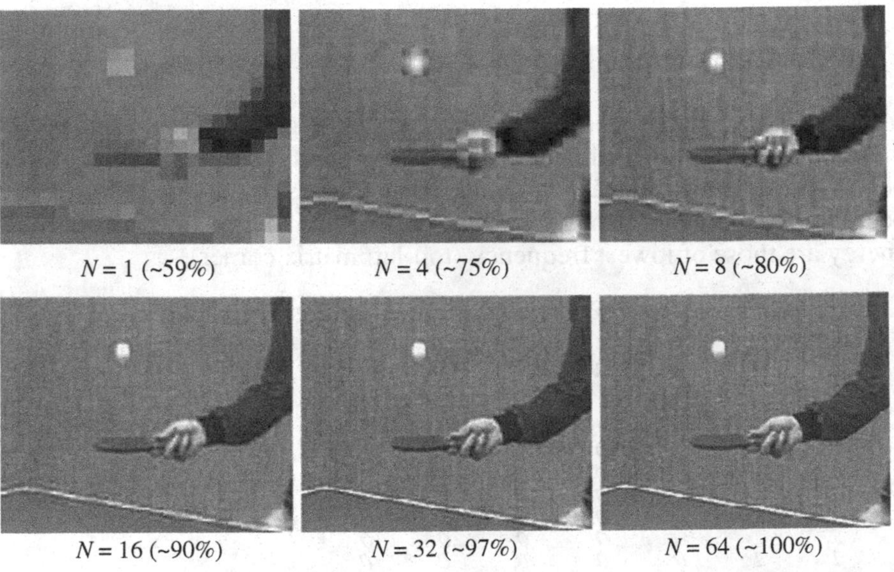

Fig. 8.22 Reconstructed images using the first N coefficients for each block with a zigzag scan (in brackets the percentage of energy corresponding to the N selected coefficients)

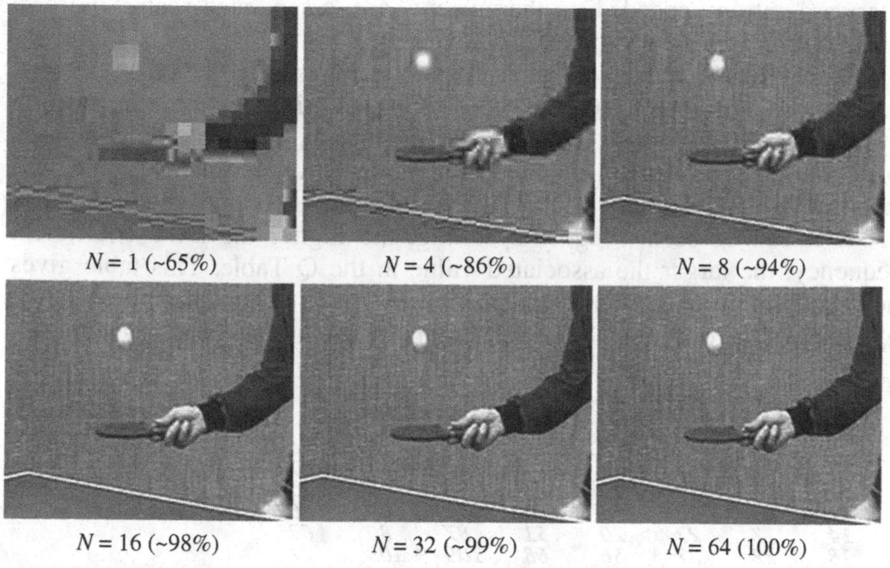

Fig. 8.23 Reconstructed images using the first N coefficients with higher energy (in brackets the percentage of energy corresponding to the N selected coefficients)

The analysis of the previous examples leads to the following comments:

- Increasing the same amount of coefficients/energy in all blocks does not translate into the same increase in visual quality for all blocks. This is due to the fact that not all DCT coefficients are equally relevant for the human visual system. This concept will be further exploited in the sequel to adapt the quantization of the DCT coefficients to the final human observer.
- In order to obtain a homogenous image quality (for instance, a given global PSNR), a different amount of coefficients/energy (and, eventually, bits) can be assigned to the various blocks. Therefore, given a bit budget for a specific image, those bits can be shared among the blocks in a content-dependent manner. That is, more/fewer bits are used for those blocks containing more/less complex information. This concept is the basis for the *rate-distortion theory,* which will be further addressed in chapter 10.

8.2.3 DCT quantization

As commented in the previous section, DCT coefficients need to be quantized in order to generate a discrete source whose symbols can be afterward entropy encoded. In the current case, a separated scalar, uniform quantization is used for each DCT coefficient.

The different quantization of each coefficient is represented by the so-called quantization table (Q Table). This is a matrix containing at position `(i,j)` the value to which the DCT coefficient `(i,j)` should be compared to generate its associated quantization index.[10] A possible Q Table was proposed by Lohscheller[11] (Lohscheller 1984) and it is presented as matrix *Q*. As it can be seen, roughly speaking, the higher the DCT coefficient frequency, the larger the associated value in the Q Table. This table gives the results of psychovisual experiments that have determined the relative relevance of the different DCT coefficients for the human visual system.

```
Q = jpegsteps,

Q =
     8    11    10    16    24    40    51    61
    12    12    14    19    26    58    60    55
    14    13    16    24    40    57    69    56
    14    17    22    29    51    87    80    62
    18    22    37    56    68   109   103    77
    24    35    55    64    81   104   113    92
    49    64    78    87   103   121   120   101
    72    92    95    98   112   100   103    99
```

The comparison between a given DCT coefficient and its associated value in the Q Table is carried out as expressed in the following code in which, for illustration purposes, we are using the same block we have used in the previous subsection (see Fig. 8.16):

```
dq = round( d./Q ),

dq =
    -2   -15   -25    -1    -1     1     0     0
    15   -15     5     6     2     0     0     0
     1     1     7     3    -2    -1     0     0
    -2    -1     0     0    -1     0     0     0
```

10 There is a difference in the way in which DCT coefficients are commonly indexed and their MATLAB implementation. DCT coefficients of an 8×8 pixel block are usually indexed from 0 to 7 in both dimensions with the first index indicating the desired column and the second indicating the desired row. In turn, the coordinates of a MATLAB matrix are indexed from 1 to 8 in both dimensions with the first index indicating the desired row and the second indicating the desired column.

11 Lohscheller tables have been adopted in the JPEG and MPEG standards as default proposals, although, in these standards, different tables can be set by the encoder and included in the bit stream. In the case of JPEG, there is a slight variation in the `Q(1,1)` value since, due to implementation features, the implemented DCT transform is not unitary.

```
  -1     0    -1     0     0     0     0     0
  -2    -1     0     0     0     0     0     0
   0     0     0     0     0     0     0     0
   0     0     0     0     0     0     0     0
```

For example, a coefficient `d(3,1)` smaller than 5 implies that the coefficient is zeroed after quantization (`Q(3,1)`=10) whereas coefficient `d(6,8)` has to be smaller than 50 in order to be zeroed (`Q(6,8)`=100).

Different coding qualities can be obtained by multiplying the values in the Q Table by a given constant, `k`, producing a new Q Table: `kQ`. The following MATLAB code presents the quantization of block `b` using `k=3` and `k=5`.

```
k=3;
k.*Q,round(d./k./Q),

ans =
    24    33    30    48    72   120   153   183
    36    36    42    57    78   174   180   165
    42    39    48    72   120   171   207   168
    42    51    66    87   153   261   240   186
    54    66   111   168   204   327   309   231
    72   105   165   192   243   312   339   276
   147   192   234   261   309   363   360   303
   216   276   285   294   336   300   309   297

ans =
    -1    -5    -8     0     0     0     0     0
     5    -5     2     2     1     0     0     0
     0     0     2     1    -1     0     0     0
    -1     0     0     0     0     0     0     0
     0     0     0     0     0     0     0     0
    -1     0     0     0     0     0     0     0
     0     0     0     0     0     0     0     0
     0     0     0     0     0     0     0     0
```

```
k=5;
k.*Q,round(d./k./Q),

ans =
    40    55    50    80   120   200   255   305
    60    60    70    95   130   290   300   275
    70    65    80   120   200   285   345   280
    70    85   110   145   255   435   400   310
    90   110   185   280   340   545   515   385
   120   175   275   320   405   520   565   460
   245   320   390   435   515   605   600   505
   360   460   475   490   560   500   515   495

ans =
     0    -3    -5     0     0     0     0     0
     3    -3     1     1     0     0     0     0
     0     0     1     1     0     0     0     0
     0     0     0     0     0     0     0     0
     0     0     0     0     0     0     0     0
     0     0     0     0     0     0     0     0
     0     0     0     0     0     0     0     0
     0     0     0     0     0     0     0     0
```

In turn, Fig. 8.24 presents the reconstruction of the *Table Tennis* image when quantizing the DCT coefficients of its blocks using `k=1`, `k=3`, and `k=5`. In our implementation, the MATLAB functions `quantizedct8` and `iquantizedct8` perform the (direct and inverse) quantization of the DCT coefficients of a given matrix. Moreover, they are responsible for clipping the DC value between [0,255] and AC values between [−128, 127][12].

```
tabledct = blkproc(table, [8 8], @dct2);
for k=[1 3 5],
    tabledctq = blkproc(tabledct, [8 8], @quantizedct8, k.*Q);
    tabledctqi = blkproc(tabledctq,[8 8],@iquantizedct8,k.*Q);
    tabler = uint8(blkproc(tabledctqi, [8 8], @idct2));
    imshow(tabler);
end;
```

$k=1$ $k=3$ $k=5$

Fig. 8.24 Reconstructed images using various quantization tables

8.2.4 Spatial decorrelation between blocks

So far, we have decorrelated the information in the image only using the DCT and this has been done grouping the data into blocks of 8×8 pixels. Since this grouping is independent of the image content, it is likely that collocated transformed coefficients in neighbor blocks will still be correlated. If such a correlation is present, an additional transformation can be applied to the DCT coefficients to further reduce it. To illustrate this correlation, the next figures present the 2-dimensional distribution of collocated DCT coefficients in consecutive blocks. To ensure that the blocks that are compared are neighbors, blocks are ordered following the zigzag scan.

---

[12] Note that in the proposed implementation, the value `Q(1,1)`=8 is used and there is no subtraction of 128 to the pixel values before computing the DCT transform. Therefore, the DC coefficient can be directly quantized in the range [0,255].

In order to create more meaningful 2-Dimensional histograms (i.e., to have more data), we use a 352×288 pixels image. This image format is usually known as common intermediate format (CIF). Moreover, we have selected the first image of the *Foreman* sequence instead of the *Table Tennis* one for this specific example. The *Table Tennis* image has a large number of blocks with similar DC values (all blocks corresponding to the wall in the background), which prevents from correctly illustrating the spatial decorrelation effect. On the contrary, the *Foreman* sequence (see Fig. 8.25) presents more variation of the DC values as blocks are less homogeneous. Figures 8.26, 8.27 and 8.28 show the distribution of the values of DCT coefficients in positions `(1,1)`, `(2,1)`, and `(4,4)` for consecutive blocks in the zigzag scan, respectively.

Fig. 8.25 *Foreman* CIF original image

```
k = 1;
im = imread('seqs/foreman/foreman_cif_000_g.bmp');
imdct = blkproc(im,[8 8],@dct2);
imdctq = blkproc(imdct,[8 8],@quantizedct8,k*Q);
pos = {[1 1],[2 1],[1 3]};
for p = 1:length(pos),
    coeffsm = imdct(pos{p}(1):8:end,pos{p}(2):8:end);
    coeffs = zigzag(coeffsm);
    figure,hold on,grid
    for i=2:length(coeffs),
        plot(coeffs(i-1),coeffs(i),'o');
    end;
end;
```

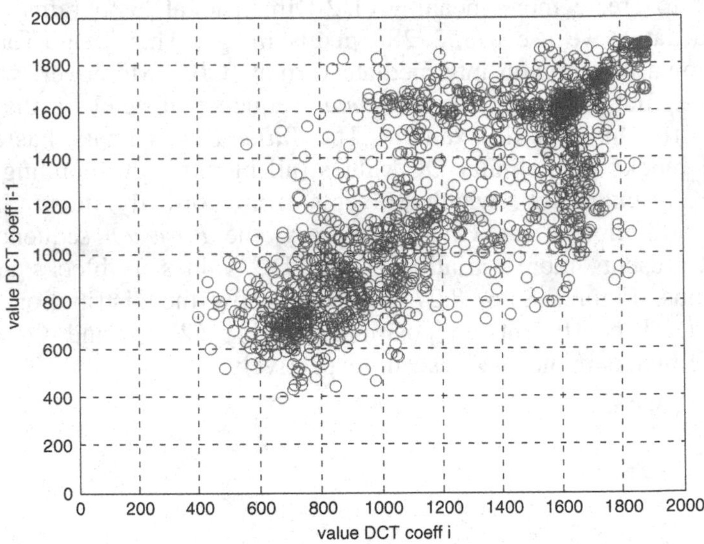

Fig. 8.26 2-Dimensional distribution of consecutive DC coefficients (zigzag scan)

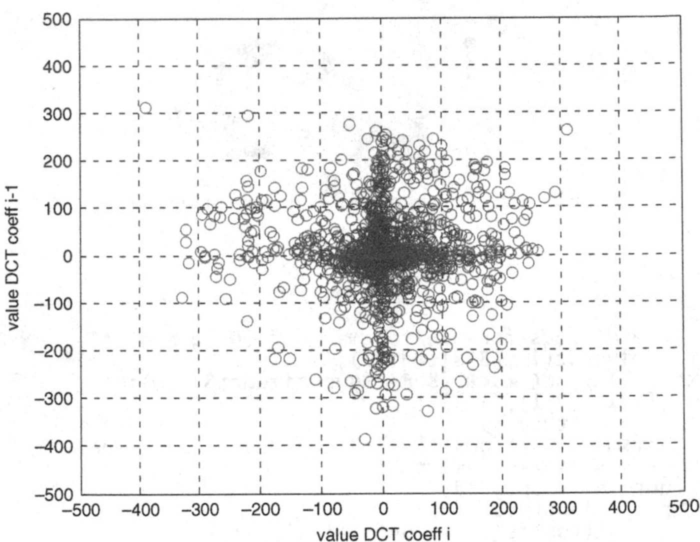

Fig. 8.27 2-Dimensional distribution of AC coefficients in position (2,1) in consecutive blocks (zigzag scan)

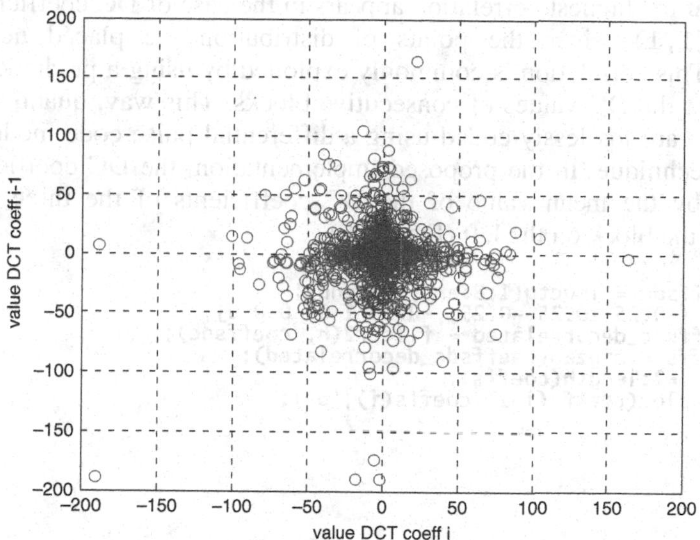

Fig. 8.28 2-Dimensional distribution of AC coefficients in position (1,3) in consecutive blocks (zigzag scan)

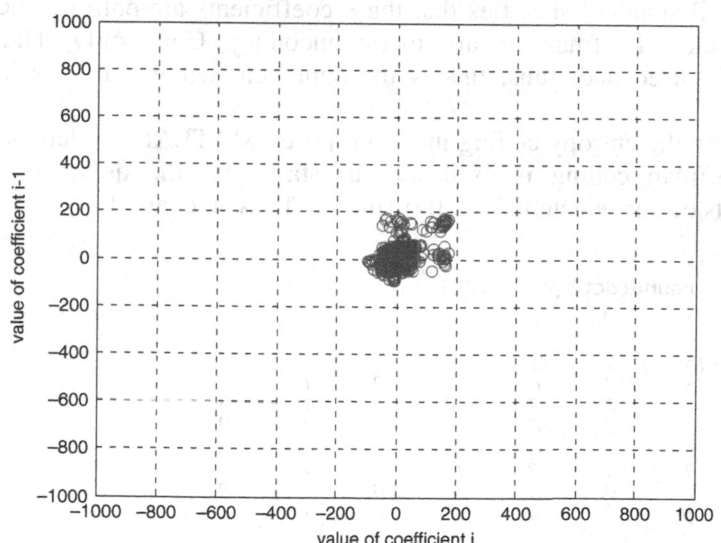

Fig. 8.29 2-Dimensional distribution of consecutive DC coefficients (zigzag scan) after decorrelation

Note that the highest correlation appears in the case of DC coefficient (at position (1,1)) since the points of distribution are placed near the diagonal. This correlation is commonly exploited by using a prediction step to represent the DC values of consecutive blocks. This way, quantized DC coefficients are losslessly coded using a differential pulse code modulation (DPCM) technique. In the proposed implementation, the DC coefficient is predicted by the mean value of the DC coefficients of the three blocks above and the block on the left (Fig. 8.29).

```
coeffsdc = imdctq(1:8:end,1:8:end);
h = [-0.25 -0.25 -0.25; -0.25 1 0; 0 0 0];
coeffsdc_decorrelated = filter2(h, coeffsdc);
coeffs = zigzag(coeffsdc_decorrelated);
for i=2:length(coeffs),
    plot(coeffs(i-1),coeffs(i),'o');
end;
```

8.2.5 Entropy coding

At this moment, we have two different sets of quantization indices or symbols: those related to the prediction error of the decorrelated quantized DC coefficients and those associated with the zigzag-ordered AC coefficients.

The JPEG standard specifies that these coefficients are entropy encoded through either Huffman or arithmetic encoding. Commonly, Huffman encoding is used and, thus, this is the approach that we discuss in this Section.

Let us see the entropy coding in the proposed MATLAB implementation where Huffman coding is used. Let us start with the quantized DCT coefficients of our selected block b (using $k = 1$) as an example.

```
k = 1;
dq = round(dct2(b)./k./Q),
```

```
dq =
   126   -15   -25    -1    -1     1     0     0
    15   -15     5     6     2     0     0     0
     1     1     7     3    -2    -1     0     0
    -2    -1     0     0    -1     0     0     0
    -1     0    -1     0     0     0     0     0
    -2    -1     0     0     0     0     0     0
     0     0     0     0     0     0     0     0
     0     0     0     0     0     0     0     0
```

As explained earlier in this chapter, the DC coefficient is decorrelated with previous DC values and encoded separately. In the case of AC coefficients, as it can be seen in the example above, there are many AC coefficients with value 0 (even with k=1). Therefore, it is important to precede the Huffman encoding of AC coefficients with a runlength encoder. Again, the AC coefficients are zigzag scanned before the runlength encoder:

```
dc = dq(1,1);
ac = zigzag(dq); ac = ac(2:end);
[nz vv] = runlength(ac,63);
```

The MATLAB function `encodemessage` is responsible for translating the quantized DC coefficients and the runlength AC coefficients into symbols using Huffman codes. The MATLAB file `"codes/imcodes.mat"` includes the previously generated Huffman codes (as seen in the previous section) corresponding for each DC coefficient (in variable `dcdcode`) and each AC coefficient (variables `nzcode` and `vvcode`). In our example, the final list of symbols for the DC coefficient (in this case, as we analyze an isolated block, the DC coefficient has not been decorrelated) is

```
load('codes/imcodes.mat');
codeddcd = encodemessage(dc+256', dcdcode),
```

codeddcd = 00001011000101

As the decorrelated DC values have a dynamic range between [–255,255], a value of 256 is added to form the index to access Huffman codes stored in variable `dcdcode`.

The runlength encoding of AC coefficients is (again, 2 is added to the runlength values to create indexes as the dynamic range is [–1,62] and 129 is added to runlength values with dynamic range [–128,128]):

```
codednz = encodemessage(nz(:)' + 2, nzcode),
```

codednz =0000000000000000011001000100110110101

```
codedvv = encodemessage(vv(:)' + 129, vvcode),
```

codedvv =
0010111110011010011000101111110011011110001001001100000010100100000011
1001100001111100000010100000101

The following code implements the entropy coding for all blocks of the input image.

```
tabledct  = blkproc(table, [8 8], @dct2);
tabledctq = blkproc(tabledct, [8 8], @quantizedct8, k.*Q);

zzdctq = double(blkproc(tabledctq, [8 8], @zigzag8));

% Separate DC from AC components of the DCT.
[zznrows zzncolumns] = size(zzdctq);
mask = true([zznrows zzncolumns]);
mask(:, 1 : 64 : end) = false;
dc = zzdctq(~mask);
```

```
ac = reshape(zzdctq(mask), ...
    [zznrows, zzncolumns - zzncolumns / 64])' - 128;

% Decorrelate DC (add 256 to directly create index)
h = [-0.25 -0.25 -0.25; -0.25 1 0; 0 0 0];
dc = reshape(dc, size(table) / 8);
dcdindex = floor(filter2(h, dc)) + 256;

% Get NZ and VV sequence of AC component.
[nz vv] = runlength(ac(:)', 63);

% Encode DC, ACNZ and ACVV components with the huffman code.
codeddcd = encodemessage(dcdindex(:)', dcdcode);
codednz = encodemessage(nz(:)' + 2, nzcode);
codedvv = encodemessage(vv(:)' + 129, vvcode);

% Header with codedacnz and codedacvv length
% get the number of bits needed to store the
% maximum possible number of runlength values
% for this image
nbits = ceil(log2(63 * prod(size(table)) / 64));
header = strcat(dec2bin(length(nz), nbits), ...
        dec2bin(length(vv), nbits));
```

After entropy coding, the final bitstream is constructed by concatenating the Huffman coding of DC and AC coefficients. In the example, we have also included the header information with the input image size

```
% Final bitstream
bitstream = strcat(header, codeddcd, codednz, codedvv);
```

In the case of *Table Tennis* image, the resulting size in bits of the final bitstream corresponds to

```
final_bits = size(bitstream,2),
```

final_bits = 22551

which yields the following figure of bits per pixel used in the representation of the decoded gray image:

```
bitspixel = final_bits / (size(table,1)*size(table,2)),
```

bitspixel = 0.8898

If we compare the obtained bits per pixel with the case of using 8 bits per pixel (when there is no compression), the final compression ratio achieved by our JPEG-oriented encoder for the *Table Tennis* image is

```
compression_ratio = 8 / bitspixel
```

```
compression_ratio = 8.993
```

In order to compare the final compressed image with the original one, the produced bitstream must be decoded. The following code shows how the bitstream can be decoded back into quantized DCT coefficients.

```
% Extract header from codedmessage and calculate number of
% ACNZ and ACVV to decode.
zzsize = size(table) .* [1/8 8];
dcsize = [zzsize(1, 1) ceil(zzsize(1, 2) / 64)];
ndc = prod(dcsize); nac = ndc * 63;
nbits = ceil(log2(nac));
nzlen = bin2dec(bitstream(1 : nbits));
vvlen = bin2dec(bitstream(nbits + 1 : 2 * nbits));
bitstream2 = bitstream(2 * nbits + 1 : end);

load('decoders/im/dcddecoder.mat');
[dcdindex bitstream2] = decodemessage(bitstream2, dcddecoder,
ndc);
load('decoders/im/nzdecoder.mat');
[nzindex bitstream2] = decodemessage(bitstream2, nzdecoder,
nzlen);
load('decoders/im/vvdecoder.mat');
[vvindex bitstream2] = decodemessage(bitstream2, vvdecoder,
vvlen);

% Calculate DCT - DC components from DC differentiated.
dcd = reshape(dcdindex, dcsize) - 256;
dc = zeros(dcsize + [1 2]);
for i = 2 : size(dc,1)
    for j = 2 : size(dc,2) - 1
        dc(i,j) = ceil(dcd(i-1,j-1) + 1/4 * (dc(i, j-1) + dc(i-
1, j-1) + dc(i-1, j) + dc(i-1, j+1)));
    end
end
dc = dc(2 : end, 2 : end - 1);

% Calculate AC coeficients from AC number of zeros (nz) and AC
% values (vv).
ac = irunlength(nzindex - 2, vvindex - 129, 63, nac);
ac = reshape(ac, [nac/zzsize(1, 1) zzsize(1, 1)])' + 128;

% Join DC and AC components into a new DCT.
zzdctq = zeros(zzsize);
mask = true(zzsize);
mask(:, 1 : 64 : end) = false;
zzdctq(~mask) = dc;
zzdctq(mask) = ac(:);

% reconstruct the image
fun = @(x) idct2(iquantizedct8(izigzag8(x),Q));
tabler = uint8(blkproc(zzdctq, [1 64], fun));
```

The PSNR is a good measure of the visual quality of the resulting compressed image (see Section 8.1.5). In the case of the *Table Tennis* image that we have used in all our examples, the final PSNR (for k=1) is

```
psnr = mypsnr(table,tabler),
```

```
psnr = 32.5078
```

8.2.6 Still image coding

For completeness, the MATLAB function `stillimagegrayencode` implements the total encoding of a single image following the concepts explained in this chapter. It accepts the matrix of the original image and the factor `k` to multiply the quantization table. The function returns the reconstructed image, the bitstream with the encoded DCT coefficients, and the peak signal-to-noise ratio (PSNR) of the compressed image.

```
[Ir,bits,psnr] = stillimagegrayencode(table,k);
```

Fig. 8.30 presents the result of applying this function to the *Foreman* CIF and *Tennis Table* QCIF original images previously used.

```
im = imread('seqs/foreman/foreman_000_cif.ras');
for k = [1 3 5],
    [Ir,bits,p] = stillimagegrayencode(im,k);
    imshow(Ir);
end;
im = imread('seqs/table/table_000_g.bmp');
for k = [1 3 5],
    [Ir,bits,p] = stillimagegrayencode(im,k);
    imshow(Ir);
end;
```

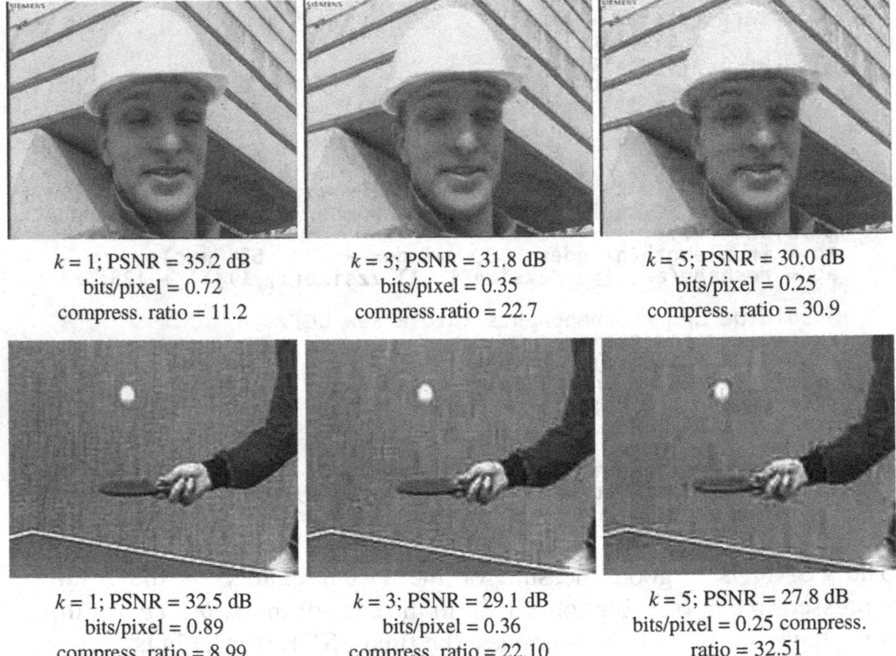

$k = 1$; PSNR = 35.2 dB bits/pixel = 0.72 compress. ratio = 11.2	$k = 3$; PSNR = 31.8 dB bits/pixel = 0.35 compress.ratio = 22.7	$k = 5$; PSNR = 30.0 dB bits/pixel = 0.25 compress. ratio = 30.9
$k = 1$; PSNR = 32.5 dB bits/pixel = 0.89 compress. ratio = 8.99	$k = 3$; PSNR = 29.1 dB bits/pixel = 0.36 compress. ratio = 22.10	$k = 5$; PSNR = 27.8 dB bits/pixel = 0.25 compress. ratio = 32.51

Fig. 8.30 Decoded gray scale images of *Foreman* and *Table Tennis* for various values of k

For color images, the same concepts apply to each of the three color channels (in the case of JPEG standard, YCbCr color space is commonly used and the chrominance channels are decimated by a factor of 2 in horizontal and vertical directions). The MATLAB function `stillimageencode` implements the encoder for color images. Fig. 8.31 shows the resulting compressed images for the same *Table Tennis* (QCIF) and *Foreman* images of previous examples (in QCIF and CIF formats), respectively:

```
table = imread('seqs/table/table_000.bmp');
foreman = imread('seqs/foreman/foreman_000.bmp');
for k=[1 3 5],
    [Ir,bits,p] = stillimageencode(table,k);
    imshow(Ir);
    [Ir,bits,p] = stillimageencode(foreman,k);
    imshow(Ir);
end;
```

$k = 1$; PSNR = 30.0 dB
bits/pixel = 0.95
compress. ratio = 25.1

$k = 3$; PSNR = 27.2 dB
bits/pixel = 0.41
compress. ratio = 59.1

$k = 5$; PSNR = 25.9 dB
bits/pixel = 0.30
compress. ratio = 78.8

$k = 1$; PSNR = 31.3 dB
bits/pixel = 1.12
compress. ratio = 41.4

$k = 3$; PSNR = 27.8 dB
bits/pixel = 0.56
compress. ratio = 42.5

$k = 5$; PSNR = 26.9 dB
bits/pixel = 0.42
compress. ratio = 57.3

$k = 1$; PSNR = 33.6 dB	$k = 3$; PSNR = 30.3 dB	$k = 5$; PSNR = 28.5 dB
bits/pixel = 0.76	bits/pixel = 0.39	bits/pixel = 0.30
compress. ratio = 31.7	compress. ratio = 61.8	compress. ratio = 80.4

Fig. 8.31 Decoded color images of *Table Tennis* QCIF (*first row*), *Foreman* QCIF (*second row*) and *Foreman CIF* (*third row*) for various values of k

8.3 Going further

What we have presented in this chapter represents the basic tools and concepts of the so-called baseline JPEG algorithm, known as well as the *JPEG sequential mode*. Nevertheless, the standard proposes some extensions that are worth commenting.

Progressive JPEG allows the compressed image to be incrementally refined through multiple scans as the compressed data arrives. It is a first version of a very useful functionality commonly known as *scalability*. There are two different implementations of progressive JPEG:

- *Successive approximation*, in which the DCT-quantized coefficients are successively refined by transmitting their more-significant bits initially and the less-significant ones subsequently.
- *Spectral selection*, in which the DCT-quantized coefficients are progressively sent, initially the DC coefficient and in successive scans the remaining AC coefficients.

In this chapter, we have assumed that the same Q Table was used for all the blocks in an image. Commonly, this is not the case and, previous to quantization, there is a step in which the block under consideration is analyzed and classified. This way, quantization can be adapted to the characteristics of the block (for instance, blocks with more/less complex information), leading to the concept of *adaptive transform coding*. If we have given a bit budget for an image, those bits can be shared among the blocks in a content-dependent manner. That is, more/fewer bits are used for those blocks containing more/less complex information. This concept is the basis for the *rate-distortion theory*, which is paramount in image and video compression.

Finally, an extension of the baseline JPEG algorithm is the so-called *motion JPEG* (usually referred to as M-JPEG), which has been for a long time a de facto standard for video coding. It is essentially identical to the baseline JPEG algorithm, with the exception that the quantization and Huffman tables cannot be dynamically selected in the encoding process but those originally given as examples in the JPEG standard have been fixed.

8.4 Conclusions

In this chapter, we have analyzed how most of digital images are compressed in the web. We have learned how spatial redundancy in images can be reduced using transforms and specifically the discrete cosine transform (DCT) on consecutive blocks of the image. We have also shown that DC coefficient information can be further decorrelated using prediction from the DC coefficients of neighbor blocks. The resulting DCT coefficients for each block are quantized and entropy-encoded to form the final compressed image.

The techniques studied in this chapter compose the basic tools used in the JPEG standard for coding gray and color images, which is the standard commonly used to compress images in the web and for storing digital photographs. For further details on the JPEG standard, the reader is referred to the standard description (JPEG 1994).

References

ISO/IEC 10918-1 or ITU-T Recommendation T.87 (1999). Information technology Lossless and near-lossless compression of continuous-tone still images

ISO/IEC 14495-1 or ITU-T Recommendation T.81 (1994). Information technology – Digital compression and coding of continuous-tone still images requirements and guidelines

Arai Y, Agui T, Nakajima M (1988) A fast DCT-SQ scheme for images. Transactions of the IEICE, 71: 11, 1095

Chiariglione L (1999) Communication standards: Götterdämmerung? In: Advances in Multimedia, Standards and Networks. Puri, Chen Eds. Marcel Dekker, Inc. New York

Cover T, Thomas A (1991) Elements of Information Theory. New York: John Wiley & Sons, Inc.

Gersho A, Gray RM (1993) Vector Quantization and Signal Compression. Boston: Kluwer Academic Publishers

Lohscheller H (1984) A subjectively adapted image communication system. IEEE Transactions on communications COM-32, 1316–1322

Noll NS, Jayant P (1984) Digital Coding of Waveforms: Principles and Applications to Speech and Video. Englewood Cliffs, NJ: Prentice Hall.

Pennebaker W, Mitchell J (1993) JPEG Still Image Data Compression Standard. New York: Van Nostrand Reinhold

Chapter 9

How are digital TV programs compressed to allow broadcasting?

F. Marqués(°), M. Menezes(*), J. Ruiz-Hidalgo(°)

(°) Universitat Politècnica de Catalunya, Spain
(*) Instituto Superior de Ciências do Trabalho e da Empresa, Portugal

The raw bit rate of a studio video sequence is 166 Mbps whereas the capacity of a terrestrial TV broadcasting channel is around 20 Mbps.

In 1982, the CCIR defined a standard for encoding interlaced analogue video signals in digital form mainly for studio applications. The current name of this standard is ITU-R BT.601 (ITU 1983). Following this standard, a video signal sampled at 13.5 MHz with a 4:2:2 sampling format (double the number of samples for the luminance component than for the two chrominance components) and quantized with 8 bits per component produces a raw bit rate of 216 Mbps. This rate can be reduced by removing the blanking intervals present in the interlaced analogue signal leading to a bit rate of 166 Mbps, which is still a figure far above the main capacity of usual transmission channels or storage devices.

Bringing digital video from its source (typically, a camera) to its destination (a display) involves a chain of processes, among which compression (encoding) and decompression (decoding) are the key ones. In these processes, bandwidth-intensive digital video is first reduced to a manageable size for transmission or storage, and then reconstructed for display. This way, video compression allows using digital video in

T. Dutoit, F. Marqués (eds.), *Applied Signal Processing*,
DOI 10.1007/978-0-387-74535-0_9,

transmission and storage environments that would not support uncompressed video.

In the last years, several image and video coding standards have been proposed for various applications such as JPEG for still image coding (see Chapter 8), H.263 for low-bit rate video communications, MPEG1 for storage media applications, MPEG2 for broadcasting and general high-quality video application, MPEG4 for streaming video and interactive multimedia applications and H.264 for high-compression requests. In this chapter, we describe the basic concepts of video coding that are common to these standards.

9.1 Background – Motion estimation

Figure 9.1 presents a simplified block diagram of a typical video compression system (Clarke 1995, Zhu et al. 2005). Note that the system can work in two different modes: *intra* and *inter* modes. In intra-mode, only the spatial redundancy within the current image is exploited (see Chapter 8), whereas inter-mode, in addition, takes advantage of the temporal redundancy among temporal neighbor images.[1]

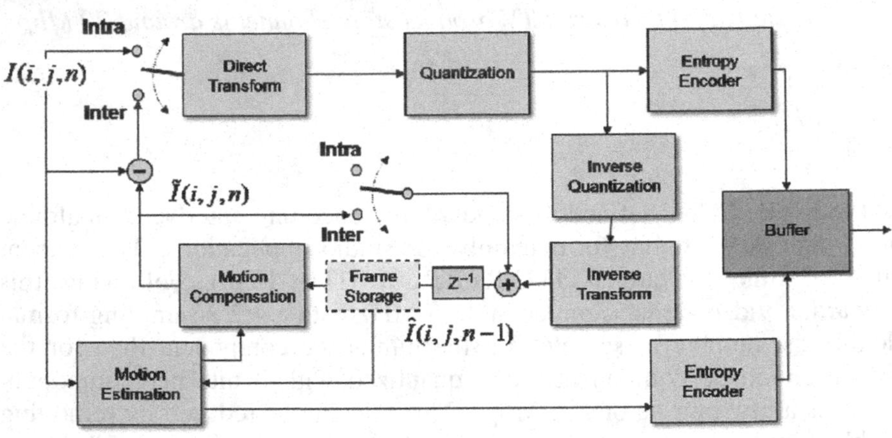

Fig. 9.1 Block diagram of video compression system

[1] In this chapter, for simplicity, we will assume that a whole picture is compressed either in intra-or inter-mode. In current standards, this decision is adopted at a more local scale within the image: at the so-called macroblock or even at block level. These concepts will be defined in the sequel (actually, the block concept has already been defined in Chapter 8).

Let us start analyzing the system in intra-mode. This is the mode used, for instance, for the first image in a video sequence. In intra-mode, the image is handled as it is described in Chapter 8. Initially, a transform is applied to it in order to decorrelate its information.[2] The image is initially partitioned into *blocks* of 8 × 8 pixels and the DCT transform is separately applied to the various blocks (Noll and Jayant 1984). The transformed coefficients are scalar quantized (Gersho and Gray 1993) taking into account the different relevancy for the human visual system of the various DCT coefficients. Quantized coefficients are zigzag scanned and entropy coded (Cover and Thomas 1991) in order to be efficiently transmitted.

The quantized data representing the intra-mode-encoded image is used in the video coding system to provide the encoder with the same information that will be available at the decoder side, i.e., a replica of the decoded image.[3] This way, a decoder is embedded in the transmitter and, through the inverse quantization and the inverse transform, the decoded image is obtained. This image is stored in the *Frame Storage* and will be used in the coding of future frames.

This image is represented as $\hat{I}(i, j, n-1)$, where the symbol "^" denotes that it is not the original frame but a decoded one. Moreover, since the system typically has already started coding the following frame in the sequence, this frame is stored as belonging to time instant n–1 (and this is the reason for including a delay in the block diagram).

Now, we can analyze the behavior of the system in inter-mode. The first step is to exploit the temporal redundancy between previously decoded images and the current frame. For simplicity, we are going to assume in this initial part of the chapter that only the previous decoded image is used for exploiting the temporal redundancy in the video sequence. In subsequent sections, we will see that the *Frame Storage* may contain several other decoded frames to be used in this step.

The previous decoded frame is used to estimate the current frame. Toward this goal, the *Motion Estimation* block computes the motion in the scene, i.e., a motion field is estimated, which assigns to each pixel in the current frame a motion vector (d_x, d_y) representing the displacement that this pixel has suffered with respect to the previous frame. The information in

---

[2] In recent standards such as H.264 (Sullivan and Wiegand 2005), several other intra mode decorrelation techniques are proposed based on prediction.

[3] Note that the quantization step very likely will introduce losses in the process and, therefore, the decoded image is not equal to the original one.

this motion field is part of the new, more compact representation of the video sequence and, therefore, it is entropy encoded and transmitted.[4]

Based on this motion field and on the previous decoded frame, the *Motion Compensation* block produces an estimate of the current image $\tilde{I}(i,j,n)$. This estimated image has been obtained applying motion information to a reference image and, thus, it is commonly known as the *motion-compensated image* at time instant *n*. In this chapter, motion compensated images are denoted by the symbol "~".

The system then subtracts the motion-compensated image from the original image, both at time *n*. The result is the so-called *motion-compensated error image* and contains all the information from the current image that has not been correctly estimated using the information in the previous decoded image. These concepts are illustrated in Fig. 9.2.

Fig. 9.2 (**a**) Original frame #12 of the *Stefan* sequence, (**b**) original frame #10 of the sequence *Stefan*, (**c**) motion field estimated between the previous images. For visualization purposes, only a motion vector for each 16 × 16 pixels area is represented. (**d**) estimation of frame #12 obtained as motion compensation of image #10 using the previous motion field, (**e**) motion-compensated error image at frame #12. For visualization purposes, an offset of 128 has been added to the error image pixels, which have been conveniently scaled afterwards

[4] As it will be discussed in Section 9.1.2, this information typically does not require quantization.

The motion-compensated error image is now handled as an original image in intra-mode (or in a still image coding system, see Chapter 8). That is, the information in the image is decorrelated (*direct transform*) typically using a DCT block transform, then the transform values are scalar quantized, (*quantization*), and finally, quantized coefficients are entropy encoded (*entropy encoder*) and transmitted.

As previously, the encoding system contains an embedded decoder that allows the transmitter to use the same decoded frames that will be used at the receiver side.[5] In this case, the reconstruction of the decoded image implies adding the quantized error to the motion-compensated image and this is the image that is stored in the *Frame Storage* to be used as reference for subsequent frames.

Figure 9.3 presents the block diagram of the decoder associated with the previous compression system. As it can be seen, previously decoded images are stored in the *Frame Storage*. They will be motion compensated using the motion information that is transmitted when coding future frames. Now the usefulness of recovering the decoded images in the encoder is even clearer. If it was not the case, the encoder and the decoder would use different information in the motion compensation process. That is, the encoder would estimate the motion information using original, data whereas the decoder will apply this motion information on decoded data, leading to different motion-compensated images and motion-compensated error images.

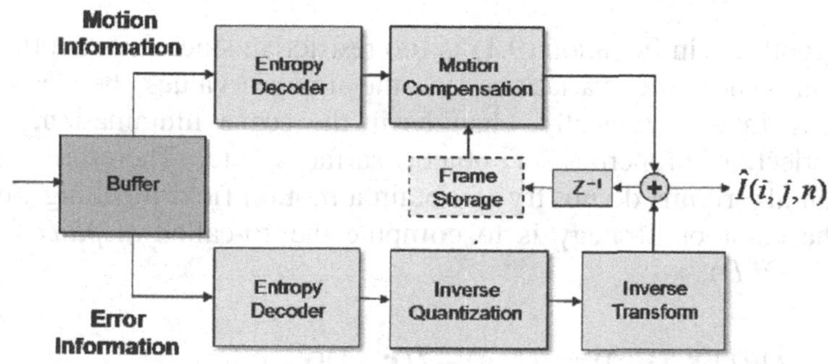

Fig. 9.3 Decoder system associated with the encoder of Fig. 9.1

[5] Such a system is commonly referred to as a *closed-loop* coder–decoder *(codec)*.

9.1.1 Motion estimation: The block matching algorithm

In the initial part of this Section, it has been made clear that a paramount step in video coding is motion estimation (and compensation). There exist a myriad of motion estimation algorithms (some of them are presented in Tekalp 1995, Fleet and Wiess 2006) that present different performance in terms of accuracy of the motion field, computational load, compactness of the motion representation, etc. Nevertheless, the so-called *Block Matching* algorithm has been shown to globally outperform all other approaches in the coding context.

Most motion estimation algorithms rely on the hypothesis that, between two consecutive (close enough) images, changes are only due to motion and therefore, every pixel in the current image $I(\mathbf{r}, n)$ has an associated pixel (a referent) in the reference image $I(\mathbf{r}, n-1)$

$$I(\mathbf{r},n) = I(\mathbf{r} - \mathbf{D}(\mathbf{r}), n-1) \tag{9.1}$$

where vector $\mathbf{r}$ represents the pixel location (x, y) and $\mathbf{D}(\mathbf{r})$ the motion field (also known as *optical flow*):

$$\mathbf{D}(\mathbf{r}) = [d_x(\mathbf{r}), d_y(\mathbf{r})] \tag{9.2}$$

The hypothesis in Equation (9.1) is too restrictive since factors other than motion influence the variations in the image values between even consecutive images, typically, changes in the scene illumination, camera noise, reflecting properties of object surfaces, etc. Therefore, motion estimation algorithms do not try to obtain a motion field fulfilling Equaion (9.1). The common strategy is to compute the so-called *displaced frame difference* (*DFD*)

$$DFD(\mathbf{r}, \mathbf{D}(\mathbf{r})) = I(\mathbf{r},n) - I(\mathbf{r} - \mathbf{D}(\mathbf{r}), n-1) \tag{9.3}$$

to define a given metric $M\{.\}$ over this image and to obtain the motion field that minimizes this metric:

$$\mathbf{D}^*(\mathbf{r}) = \arg\min_{\mathbf{D}(\mathbf{r})} M\{DFD(\mathbf{r}, \mathbf{D}(\mathbf{r}))\} \tag{9.4}$$

Note that the image that is subtracted from the current image in Equation (9.3) is obtained by applying the estimated motion vectors **D(r)** to the pixels of the reference image. This is what, in the context of video coding, we have called motion-compensated image:

$$\tilde{I}(\mathbf{r},n) = I(\mathbf{r}-\mathbf{D}(\mathbf{r}),n-1) \tag{9.5}$$

Consequently, the *DFD* is nothing but the motion-compensated error image and the minimization process expressed in Equation (9.4) looks for the minimization of a metric defined over an estimation error. Therefore, a typical choice for the selection of that metric is the energy of the error:[6]

$$\mathbf{D}^*(\mathbf{r}) = \arg\min_{\mathbf{D}(\mathbf{r})} \sum_{\mathbf{r}\in R} \left\| DFD(\mathbf{r},\mathbf{D}(\mathbf{r})) \right\|^2 \tag{9.6}$$

where R is, originally, the region of support of the image.

So far, we have not imposed any constraints on the motion vector field and all its components are independent. This pixel independency is neither natural, because neighbor pixels are likely to present similar motion, nor useful in the coding context, because it leads to a different displacement vector for every pixel, which results in a too large amount of information to be coded.

Parametric vector fields impose a given motion model to a specific set of pixels in the image. Common motion models are translational, affine, linear projective, or quadratic. If we assume that the whole image undergoes the same motion model, the whole motion vector field can be parameterized **D(r, p)**, where **p** is the vector containing the parameters of the model, and the *DFD* definition depends now on these parameters **p**:

$$DFD(\mathbf{r},\mathbf{p}) = I(\mathbf{r},n) - I(\mathbf{r}-\mathbf{D}(\mathbf{r},\mathbf{p}),n-1) \tag{9.7}$$

The minimization process of Equation (9.6) aims now at obtaining the optimum set of parameters:

$$\mathbf{D}^*(\mathbf{r},\mathbf{p}) = \arg\min_{\mathbf{p}} \sum_{\mathbf{r}\in R} \left\| DFD(\mathbf{r},\mathbf{p}) \right\|^2 \tag{9.8}$$

[6] Although this is a useful choice when theoretically deriving the algorithms, when actually implementing them, the square error (L2 norm) is commonly replaced by the absolute error (the L1 norm), given its lower computational load.

The perfect situation would be to have a parametric motion model assigned to every different object in the scene. This solution requires the segmentation of the scene into its motion homogeneous parts. However, motion-based image segmentation is a very complicated task (segmentation is often named an ill-posed problem). Moreover, if the image is segmented, the use of the partition for coding purposes would require transmitting the shapes of the arbitrary regions and this boundary information is extremely difficult to compress.

The adopted solution in video coding is to partition the image into a set of square blocks (usually referred to as *macroblocks*) and to estimate the motion separately within each one of these macroblocks. Therefore, a fixed partition is used, which is known beforehand by the receiver and does not require to be transmitted. Since this partition is independent of the image contents, data contained in each macroblock may not share the same motion. However, given the typical macroblock size (e.g., 16×16 pixels), motion information within a macroblock can be considered close to homogeneous. Furthermore, the imposed motion model is translational, i.e., all pixels in a macroblock are assumed to undergo the same motion which, for each macroblock, is represented by a single displacement vector.

As it has been previously said, the motion (displacement) of every macroblock is separately determined. Therefore, the global minimization problem is divided into a set of local minimization ones where, for the *ith* macroblock MB_i, the optimum parameters $\mathbf{p}_i^*=[d_x^*, d_y^*]_i$ are obtained:

$$\mathbf{p}_i^* = \left[d_x^*, d_y^*\right]_i = \arg\min_{\mathbf{p}} \sum_{\mathbf{r}\in MB_i} \left\|DFD(\mathbf{r},\mathbf{p})\right\|^2 \tag{9.9}$$

The common implementation of this minimization process is the so-called *Block Matching* algorithm. In the Block Matching, a direct exploration of the solution space is performed. In our case, the solution space is the space containing all possible macroblock displacements. This way, for each macroblock in the current image, a *search area* is defined in the reference image. The macroblock is placed at various positions within the search area in the reference image. Those positions are defined by the *search strategy* and the *quantization step* and every position corresponds to a possible displacement, i.e., a point in the solution space. At each position, the pixel values overlapped by the displaced macroblock are compared (matched) with the original macroblock pixel values. The way to perform this comparison is defined by the selected *metric*. The vector representing the displacement leading to the best match (lowest metric value) is the motion vector assigned to this macroblock. The final result of the Block

Matching algorithm is a set of displacements (motion vectors), each one associated with a different macroblock of the current image.

Therefore, several aspects have to be fixed in a concrete implementation of the Block Matching, namely[7]

- the *metric* that defines the best match, while the square error is the optimum metric in the case of assessing the results in terms of PSNR (see Chapter 8), it is common to implement the absolute error, given its lower computational complexity;
- the *search area* in the reference image, which is related to the maximum allowed displacement, i.e., to the maximum allowed speed of objects in the scene (see Section 9.2.2);
- the *quantization step* to represent the parameters of the motion model, in our case, the coordinates of the displacement, which is related to the accuracy of the motion representation;
- the *search strategy* in the parameter space, which defines which possible solutions (i.e., different displacements) are analyzed in order to select the motion vector.

The size of the search area is application dependent. For example, it is clear that the maximum displacement expected by objects in the scene is very different for a sport video than for an interview program. A possible simplification is to fix the maximum displacement in any direction close to the size of a macroblock side. This leads to space solutions covering zones of, for example, size $[-15,15] \times [-15,15]$.

The quantization step is related to the precision used to represent the motion parameters. Objects in the scene do not move in terms of pixels and therefore, to describe their motion by means of an integer number of pixels, is to reduce the accuracy of the motion representation. In current standards, techniques to estimate the motion with a quantization step of ½ and even ¼ of pixels are implemented.[8] The quantization step samples the space solution and defines a finite set of possible solutions. For instance, if we fix the quantization step to 1 pixel, the previous space solution of size $[-15,15] \times [-15,15]$ leads to $31 \times 31=961$ possible solutions i.e., 961 different displacement vectors that have to be tested to find the optimum one. Note that if we use a quantization step of 1/N of pixel, we are increasing the number of possible solutions by a factor of N^2.

---

[7] We could add here other aspects such as the macroblock shape and size but, as previously commented, we are assuming in the whole chapter that square macroblocks of size 16×16 pixels are used.

[8] Subpixel accuracy requires interpolation of the reference image. This kind of techniques, although very much used in current standards mainly to reach high-quality decoded images, is out of the scope of this chapter.

A search strategy is necessary to reduce the amount of computations required by the Block Matching algorithm. Following with the previous example, let us see the computational load of an exhaustive analysis of the 961 possible solutions.[9] For each possible displacement (solution), the pixels in the macroblock have to be compared with the pixels in the reference subimage overlapped by the macroblock at this position. Since we are assuming macroblocks of size 16×16, this leads to $(16 \times 16 \times 31 \times 31)$ 246.016 comparisons between pixels, and this is for a single macroblock! Suboptimal search strategies (Ghanbari 2003) have been proposed to reduce the amount of solutions to be analyzed per macroblock.[10] These techniques rely on the hypothesis that the metric (the error function) to be minimized is a (close to) U-convex function and, therefore, by analyzing a few solutions, the search algorithm can lead to the optimum one (see Fig. 9.4).

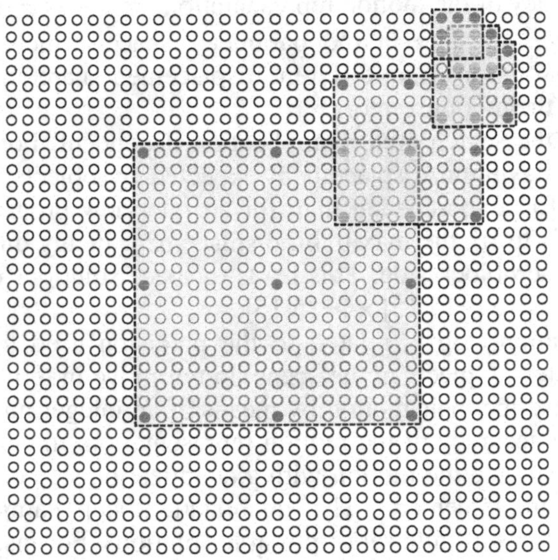

Fig. 9.4 Example of suboptimal search strategy: the so-called *nstep search* strategy. Nine possible solutions are analyzed corresponding, initially, to the center of the solution space and eight other solutions placed in the side centers and corners of a square of a given size (usually, half the solution space). The solution leading to a minimum is selected and the process is iterated by halving the size of the square and centering it in the current solution, until the final solution is found

[9] This is commonly referred to as *full-search* strategy.

[10] There exists another family of techniques that, rather than (or in addition to) reducing the amount of solutions to be analyzed, aims at reducing the amount of comparisons to be performed at each analyzed solution.

9.1.2 A few specificities of video coding standards

The complexity of current video standards is extremely high, incorporating a large variety of techniques for decorrelating the information and allowing the selection of one technique or another at, even, the subblock level (see, for instance, Sullivan and Wiegand 2005). Analyzing all these possibilities is out of the scope of this chapter and, instead, we are going to concentrate on a few additional concepts that (i) are basically shared by all video standards, (ii) complete the description of a simple video codec, and (iii) help understanding the potential of the motion estimation process. This way, we present three different types of frames that are defined and the concept of Group of Pictures (GOP), we discuss how the motion information is decorrelated and entropy coded, and, finally, we comment on the main variations that the specific nature of the motion-compensated error image introduces in its coding.

I-Pictures, P-Pictures, and B-Pictures

The use of motion information to perform temporal prediction improves the compression efficiency, as it will be shown in Section 9.2. However, coding standards are required not only to achieve high-compression efficiency but also to fulfill other functionalities. Among these additional features, a basic one is *random access*. Random access is defined as "the process of beginning to read and decode the coded bitstream at an arbitrary point" (MPEG 1993). Actually, random access requires that any picture can be decoded in a limited amount of time, which is an essential feature for video on a storage medium[11]. This requirement implies having a set of access points in the bitstream, associated with specific images, which are easily identifiable and can be decoded without reference to previous segments of the bitstream. The way to implement these requirements is by breaking the temporal prediction chain and introducing some images encoded in intra-mode, the so-called intra-pictures or *I-Pictures*. The spacing of two I-Pictures per second depends on the application. Applications requiring random access may demand short spacing, which can be achieved without significant loss of compression rate. Other applications may even use I-Pictures to improve the compression where motion compensation reveals ineffective, for instance, in a scene cut. Nevertheless, since I-Pictures will be used as reference for future frames in the sequence, they are coded with good quality, i.e., with moderate compression.

---

[11] Think, for example, about the functionalities of fast-forward and fast-reverse playback that we commonly use in a DVD player.

By relying on I-Pictures, new images can be temporally predicted and encoded. This is the case of predictive coded pictures (*P-Pictures*), which are coded more efficiently using as reference image a previous I-Picture or P-Picture, commonly, the nearest one. Note that P-Pictures are used as reference for further prediction and, therefore, their quality has to be high enough for that purpose.

Finally, a third type of images is defined in usual video coding systems, which are the so-called bidirectionally predictive coded pictures (*B-Pictures*). B-Pictures are estimated by combining the information of two reference images, one preceding it and the other following it in the video sequence (see Fig. 9.5). Therefore, B-Pictures use both past and future reference pictures for motion compensation.[12] The combination of past and future information has several implications.

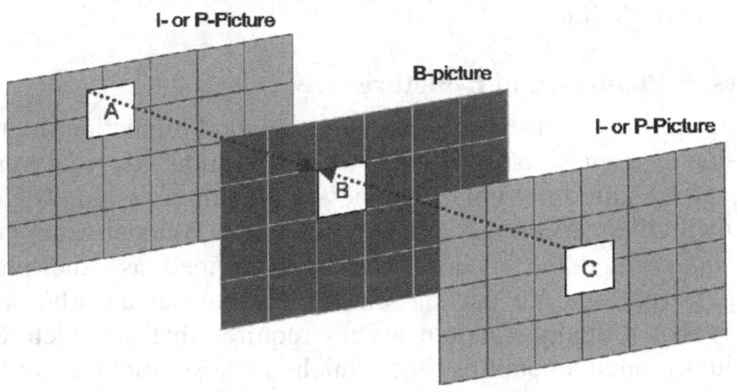

Fig. 9.5 Illustration of the motion prediction in B-Pictures

First, in order to allow using future frames as references, the image transmission order has to be different from the displaying order. This concept will be further discussed in the sequel. Second, B-Pictures may use past, future, or combinations of both images in their prediction. In the case of combining both references, the selected subimages in the past and future reference images are linearly combined to produce the motion-compensated estimation.[13] This selection of references leads to an increase in motion compensation efficiency. In the case of combining past and future references, the linear combination may imply a noise reduction. Moreover, in the case of using only a future reference, objects appearing in the scene

[12] Commonly, estimation from past references is named *forward prediction,* whereas estimation from future references is named *backward prediction*.

[13] Linear weights are inversely proportional to the distance between the B-Picture and the reference image.

can be better compensated using future information (see Section 9.2.6 for a further discussion on that topic). In spite of the double set of motion information parameters that they require, B-Pictures provide the highest degree of compression.

Group of Pictures (GOP)

The organization of the three types of pictures in a sequence is very flexible and the choice is left to the encoder. Nevertheless, pictures are structured in the so-called *group of pictures* (*GOP*), which is one of the basic units in the coding syntax of video coding systems. GOPs are intended to assist random access. The information on the length and organization of a GOP is stored in its header, allowing, therefore, easily identifiable access points in the bitstream. Let us use a simplified version of a GOP in order to illustrate its usage and usefulness.

First, let us clarify that given their usually lower quality B-Pictures are not (actually, seldom) used as reference for prediction. That is, the reference images for constructing a P-Picture or a B-Picture are either I-Pictures or P-Pictures. Now, let us fix the following GOP structure:

I(1) B(2) B(3) P(4) B(5) B(6) I(7) B(8) B(9) P(10) B(11) B(12) I(13) …

Frame I(1) is coded in intra-mode and therefore does not need any reference and can be directly transmitted. However, frames B(2) and B(3) are B-Pictures and they need a past and a future reference to be encoded and transmitted before they can be processed. Therefore, prior to encoding B(2) and B(3), frame P(4) is encoded as a P-Image, having the previous I(1) as reference. Once P(4) has been transmitted, B(2) and B(3) can be encoded as B-Images and transmitted. A similar situation happens now for B(5) and B(6), since they also require as well a future reference. In that case, the GOP structure imposes I(7) to be an I-Picture and this is first encoded and transmitted.[14] After that, B(5) and B(6) can be encoded as B-Pictures, having P(4) and I(7) as references.

In order to allow bidirectional prediction, a reordering of the images in transmission has to be imposed. This way, the previous GOP structure forces the following ordering in transmission:

I(1) P(4) B(2) B(3) I(7) B(5) B(6) P(10) B(8) B(9) I(13) B(11) B(12) …

---

[14] Note that the GOP structure could have imposed P(7). In that case, the only change would have been that P(7) should be predicted from P(4).

These concepts are further illustrated in Fig. 9.6, where images are associated with their order in the transmission process. For the sake of clarity, only arrows showing the temporal dependency among a few images are shown.

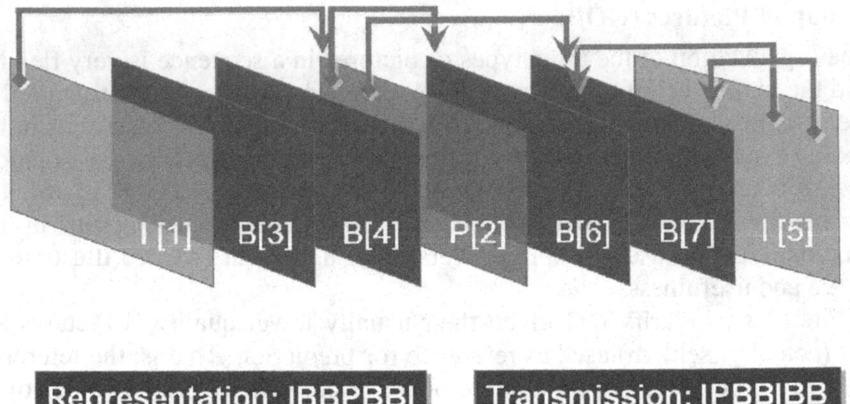

Fig. 9.6 Example of GOP structure. Frames are presented in display order but numbers associated with each frame represent its order of transmission

A final remark has to be made regarding the reordering imposed by B-Pictures. Note that this technique demands much more processing and storing at the receiver side. First, more complicated motion estimation strategies are defined in which combinations of past and future estimations are performed. Second, the receiver can no longer decode, display, and remove from memory a given image but to allow decoding B-Pictures, it has to keep decoded I-Pictures and B-Pictures into memory before displaying them.

Coding of motion vectors

As has been said (and will be further illustrated in Section 9.2), motion information is very relevant for the performance of the video coding system. Due to that, no quantization of the motion vectors is performed so that no losses are introduced in this information.

As in the case of the DC coefficients in neighbor blocks in the JPEG standard (see Chapter 8), motion vectors in neighbor macroblocks are usually quite correlated. Therefore, a differential coding is performed in which a prediction based on previously decoded macroblocks is used. Depending on the standard and on the type of image, the way to perform this prediction varies.

DCT coefficient quantization of the motion-compensated error image

When presenting the procedure for quantizing the DCT coefficients in still image coding (see Chapter 8), it was commented that a specific quantization table was used to take into account the different relevance of each coefficient for the human visual system. In that case, the study of the human visual system led to the use of the so-called Lohscheller tables, which roughly speaking, impose a stronger quantization step to higher frequency coefficients.

In the case of coding the motion-compensated error image, this strategy is no longer useful. Note that the presence of high-frequency components in the error image may not be related to high-frequency information in the original image but, very likely, to poor motion compensation. Therefore, a flat default matrix is commonly used (all matrix components set to 16).

9.2 MATLAB proof of concept

In the following section, we are going to illustrate the main concepts behind compression of image sequences. Image sequences present temporal as well as spatial redundancy. In order to correctly exploit the temporal redundancy between temporally neighbor images, the motion present in the scene has to be estimated. Once the motion is estimated, the information in the image used as reference is motion compensated to produce a first estimation of the image to be coded. Motion estimation is performed by dividing the image into nonoverlapping square blocks and estimating the motion of each block independently.

9.2.1 Macroblock processing

Typically, for motion estimation, images are partitioned into blocks of 16×16 pixels,[15] which will be referred to as *macroblocks* (see Fig. 9.7). The macroblock partition is hierarchical with respect to the block partition (see Chapter 8) and, therefore, every macroblock contains four blocks. As in the block case, images are padded to allow an exact partition in terms of macroblocks.

[15] Current standards allow finer partitions using both smaller square blocks (e.g., 8×8) and smaller rectangular blocks (e.g., 16×8, 8×16)

```
for i=1:3,
    table{i} =
imread(sprintf('seqs/table/gray/table_%03d_g.bmp',i-1));
    imshow(table{i});
    addgridtofigure(size(table{i}),[16 16]);
end;
```

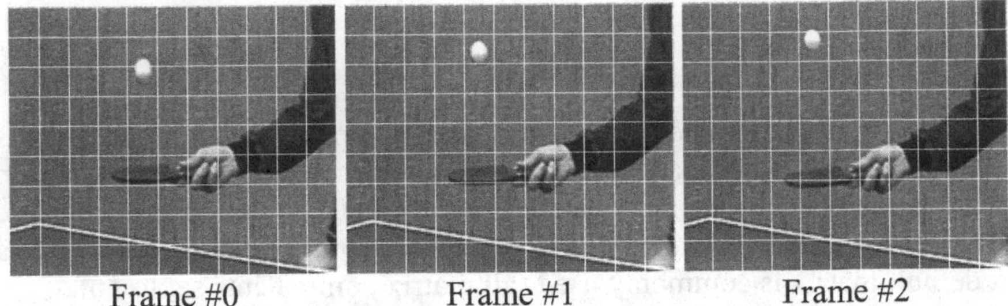

Fig. 9.7 Original frames of the *Table Tennis* sequence and macroblock partitions

In Section 9.1, we have discussed that usual coding systems work in closed loop, i.e., the encoder uses decoded data (instead of the original one) to perform the prediction steps. This ensures that the transmitter and the receiver work in the same conditions. However, for the sake of simplicity, in the following sections, we are going to use original data in the prediction steps. Results in closed loop will be presented in Section 9.2.9 when the complete coding system will be analyzed.

9.2.2 Block matching motion estimation

Now, we will analyze how our basic motion compensation unit (i.e., a generic 16×16 pixel macroblock) is processed. Among all the different techniques for estimating the motion associated with a macroblock with respect to a given reference image, the Block Matching algorithm has been shown to present very good coding properties.

The *Block Matching* algorithm (BM) works independently for each macroblock of the current image and for each macroblock, it looks for the best representation in the reference image, assuming that the whole macroblock has undergone only a translational movement (see Section 9.1.1). Therefore, the selected macroblock is placed at various positions in the reference image (those positions defined by the search area and the search strategy) and the pixel values overlapped by the displaced macroblock are compared (matched) with the original macroblock pixel values. The vector representing the displacement leading to the best match is the motion vector assigned to the selected macroblock. The final result of the BM algorithm is a set of displacements (motion vectors), each one associated with a different macroblock of the current image.

As commented in Section 9.1.1, several aspects have to be fixed in a concrete implementation of the BM, namely (i) the metric that defines the best match; (ii) the search area in the reference image, which is related to the maximum allowed speed of objects in the scene; (iii) the quantization step to represent the coordinates of the motion vector, which is related to the accuracy of the motion representation; and (iv) the search strategy in the parameter space, which defines how many possible solutions (i.e., different displacements) are analyzed in order to select the motion vector.

The MATLAB function `estimatemotion` performs the BM of a given image with respect to a reference image. In it, the previous aspects are implemented as follows: (i) the selected *metric* is the absolute difference; (ii) the *search area* can be set by the fourth parameter (actually, this parameter does not directly set the search area but the maximum displacement allowed to the motion vector, so a value of [16,12] indicates that the motion vector can have values from (–16, –12) to (16,12)); (iii) the *quantization step* has been set to 1 pixel; and (iv) as *search strategy* both a full-search and an nstep-search (see Fig. 9.4) within the search area are implemented.

```
t = 2;
[mvv mvh] = estimatemotion(table{t},table{t-1},[16 16],...
    [15 15],'fullsearch');
```

Figure 9.8 presents the result of computing the BM between frames #1 and #0 of the *Table Tennis* sequence. BM motion vectors are plotted in white over each of the macroblocks of frame #1 (motion vector field). Since an exhaustive search is performed, the estimated motion vectors are the optimum ones under the selected metric, search area, and quantization step.

```
plotmv(table{t},mvv,mvh,[16 16]);
```

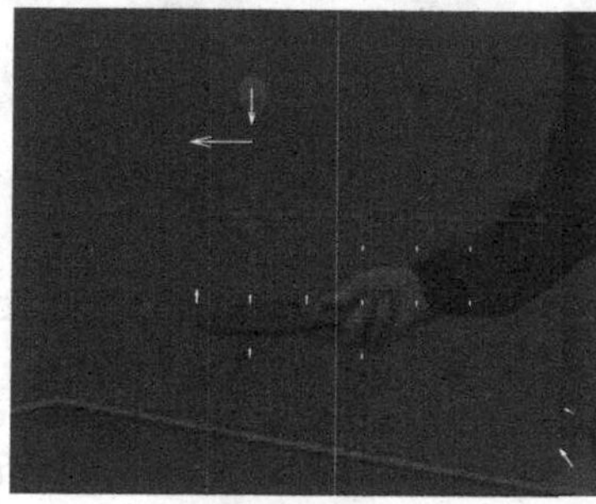

Fig. 9.8 Motion vectors for block matching estimation between frame #1 and frame #0 of the *Table Tennis* sequence

In order to illustrate how these vectors are computed, let us analyze the specific case of two different macroblocks. In Fig. 9.9, we show the information associated with the first of these macroblocks, which is placed five blocks from the left and two from the top.

```
posr = 2; posc = 5;
f = table{t}*0.2;
bs = 16;
f(bs*(posr-1)+1:bs*(posr-1)+bs,bs*(posc-1)+1:bs*(posc-1) ...
  +bs) = table{t}(bs*(posr-1)+1:bs*(posr-1)+bs,bs*(posc-1) ...
  +1:bs*(posc-1)+bs);
imshow(f);
```

Fig. 9.9 Macroblock at position [2,5] (highlighted)

Figure 9.10 shows a magnification of the selected macroblock.

```
b = table{t}(bs*(posr-1)+1:bs*(posr-1)+bs, ...
   bs*(posc-1)+1:bs*(posc-1)+bs);
```

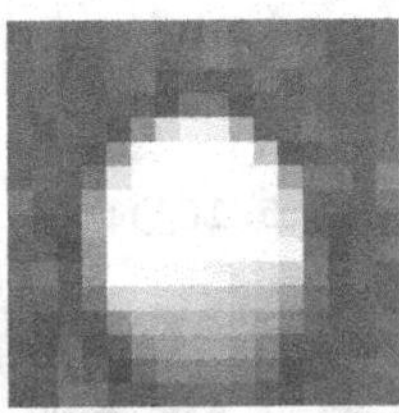

Fig. 9.10 Magnification of the macroblock at position [2,5]

Figure 9.11 shows the search area in the reference image (frame #0). The search area is centered at the same position of the macroblock under analysis and, in this case, the search area has been set to 46×46 pixels (parameter `sa` set to [15,15]). The macroblock from the current image (frame #1) is compared to all possible 16×16 subimages within the search area in the reference image (frame #0) (i.e., 31×31=961 possible positions for each macroblock, see Section 9.1.1).

```
sa = [15 15];
f = table{t-1}*0.2;
f(bs*(posr-1)+1-sa(1):bs*(posr-1)+bs+sa(1),...
   bs*(posc-1)+1-sa(2):bs*(posc-1)+bs+sa(2)) = ...
   table{t}(bs*(posr-1)+1-sa(1):bs*(posr-1)+bs+sa(1),...
   bs*(posc-1)+1-sa(2):bs*(posc-1)+bs+sa(2));
imshow(f);
```

Fig. 9.11 Search area at frame #0 for the macroblock at position [2,5]

Figure 9.12 shows a magnification of the search area.

```
br = table{t}(bs*(posr-1)+1-sa(1):bs*(posr-1)+ ...
   bs+sa(1),bs*(posc-1)+1-sa(1):bs*(posc-1)+bs+sa(2));
figure,imshow(kron(br,uint8(ones(8))));
```

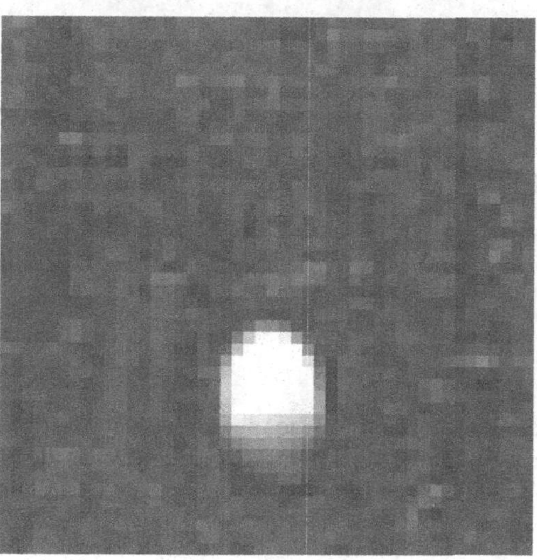

Fig. 9.12 Magnified search area in image #0 for the macroblock at position [2,5]

Figure 9.13 shows the error prediction surface, i.e., the function containing the metric values of the comparison between the macroblock under analysis and all possible 16×16 subimages within the search area. Darker values of the error function (lower points in the 3D surface) correspond to lower error values between the macroblock and all possible subimages and, thus, better candidates to be the reference for the current macroblock.

```
% function estimate_macroblock does the same as estimatemotion
% but for only 1 macroblock
[bcomp, bvv, bvh, errorf] = estimate_macroblock(b, ...
   table{t-1}, [posr posc], [16 16], ...
   [15 15],'fullsearch');
figure,mesh(-15:15,-15:15,errorf);
colormap('default'); colorbar; view(-20,22);
hold on; pcolor(-15:15,-15:15,errorf);
axis([-15 15 -15 15]);
```

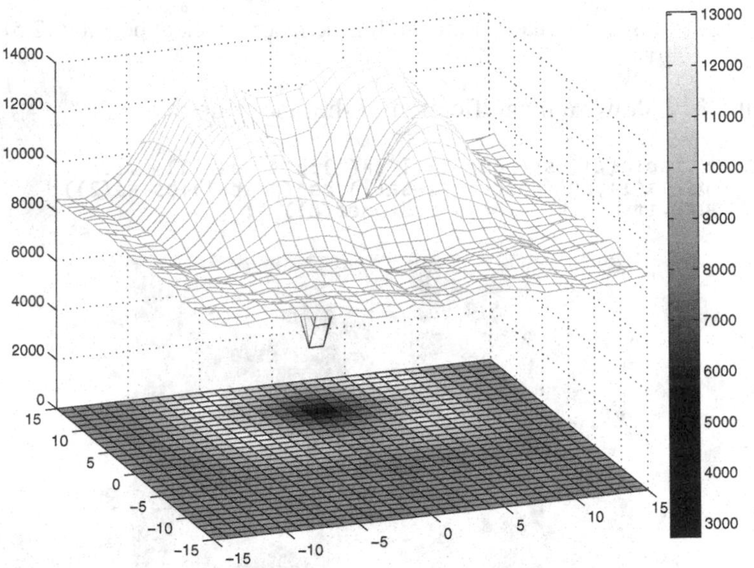

Fig. 9.13 Error prediction function for the macroblock at position [2,5]. The surface represents the error function as a 3D curve. For clarity, at the bottom (on plane z=0), the same error function is represented as a 2D image

In this case, even if the error function presents several local minima, there is a global minimum (which corresponds to the movement of the ball between frames #0 and #1) at position [0,8] (motion vector). That is, the best possible reference subimage (within the search area) for this macroblock is situated 8 pixels below and 0 pixels to the right of the original macroblock position. As the selected search strategy (full-search) performs an exhaustive search through the entire search area, it is guaranteed that the global minimum is always found. Other search strategies such as the nstep-search (See Section 9.1.1), which visits less positions in the search area, may be able to obtain similar (or even the same) results as the exhaustive search. For instance, for the case of the macroblock at position [2,5], an nstep-search also finds the global minimum, as shown in Fig. 9.14.

```
[bcomp2, bvv2, bvh2, errorf2] = estimate_macroblock(b, ...
    table{t-1}, [posr posc], [16 16], [15 15],'nstep');
figure,surf(-15:15,-15:15,errorf2);
colormap('default'); colorbar;
view(-20,22);
hold on; pcolor(-15:15,-15:15,errorf2);
axis([-15 15 -15 15]);
```

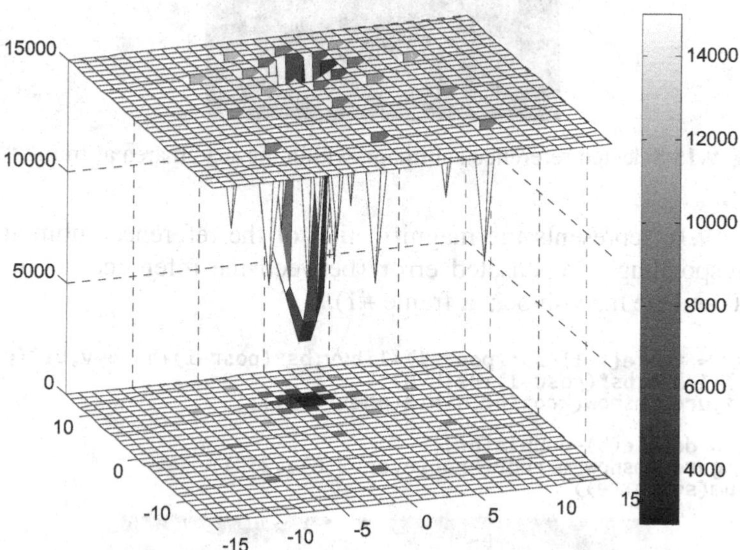

Fig. 9.14 Error prediction function for the selected macroblock (at position [2,5]) using an nstep-search strategy. The surface represents the error function as a 3D curve where the highest values (in that case 15,000) are unvisited positions in the search. At the bottom (on plane z=0), the error function is represented as a 2D image (white pixels correspond to unvisited positions due to the search strategy)

In this example, both search strategies implemented in the MATLAB function `estimatemotion` (full-search and nstep-search) lead to the same result, a motion vector of [0,8]. Figure 9.15 shows the selected reference subimage applying a motion vector of [0,8] within the search range.

```
f = table{t-1}*0.2;
f(bs*(posr-1)+1-sa(1):bs*(posr-1)+bs+sa(1),bs*(posc-1)+1-
sa(2):bs*(posc-1)+bs+sa(2)) = ...
    0.45*table{t-1}(bs*(posr-1)+1-sa(1):bs*(posr-
1)+bs+sa(1),bs*(posc-1)+1-sa(2):bs*(posc-1)+bs+sa(2));
f(bs*(posr-1)+1+bvv:bs*(posr-1)+bs+bvv,bs*(posc-
1)+1+bvh:bs*(posc-1)+bs+bvh) = ...
    table{t-1}(bs*(posr-1)+1+bvv:bs*(posr-1)+bs+bvv,bs*(posc-
1)+1+bvh:bs*(posc-1)+bs+bvh);
```

Fig. 9.15 Selected reference subimage within the search area at frame #0

Figure 9.16 represents the magnification of the reference subimage and the corresponding compensated error (between the reference subimage at frame #0 and the macroblock at frame #1).

```
br = table{t-1}(bs*(posr-1)+1+bvv:bs*(posr-1)+bs+bvv,bs*(posc-
1)+1+bvh:bs*(posc-1)+bs+bvh);
figure,imshow(kron(br,uint8(ones(8))));

e = double(b)-double(br);
figure,imshow_merror(kron(e,ones(8)),250);
sum(sum(e.*e)),
```

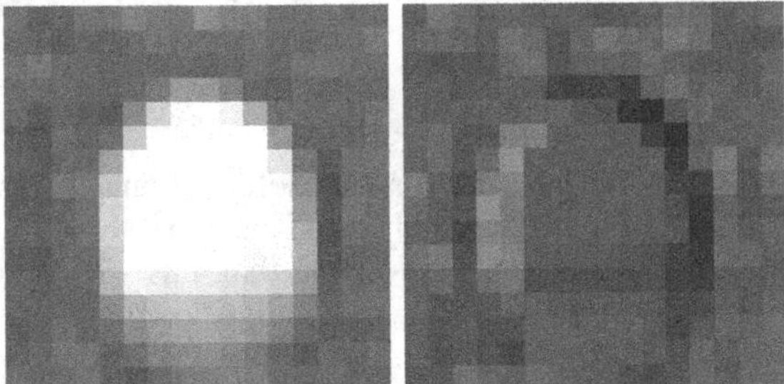

Reference subimage

Compensated error
Energy = 65.728

Fig. 9.16 Magnification of the selected reference subimage at frame #0 and the corresponding compensated error

Through this chapter, in order to present error images, an offset of 128 is added to all pixel values and, if necessary, they are clipped to the [0,255] range. This way, zero error is represented by a 128 value.

A second example is analyzed to further illustrate the algorithm. Figure 9.17 and Fig. 9.18 show the information associated with a macroblock from frame #1 placed at a different position. In this case, the position of the macroblock under analysis is six macroblocks from the top and six from the left.

```
posr = 6; posc = 6;
f = table{t}*0.2;
bs = 16;
f(bs*(posr-1)+1:bs*(posr-1)+bs,bs*(posc-1)+1:bs*(posc-1)+bs) =
    table{t}(bs*(posr-1)+1:bs*(posr-1)+bs, ...
    bs*(posc-1)+1:bs*(posc-1)+bs);
```

Fig. 9.17 Macroblock at position [3,5] (highlighted)

Figure 9.18 shows a magnification of the macroblock under analysis.

```
b = table{t}(bs*(posr-1)+1:bs*(posr-1)+bs, ...
    bs*(posc-1)+1:bs*(posc-1)+bs);
imshow(kron(b,uint8(ones(8))));
```

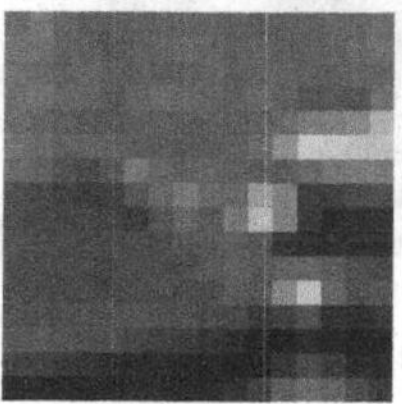

Fig. 9.18 Magnification of the macroblock at position [3,5]

The search area at frame #0 corresponding to the macroblock under analysis can be seen in Figs. 9.19 and 9.20.

```
sa = [15 15];
f = table{t-1}*0.2;
f(bs*(posr-1)+1-sa(1):bs*(posr-1)+bs+sa(1),...
    bs*(posc-1)+1-sa(2):bs*(posc-1)+bs+sa(2)) = ...
    table{t-1}(bs*(posr-1)+1-sa(1):bs*(posr-1)+bs+sa(1), ...
    bs*(posc-1)+1-sa(2):bs*(posc-1)+bs+sa(2));
```

Fig. 9.19 Search area at frame #0 for the macroblock at position [3,5]

```
br = table{t-1}(bs*(posr-1)+1-sa(1):bs*(posr-
1)+bs+sa(1),bs*(posc-1)+1-sa(1):bs*(posc-1)+bs+sa(2));
figure,imshow(kron(br,uint8(ones(8))));
```

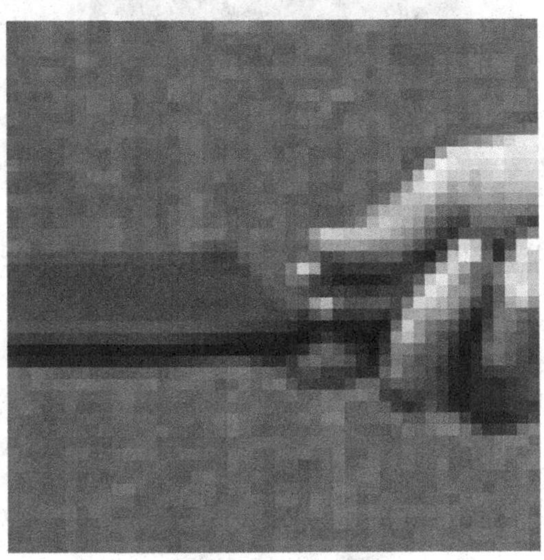

Fig. 9.20 Magnification of the search area for the macroblock at position [3,5]

As in the previous example, we show the error surface (see Figs. 9.21 and 9.22) corresponding to both full-search and nstep-search strategies.

```
[bcomp, bvv, bvh, errorf] = estimate_macroblock(b, ...
   table{t-1}, [posr posc], [16 16], [15 15],...
   'fullsearch');bvv,bvh,
figure,mesh(-15:15,-15:15,errorf);
colormap('default'); colorbar; view(-20,22);
hold on; pcolor(-15:15,-15:15,errorf);
axis([-15 15 -15 15]);
```

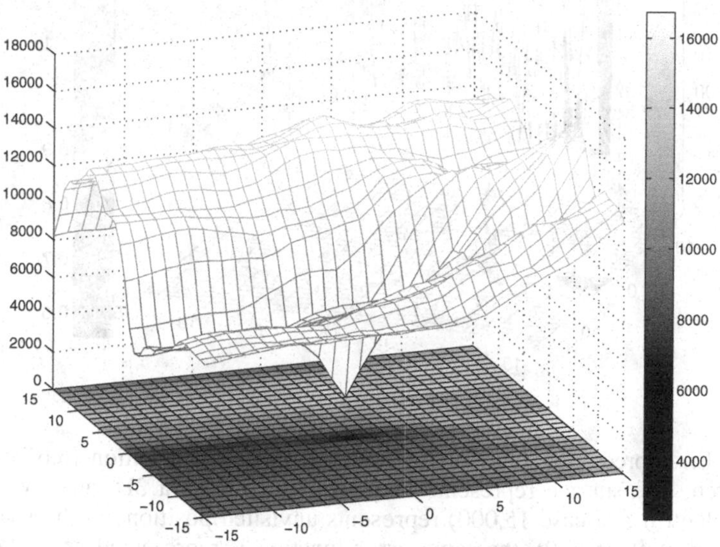

Fig. 9.21 Error prediction function for the macroblock at position [6,6]. The surface represents the error function as a 3D curve. For clarity, at the bottom (on plane z=0), the same error function is represented as a 2D image

```
[bcomp2, bvv2, bvh2, errorf2] = estimate_macroblock(b, ...
   table{t-1}, [posr posc], [16 16], [15 15],'nstep');
figure,surf(-15:15,-15:15,errorf2);
colormap('default'); colorbar;
view(-20,22);
hold on; pcolor(-15:15,-15:15,errorf2);
axis([-15 15 -15 15]);
```

In this example, the full-search obtains a motion vector of [0,−2], while the nstep-search obtains a motion vector of [−14,−1]. Note that, in this case, the nstep-search strategy does not lead to the optimum result. In particular, for the nstep-search, the initial step in the search is 8 pixels. The nine solutions that are evaluated overlooked the global minimum and following steps of the algorithm get trapped in a distant local minimum.

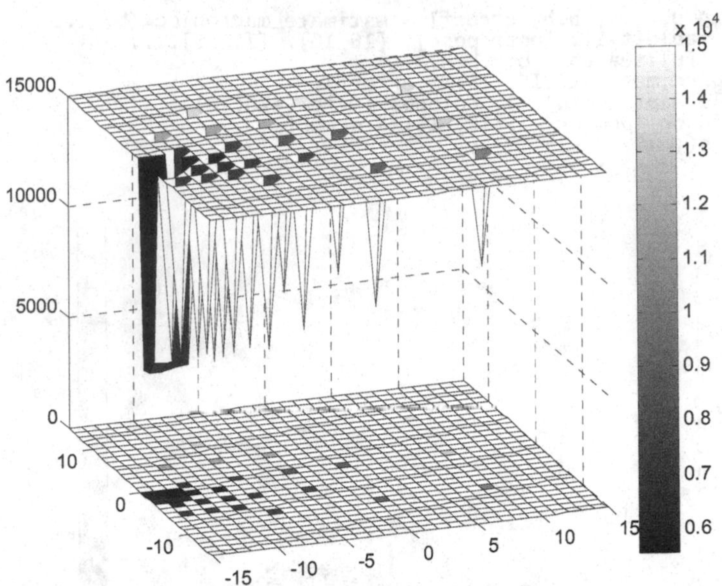

Fig. 9.22 Error prediction function for the macroblock at position [6,6] using an nstep-search. The surface represents the error function as a 3D curve where the highest value (in that case 15,000) represents unvisited positions in the search. At the bottom (on plane z=0), the same error function is represented as a 2D image (white pixels correspond to positions not visited in the search)

Figure 9.23 shows the selected reference subimage leading to a motion vector of components [0,−2] (computed with the full-search approach).

```
f = table{t-1}*0.2;
f(bs*(posr-1)+1-sa(1):bs*(posr-1)+bs+sa(1),...
   bs*(posc-1)+1-sa(2):bs*(posc-1)+bs+sa(2)) = 0.45* ...
    table{t-1}(bs*(posr-1)+1-sa(1):bs*(posr-1)+bs+sa(1),...
   bs*(posc-1)+1-sa(2):bs*(posc-1)+bs+sa(2));
f(bs*(posr-1)+1+bvv:bs*(posr-1)+bs+bvv,...
   bs*(posc-1)+1+bvh:bs*(posc-1)+bs+bvh) = ...
   table{t-1}(bs*(posr-1)+1+bvv:bs*(posr-1)+bs+bvv,...
   bs*(posc-1)+1+bvh:bs*(posc-1)+bs+bvh);
```

Figure 9.24 represents the magnification of the reference subimage and the corresponding compensated error.

```
br = table{t-1}(bs*(posr-1)+1+bvv:bs*(posr-1)+bs+bvv,...
   bs*(posc-1)+1+bvh:bs*(posc-1)+bs+bvh);
figure,imshow(kron(br,uint8(ones(8))));
e = double(b)-double(br);
figure,imshow_merror(kron(e,ones(8)),70);
sum(sum(e.*e)),
```

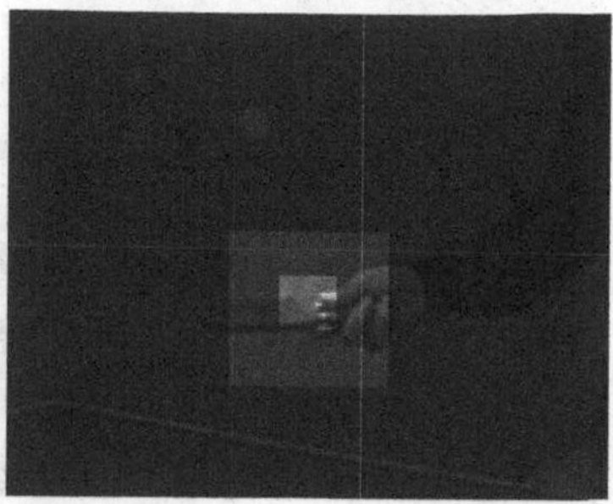

Fig. 9.23 Selected reference subimage (leading to a motion vector of components [0,−2]) within the search range at frame #0

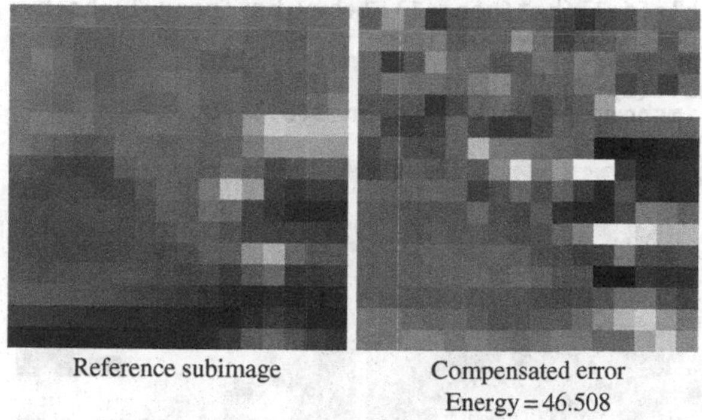

Fig. 9.24 Magnification of the selected reference subimage (with a motion vector [0,−2]) and the corresponding compensated error

In the case of the suboptimal strategy (nstep-search), Fig. 9.25 shows the selected reference subimage (leading to a motion vector of [−14, −1]) and Fig. 9.26 presents the magnification of the selected reference image and of the corresponding compensated error. As the selected reference subimage is not the optimum one (within the search range), the final compensated error energy is greater in the case of nstep-search.

```
f = table{t-1}*0.2;
f(bs*(posr-1)+1-sa(1):bs*(posr-1)+bs+sa(1),...
    bs*(posc-1)+1-sa(2):bs*(posc-1)+bs+sa(2)) = 0.45* ...
      table{t-1}(bs*(posr-1)+1-sa(1):bs*(posr-1)+bs+sa(1),...
    bs*(posc-1)+1-sa(2):bs*(posc-1)+bs+sa(2));
f(bs*(posr-1)+1+bvv2:bs*(posr-1)+bs+bvv2,...
    bs*(posc-1)+1+bvh2:bs*(posc-1)+bs+bvh2) = ...
    table{t-1}(bs*(posr-1)+1+bvv2:bs*(posr-1)+bs+bvv2,...
    bs*(posc-1)+1+bvh2:bs*(posc-1)+bs+bvh2);
```

Fig. 9.25 Selected reference subimage (leading to a motion vector of components [–14,–1]) within the search range at frame #0

```
br = table{t-1}(bs*(posr-1)+1+bvv:bs*(posr-1)+bs+bvv,...
   bs*(posc-1)+1+bvh:bs*(posc-1)+bs+bvh);
figure,imshow(kron(br,uint8(ones(8))));
e = double(b)-double(br);
figure,imshow_merror(kron(e,ones(8)),70);
sum(sum(e.*e)),
```

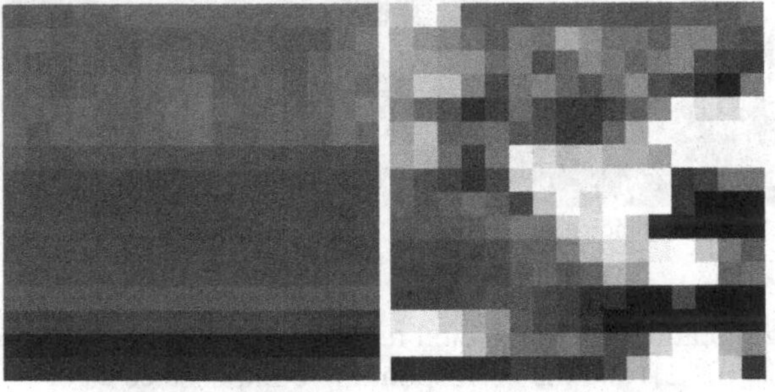

Reference subimage　　Compensated error Energy = 252.708

Fig. 9.26 Magnification of the selected reference subimage (with a motion vector [–14,–1]) and the corresponding compensated error

The use of suboptimal search strategies can lead to the selection of reference subimages, which are not the best representation of the macroblock under study. This translates into an increase in the prediction error and, therefore, a decrease in the coding performance.

Note that even in the case of using an exhaustive search, the resulting motion vectors may not represent the real motion in the scene. Vectors are computed for a fixed partition and obtained by minimization of a given metric in a specific neighborhood. These constraints impose a scenario that may not be coincident with the real one. This problem is common in areas with very similar texture that may lead to different solutions with close values. Another situation that produces such kind of results is that of having two objects (or portions of objects) in the same macroblock undergoing different motions.

9.2.3 Motion compensation

The estimation of the original frame #1 is obtained by motion compensating the reference image (in this example, the frame #0) using the motion vectors from the BM algorithm, which created the so-called motion-compensated image (see Fig. 9.27). In our implementation, the MATLAB function `applymotion` is responsible for compensating the reference image with the motion vectors computed using the previous function `estimatemotion`.

```
t = 2;
[mvv mvh] = estimatemotion(table{t},table{t-1},[16 16],...
    [15 15],'fullsearch');
tablec{t} = applymotion(table{t-1},mvv,mvh,[16 16]);
imshow(tablec{t});
```

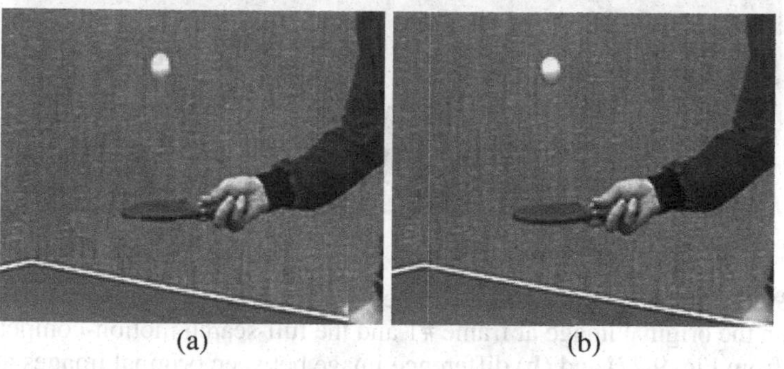

Fig. 9.27 (**a**) Motion-compensated image using frame #0 as reference and full-search motion vectors (see Fig. 9.8) and (**b**) original image at frame #1

Figure 9.28 a shows the error image between the original and the motion-compensated image (*compensation error*).

```
error{t} = double(tablec{t})-double(table{t});
imshow_merror(error{t},250);
```

As can be seen, the errors in the compensated image are located in the moving objects of the scene. However, as the computed motion vectors compensate somehow this motion (even if it is not translational), the final compensation error is relatively small.

```
Ee{t} = sum(sum(error{t}.*error{t}));
Ee{t},
```

```
ans = 932406
```

To have an idea of the improvement obtained by using motion compensation with BM, the direct difference between images #1 and #0 is presented in Fig. 9.28b. This result can be interpreted as motion-compensating image #0 with a motion vector field in which all motion vectors are `[0,0]`. It can be observed that the compensation error without motion compensation is larger than in the previous case, and the energy of the error would increase around 420%.

```
error_nocompen{t} = double(table{t})-double(table{t-1});
imshow_merror(error_nocompen{t},250);
Eenc{t} = sum(sum(error_nocompen{t}.*error_nocompen{t}));
Eenc{t},

Eenc = 4836677
```

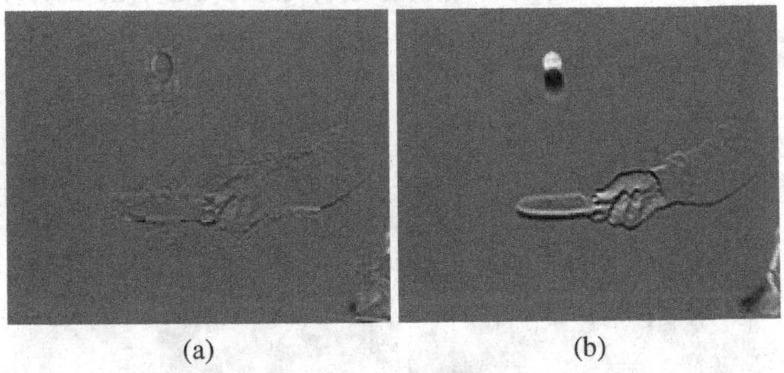

(a) (b)

Fig. 9.28 Comparison of motion-compensated images: (**a**) compensated error between the original image at frame #1 and the full-search motion-compensated image (from Fig. 9.27) and (**b**) difference image between original images at frame #1 and frame #0 (without motion compensation

Finally, Fig. 9.29 presents the motion-compensated image (and the original image), which is obtained by estimating the motion vectors using the nstep-search, sub-optimal strategy. In addition, Fig. 9.30 shows the corresponding compensation error. In that case, it can be seen that the motion-compensated image is not able to estimate the original image as well as with the full-search and therefore, the energy of the error is greater than in the full-search case. Also, some of the errors in the nstep-search strategy can be easily seen in the compensated image. For instance, the macroblock in position [6,6] (analyzed in Section 9.2.2) does not correctly reconstruct the racket or the player's hand due to the nstep-search strategy failing to obtain the optimum reference subimage in the search area.

```
t=2;
[mvv2 mvh2] = estimatemotion(table{t},table{t-1},[16 16],...
    [15 15],'nstep');
tablec2{t} = applymotion(table{t-1},mvv2,mvh2,[16 16]);
```

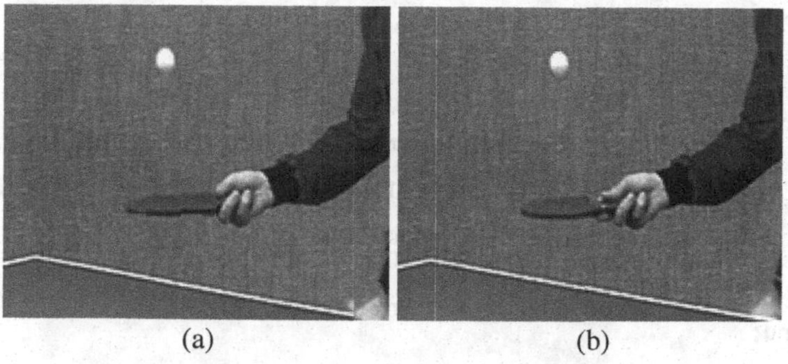

(a) (b)

Fig. 9.29 (**a**) Motion-compensated image using frame #0 as reference and nstep-search motion vectors and (**b**) original image at frame #1

```
error2{t} = double(table{t})-double(tablec2{t});
figure,imshow_merror(error2{t},250);

Ee2{t} = sum(sum(error2{t}.*error2{t}));
Ee2{t},

ans = 1212642
```

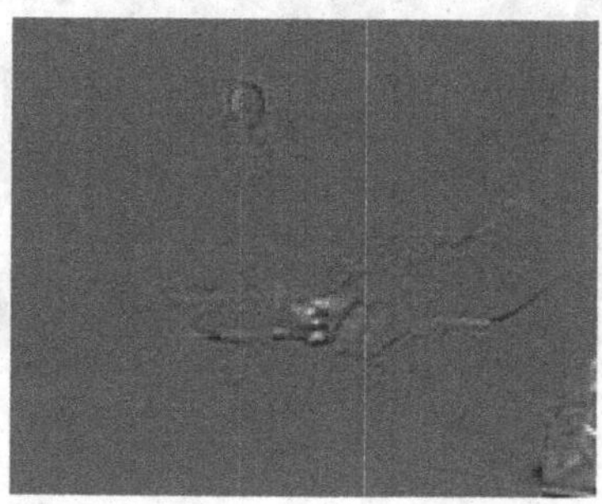

Fig. 9.30 Compensated error between the original image at frame #1 and the nstep-search, motion-compensated image

9.2.4 Selection of search area

Let us now analyze the effect of varying the size of the search area in the BM process. As we have previously commented, the search area is related to the maximum speed that we expect from the objects in the scene. In the case of the *Table Tennis* sequence, the fastest object is the ball. We use frames #1 and #0 of the same sequence. Figure 9.31 presents, on the top row, the motion-compensated images using a search area with values 24×24, 32×32, and 46×46. On the second row, the corresponding compensation errors are presented.

```
t = 2;
n = 1;
for i=[4 8 15],
    [vv vh] = estimatemotion(table{t},table{t-1},[16 16],...
        [i i]);
    mc = applymotion(table{t-1},vv,vh,[16 16]);
    imshow(mc);
    e = double(table{t}) - double(mc);
    figure,imshow_merror(e,250);
    ee = sum(sum(e.^2));
    disp(['Error energy = ' int2str(ee)]);
end;
```

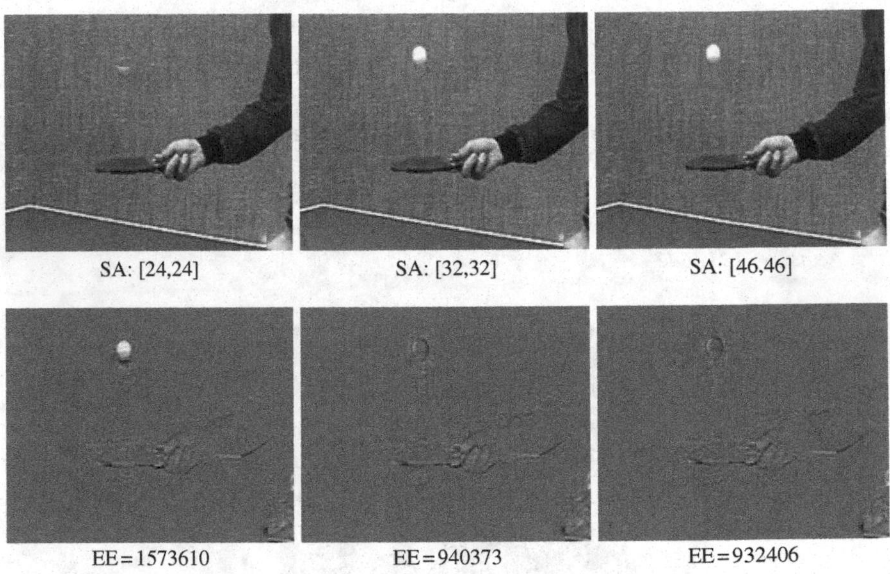

Fig. 9.31 Compensated image obtained when varying the size of the search area (SA) and the associated error compensation image with the resulting global error energy (EE)

As can be seen, a search area as small as 24×24 (corresponding to a maximum motion vector displacement of [±4, ±4]) does not allow correctly compensating the ball since its motion has displaced it outside the search area. This can be easily seen in the first column of Fig. 9.31, where the ball does not appear in the motion-compensated image (so the error in the area of the ball is very high). However, when using a search area of 32×32 or

46×46, the BM is capable of following the ball movement and the corresponding motion-compensated images provide smaller errors in the area of the ball.

Of course, the use of larger search areas implies a trade-off: the quality of the motion vectors is increased at the expenses of analyzing a larger amount of possible displacements and thereby, increasing the computational load of the motion estimation step.

9.2.5 Selection of reference image

Let us now analyze the impact of using a reference image different from the previous one. In order to do that, we will select frame #30 as the image to be estimated and successively use frames #28, #26, and #24 as reference images. The original images at those time instants can be seen in Fig. 9.32.

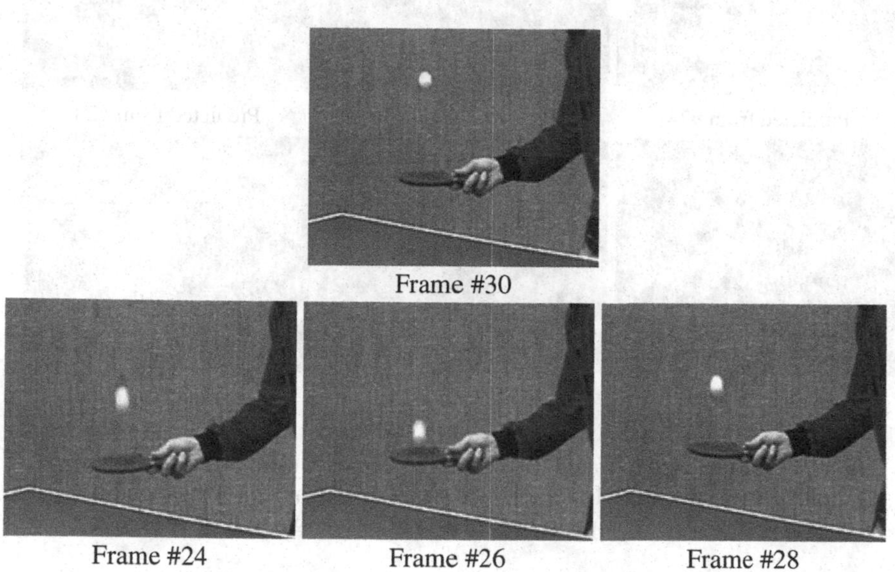

Frame #30

Frame #24 Frame #26 Frame #28

Fig. 9.32 Original images used in the analysis of the time difference between the image under study and the image reference

```
for i=25:2:31,
    table{i} = ...
        imread(sprintf('seqs/table/gray/table_%03d_g.bmp',
i-1));
    imshow(table{i});
end;
```

Figure 9.33 presents the compensated images and the associated compensation error images when using frames #24, #26, and #28, respectively as reference image.

```
t = 31;
for p=29:-2:25,
    [mvv,mvh] = estimatemotion(table{t},table{p},[16 16],...
        [16 16]);
    mc = applymotion(table{p},mvv,mvh,[16 16]);
    imshow(mc);
    e = double(table{t}) - double(mc);
    ee = sum(sum(e.^2));
    imshow_merror(e,250);
end;
```

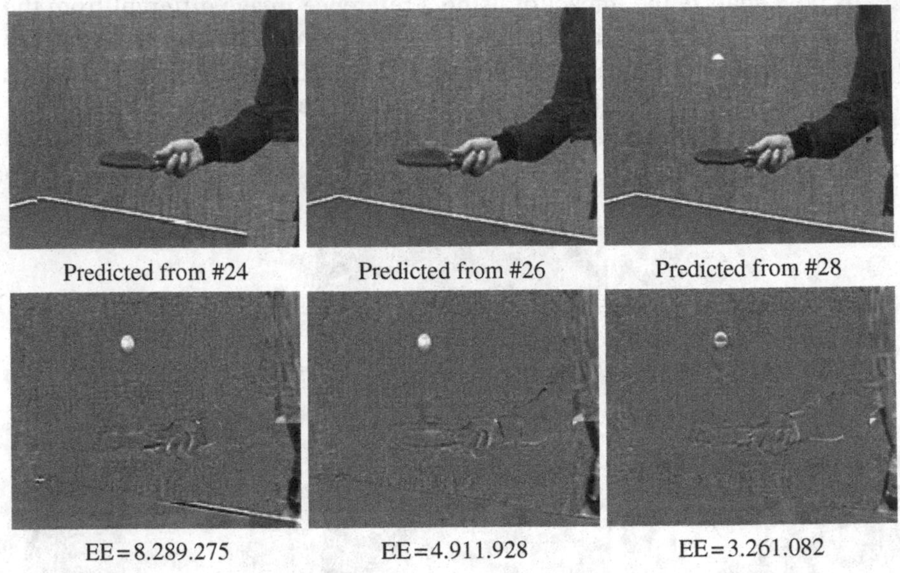

Fig. 9.33 Motion-compensated images in the *first row* using different reference frames to compensate frame #30. At the *bottom row*, the corresponding compensation error and its energy

As can be seen, the further away the two images, the higher the energy of the compensation error. This is due to the fact that distant images are less correlated and, therefore, the prediction step cannot exploit in an efficient manner the temporal correlation manner. Moreover, there is an object, the player's arm at the bottom right corner, in frame #30, which is not present in frames #24 and #26. Therefore, when using these frames as references, this object does not have any referent in the reference image and the motion compensation leads to a compensation error with higher energy.

9.2.6 Backward motion estimation

The problem of having an object in the current frame without any referent in the previous images happens every time an object appears in the scene. It is clear that, in order to obtain a referent for this object, we should look into the future frames: i.e., use a noncausal prediction. At this stage, we are going to analyze only the possibility of using past as well as future frames as reference images. We postpone the illustration of how this noncausal prediction can actually be implemented in the coding context to Section 9.2.9 (see Section 9.1.2).

Here, we analyze the possibility of estimating the current frame using the following frame in the sequence as reference image. If noncausal prediction can be used (i.e., motion-compensated future frames are possible), the information in past and future images can be used separately or combined in order to obtain a better motion-compensated image. These concepts are illustrated in the next figures.

In the first row of Fig. 9.34, we present the current image (frame #30) with the motion vector fields with respect to the previous (Fig. 9.34a) and the following (Fig. 9.34b) frames. In the second row, we present the motion-compensated images using these motion vector fields. Finally, in the last row, the associated compensation errors are shown.

```
for i=30:32,
    table{i} = ...
       imread(sprintf('seqs/table/gray/table_%03d_g.bmp',i-1));
end;

t=31;

% forward estimation
[fmvv,fmvh] = estimatemotion(table{t},table{t-1},[16 16],...
   [16 16],'fullsearch');
fmc = applymotion(table{t-1},fmvv,fmvh,[16 16]);
fe = double(table{t}) - double(fmc);
fee = sum(sum(fe.^2));

% backward estimation
[bmvv,bmvh] = estimatemotion(table{t},table{t+1},[16 16], ...
   [16 16],'fullsearch');
bmc = applymotion(table{t+1},bmvv,bmvh,[16 16]);
be = double(table{t}) - double(bmc);
bee = sum(sum(be.^2));
```

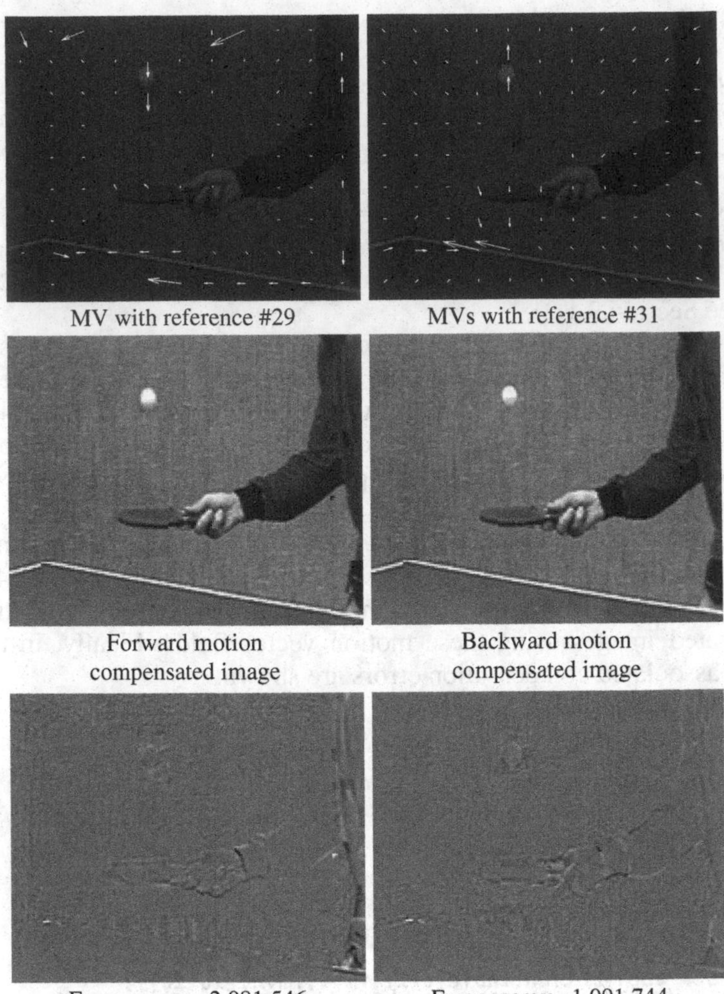

Fig. 9.34 Motion estimation and compensation using forward (#29) and backward (#31) references for frame #30

From the above example it is clear that backward motion estimation can reduce the compensation error especially when new objects appear in the scene. In this example, this is the case of the player's arm as well as a part of the background (note that the camera is zooming out). However, in order to further reduce the compensation error, both motion-compensated images (forward and backward) could be combined. Following this new idea, Fig. 9.35 shows the motion-compensated image and the compensation error where, for each macroblock, the estimation is computed by linearly combining the two reference subimages with equal weights (w=0.5).

```
bdmc = uint8(0.5 * (double(fmc) + double(bmc)));
bde = double(table{t}) - double(bdmc);
bdee = sum(sum(bde.^2));
imshow(bdmc);imshow_merror(bde,250);
```

Bi-directional motion compensated image

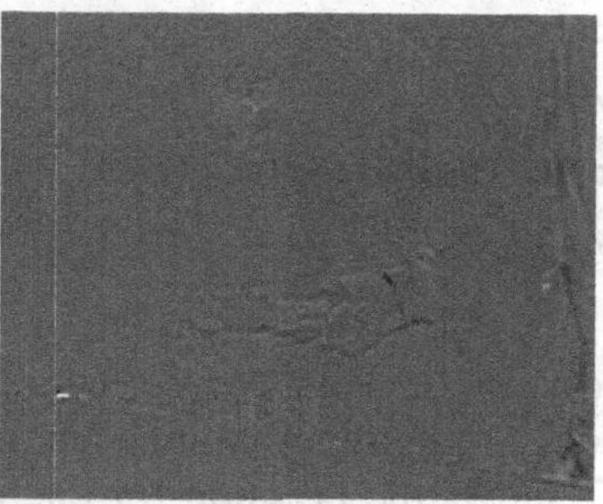

Error Energy = 936.236

Fig. 9.35 Motion-compensated image and compensation error using combined forward and backward prediction

In practice, the combination between forward and backward predictions is done in a macroblock level. For each macroblock, a decision (based on minimum error) has been made to select a forward, backward, or combined reference. Figure 9.36 shows the motion-compensated image and the compensation error where each macroblock selects the best prediction.

```
[bimc,bie,bim] = biestimatemotion(table{t}, ...
    table{t-1},table{t+1},[16 16],[15 15],'fullsearch');
imshow(bimc);
biee = sum(sum(double(bie).^2));
imshow_merror(bie,250);
```

Bi-directional motion compensated image

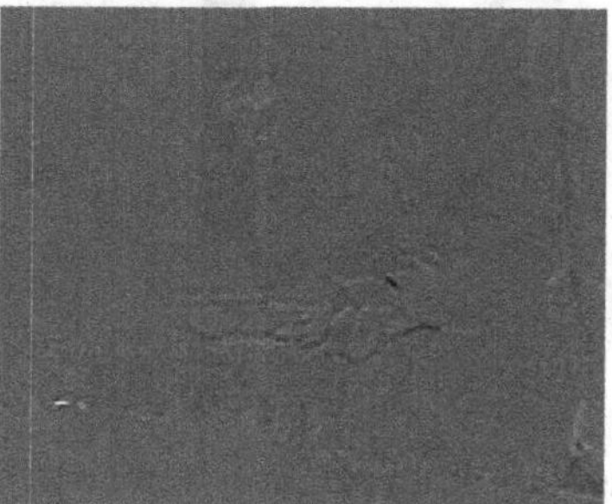

Error Energy = 740.191

Fig. 9.36 Motion-compensated image and compensation error using a selection among forward, backward, or combined prediction

For this example, combining both forward and backward predictions reduces the energy of the compensation error up to 64 and 32%, respectively, with respect to only forward or backward estimation.

```
round((fee-biee)/fee*100),

ans = 64

round((bee-biee)/bee*100),

ans = 32
```

In order to be able to see which decision has been taken for each macroblock (among forward, backward, and combined estimations), Fig. 9.37 shows the decision for each macroblock encoded in gray values. Note that backward estimation (black macroblocks) has been selected in almost all the macroblocks of the perimeter of the image. This is due to the fact that, in these images, the camera performs a zoom out and therefore, new information comes into the scene at each image. This way, a correct referent for the information in this zone in frame #30 has to be looked for in future frames (frame #31, in this case). Note, as well, that for the largest amount of macroblocks, the best estimations are built up by combining forward or backward information (macroblocks in gray). Finally, there exist a few macroblocks that are better motion compensated using only information from the past (macroblocks in gray).

```
kbim = kron(bim, ones(16));
imshow((kbim-1)/4);
addgridtofigure(size(kbim),[16 16]);
```

Fig. 9.37 Forward, backward, and combined estimation decision: a *black* macroblock means backward prediction has been selected, a *gray* one forward prediction, and a *white* one combined prediction

9.2.7 Coding of the compensation error

In all previous examples, it can be observed that although motion-compensated images show good quality, there is still relevant information on the compensation error. The compensation error can be understood as a still image that presents some correlation among adjacent pixels. Even though that correlation is much lower than in original images, the compensation error information is coded using the same strategy used when coding still images (see Chapter 8). This way, the motion compensation error image is partitioned into blocks of 8×8 pixels and each block is separately transformed (using a DCT transformation), quantized, and entropy coded.

The same transform (DCT) and the quantization table are used to code each compensation error block (as seen in the still image coding case in Chapter 8). Moreover, different k values can be used to obtain different qualities in the reconstructed compensation error. In our implementation, the MATLAB functions `quantizedct8diff` and `iquantizedct8diff` perform the (direct and inverse) quantization of the DCT coefficients for a compensation error block. As commented in Section 9.1.2, all values in the quantization table Q are set to 16.

To illustrate the transformation and quantization steps, Fig. 9.38 presents the reconstruction of the compensation error when quantizing the DCT coefficients of its blocks using k=1, k=3, and k=5. The compensation error corresponds to the block matching between frames #28 and #30 (using an exhaustive search) of the *Table Tennis* sequence (see Fig. 9.33, right column).

```
t = 31;
[mvv,mvh] = estimatemotion(table{t},table{t-2},[16 16],...
    [15 15],'fullsearch');
mc = applymotion(table{t-2},mvv,mvh,[16 16]);
e = double(table{t}) - double(mc);
Q = 16*ones(16,16);
edct = double(blkproc(e, [8 8], @dct2));
for k=[1 3 5],
    edctq = blkproc(edct, [8 8], @quantizedct8diff, k.*Q);
    edctqi = blkproc(edctq,[8 8],@iquantizedct8diff,k.*Q);
    er = uint8(blkproc(edctqi, [8 8], @idct2));
    figure,imshow_merror(er,250);
end;
```

As can be seen, increasing the values in the quantization tables reduces the information in the decoded compensation error and so does the quality of the reconstructed image (obtained by adding the compensated image and the compensation error). In this example, this effect can be observed in the better representation of the texture of the background in the first image.

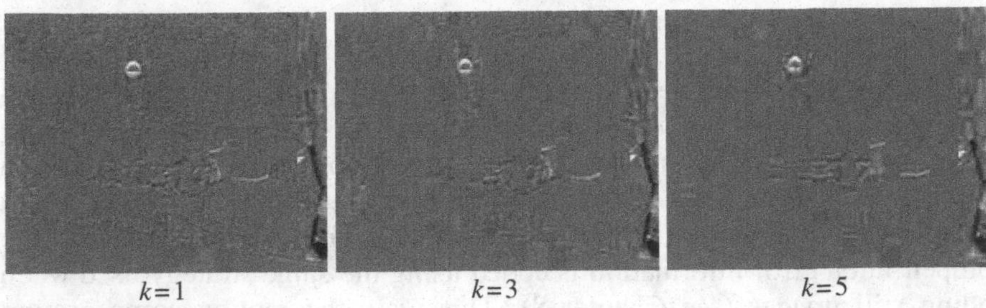

$k=1$ $k=3$ $k=5$

Fig. 9.38 Reconstructed compensation error using various quantization tables

Figure 9.39 further illustrates this concept presenting the reconstructed images.

```
for k=[1 3 5],
    edctq = blkproc(edct, [8 8], @quantizedct8diff, k.*Q);
    edctqi = blkproc(edctq,[8 8],@iquantizedct8diff,k.*Q);
    er = blkproc(edctqi, [8 8], @idct2);
    tabler = mc + er;
    figure,imshow(tabler);
    disp(['k = ' int2str(k) '; ...
        psnr = ' num2str(mypsnr(table{t},tabler))]);
end;
```

$k=1$ PSNR=36.7 dB $k=3$ PSNR=32.0 dB $k=5$ PSNR=30.7 dB

Fig. 9.39 Reconstructed images using various quantization tables and the resulting PSNR values from the original image (frame #30)

9.2.8 Entropy coding

As in the still image coding case, the quantized DCT coefficients associated which each block of the compensation error are entropy coded. In addition, the motion vectors associated with each macroblock of the image are encoded to form the final bitstream (see Section 9.1.2).

For the DCT coefficients, the same strategy as in still image coding is used. DC coefficients are decorrelated and encoded separately, while AC coefficients are preceded with a runlength encoder. However, in the case of coding the compensation error, a different set of predefined Huffman codes is used. The MATLAB file "`codes/immotiondiffcodes.mat`" includes these Huffman codes corresponding to different values of DC coefficients (in variable `dcdcode`) and AC coefficients (variables `nzcode` and `vvcode`). In our implementation, the MATLAB function `encodeimage` quantizes and encodes the corresponding DCT coefficients.

Finally, motion vectors are also entropy coded. In our implementation, the MATLAB function `encodemv` is responsible for coding the motion vectors. This function imports the file "`codes/mvcodes.mat`", which includes the Huffman codes corresponding to the various motion vector values and returns the corresponding bitstream.

The following code shows how motion vectors and compensation error are encoded into a single bitstream. Applying it to the motion vectors estimated between frames #28 and #30 of the *Table Tennis* sequence and the corresponding compensation error results in

```
t = 31;
[mvv,mvh] = estimatemotion(table{t},table{t-1},[16 16],...
    [15 15],'fullsearch');
[mvv,mvh] = estimatemotion(table{t},table{t-2},[16 16],...
    [15 15],'fullsearch');
mc = applymotion(table{t-2},mvv,mvh,[16 16]);
e = double(table{t}) - double(mc);
edct = double(blkproc(e, [8 8], @dct2));
mvcm = encodemv(mvv,mvh);
diffcm = encodeimage(edct,Q,'motiondifference');
bitstream = strcat(mvcm, diffcm);
final_bits = size(bitstream,2),

final_bits = 25546
```

which yields the following figure of bits per pixel used in the representation of the decoded gray image:

```
bitspixel = final_bits / (size(table{t},1)*size(table{t},2)),

bitspixel = 1.0080
```

In order to compare the final compressed frame with the original one, the produced bitstream must be decoded. The following code (using functions `decodemv` and `decodeimage`) shows how the bitstream can be decoded back into the reconstructed image. As entropy coding is a lossless process, the PSNR of the resulting reconstructed image is the same as in the previous section (see Fig. 9.39 with k=1).

```
[mvv2, mvh2, bitstream] = decodemv(bitstream, ...
   size(table{t}),[16 16]);
[eQ bitstream] = decodeimage(bitstream, ...
   size(table{t}),Q,'motiondifference');
mc = applymotion(table{t-2},mvv,mvh,[16 16]);tabler =
uint8(limit(double(mc) + double(eQ),0,255));
mypsnr(table{t},tabler),

psnr = 36.6910
```

9.2.9 Video coding

In this section, we are going to illustrate the entire video coding chain with different test sequences. In these experiments, we will be working with color images. As in the still image coding case (see Chapter 8), the YCbCr color space is used and, to exploit the limitations of the human visual system, chrominance components are subsampled by a factor of 2 in the horizontal and vertical directions. The coding of chrominance components is done following the same steps as the luminance (grayscale) component. As it is common, the motion estimation is performed only in the luminance component and subsampled by 2 to be applied to the chrominance components.

In our implementation, the MATLAB function `mpegencode` is responsible for performing the motion estimation, computing and coding the prediction error, closing the coding loop, and selecting a frame type for the frame in the sequence. In order to force the use of I frames at the beginning of each GOP, the function is called separately for each GOP in the sequence. As this function does not implement any rate control, the quantization tables used in the coding of the compensation error have been multiplied by a factor of 2 to ensure that the resulting PSNR for P and B frames is similar to that of I frames.

The following MATLAB code is used to encode the first 40 frames (five GOPs of eight frames) of the *Table Tennis* sequence. Five different experiments are performed: using a GOP structure with no B frames (B=0), using one B frame between P frames (B=1), using two B frames between P frames (B=2), using three B frames between P frames (B=3), and finally, using four B frames between P frames (B=4).

```
ngops = 5;
gopsize = 8;
for ng = 1:ngops,
   [cm,bits{1},type{1},psnr{1}] = mpegencode( ...
      'results/table_color','seqs/table/color', ...
      1+(ng-1)*gopsize,ng*gopsize, 1,gopsize);
   [cm,bits{2},type{2},psnr{2}] = mpegencode(...
      'results/table_color','seqs/table/color', ...
      1+(ng-1)*gopsize,ng*gopsize+1,2,round(gopsize/2));
   [cm,bits{3},type{3},psnr{3}] = mpegencode(...
```

```
        'results/table_color','seqs/table/color', ...
        1+(ng-1)*gopsize,ng*gopsize+1,3,3);
    [cm,bits{4},type{4},psnr{4}] = mpegencode(...
        'results/table_color','seqs/table/color', ...
        1+(ng-1)*gopsize,ng*gopsize+1,4,2);
    [cm,bits{5},type{5},psnr{5}] = mpegencode(...
        'results/table_color','seqs/table/color', ...
        1+(ng-1)*gopsize,ng*gopsize+1,5,2);

    for i=1:5,
        table.bits{i}(1+(ng-1)*gopsize:ng*gopsize) = ...
            bits{i}(1+(ng-1)*gopsize:ng*gopsize);
        table.type{i}(1+(ng-1)*gopsize:ng*gopsize) = ...
            type{i}(1+(ng-1)*gopsize:ng*gopsize);
        table.psnr{i}.y(1+(ng-1)*gopsize:ng*gopsize) = ...
            psnr{i}.y(1+(ng-1)*gopsize:ng*gopsize);
        table.psnr{i}.cb(1+(ng-1)*gopsize:ng*gopsize) = ...
            psnr{i}.cb(1+(ng-1)*gopsize:ng*gopsize);
        table.psnr{i}.cr(1+(ng-1)*gopsize:ng*gopsize) = ...
            psnr{i}.cr(1+(ng-1)*gopsize:ng*gopsize);
    end;
end;
```

Figure 9.40 and 9.41 show the results of the five experiments for the sequence *Table Tennis*. In Fig. 9.40, the number of bits needed to encode each frame (for all five experiments) is represented and the corresponding PSNR (for the luminance component) is shown in Fig. 9.41. Furthermore, the average number of bits needed to represent the 40 frames and the average PSNR for all 40 frames are shown in Table 9.1.

Let us discuss the coding efficiency concept by analyzing Figs. 9.40 and 9.41. Coding efficiency is measured in terms of achieving the same or better quality with the same or lower bit rate. First, note that, as already commented in Section 9.1.2, I-Pictures require many more bits than P-Pictures and B-Pictures, while achieving similar PSNR (the largest difference being 2 dBs). A similar conclusion can be driven with respect to P-Pictures and B-Pictures: due to the better compensation obtained in B-Pictures, slightly better qualities can be obtained with a lower number of bits. Note, as well, that when the number of B frames in the GOP increases, so does the number of bits necessary to encode the following P frame (for roughly the same PSNR). As we have discussed in Section 9.2.5, a larger number of B-Pictures translates into a P image, which is more decorrelated with respect to its reference and, therefore, more bits are necessary to code the compensation error. Nevertheless, for the *Table Tennis* sequence, it can be seen that using B frames in the GOP structure really improves the coding efficiency of the video codec.

Finally, another aspect that can be commented in Fig. 9.41 is the degradation that the prediction chain introduces. This can be very well observed in the curve associated with B=0, where the PSNR decays at each new predicted frame, even though the number of bits is progressively increased.

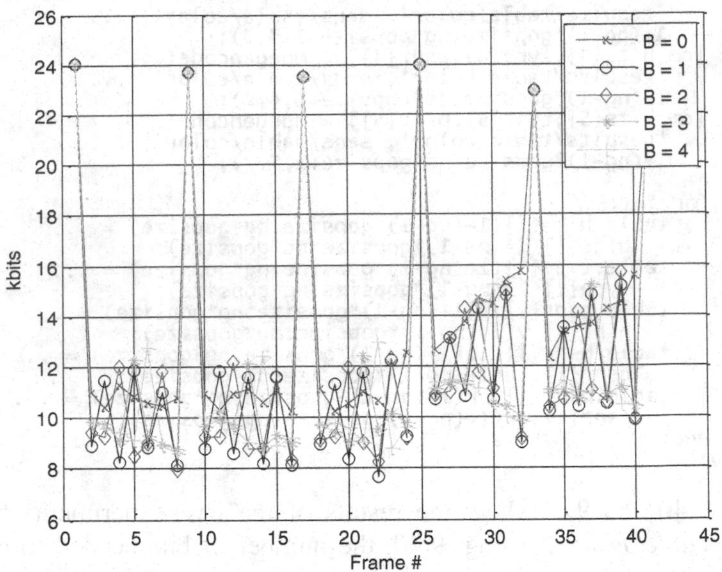

Fig. 9.40 Number of kilobits needed to encode each frame of the *Table Tennis* sequence for different number of B frames in the GOP structure

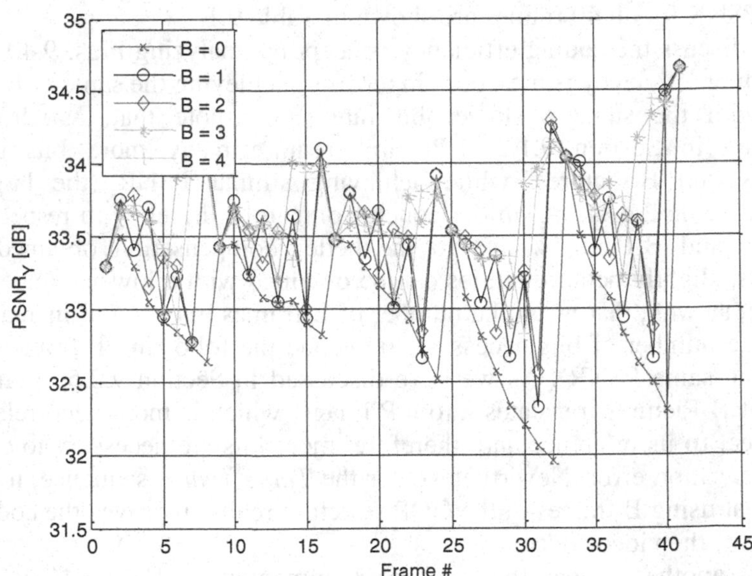

Fig. 9.41 Resulting PSNR (luminance component) for each frame of the *Table Tennis* sequence for different number of B frames in the GOP structure

Table 9.1 Average number of bits needed to represent the 40 frames and mean PSNR for all 40 frames in the *Tennis Table* sequence

	B=0	B=1	B=2	B=3	B=4
kbits	13.77	12.58	12.48	12.38	12.42
PSNR (Y)	32.97	33.42	33.49	33.69	33.62

Figure 9.42 represents the average PSNR and kilobits per second needed to encode the 40 frames of the *Table Tennis* sequence for each GOP structure and at different qualities. Qualities are selected by using a k factor, which multiplies all values in the quantization tables (both for intra and inter predictions).

Figure 9.42 shows that, for this sequence, using B frames generally improves the coding efficiency at high/normal qualities. However, for lower qualities, it is more efficient to encode using a lower number of B frames (ideally 0). This is due to the fact that, at lower qualities, reference frames in the closed loop are of very low quality and therefore the motion-compensated images are not very similar to the frame being encoded. In this case, it is more efficient to use reference frames as close as possible to the frame being coded.

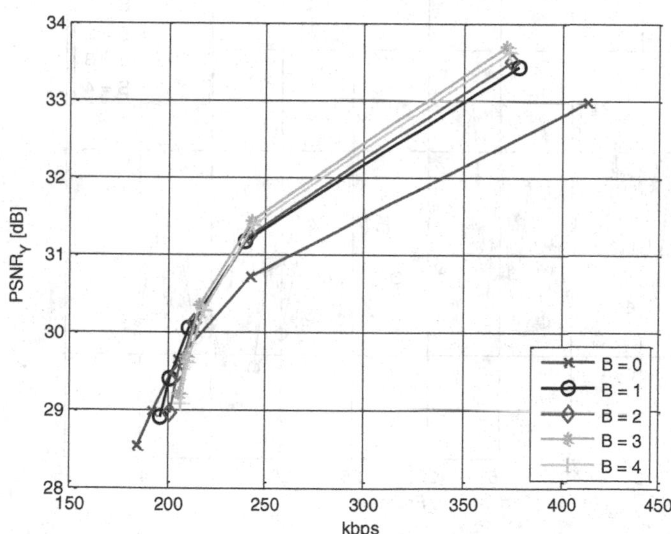

Fig. 9.42 Relation between mean PSNR and bit rate for 40 frames of the *Table Tennis* sequence for different number of B frames in the GOP structure

Figures 9.43, 9.44, and 9.45 show the same experiments conducted above but, this time, for 40 frames of the *Foreman* sequence. In this case, a specific part of the sequence is selected where there is a high amount of motion present. Similar conclusions can be driven as in the previous example. Nevertheless, as the motion estimation/compensation is not able to reduce the compensation error as much as in the previous test sequence, the more efficient GOP structure (at high/medium qualities) consists in using only 1 B frame between P frames (B=1). Figure 9.45 shows (as in the previous test sequence) that, at lower qualities, the use of B frames reduces its coding efficiency. As previously, the average number of bits needed to represent the 40 frames and the average PSNR for all 40 frames are shown in Table 9.2.

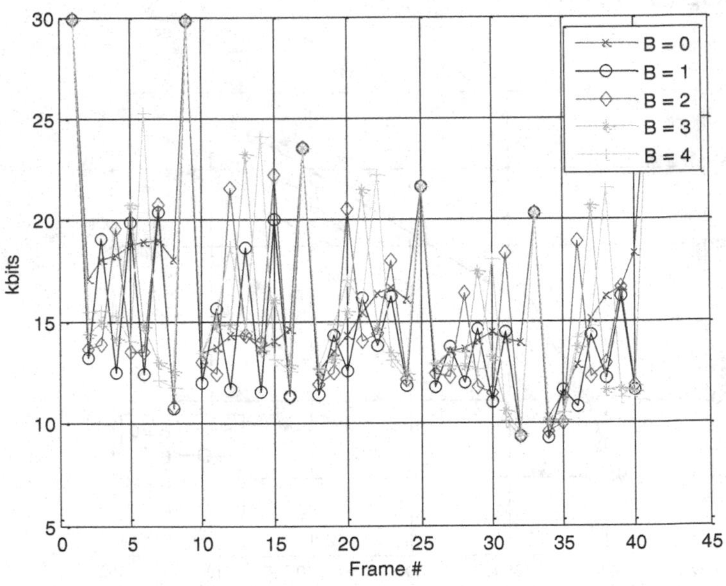

Fig. 9.43 Number of kilobits needed to encode each frame of the *Foreman* sequence for different number of B frames in the GOP structure

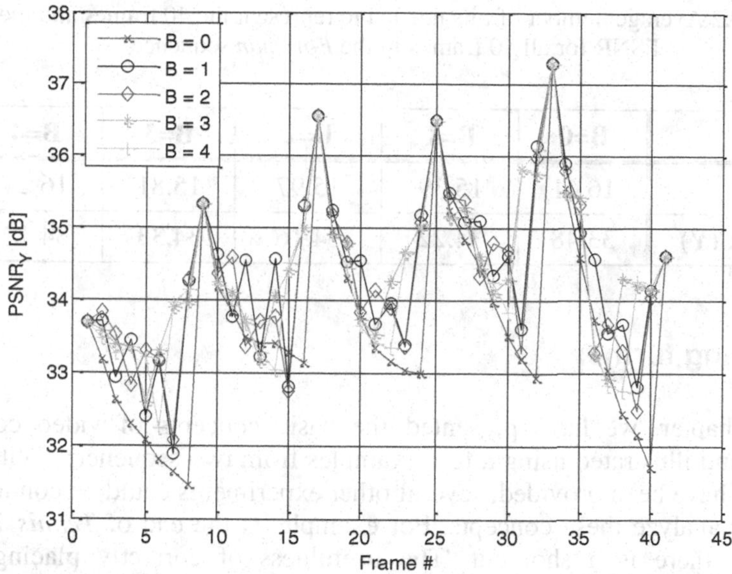

Fig. 9.44 Resulting PSNR (luminance component) for each frame of the *Foreman* sequence for different number of B frames in the GOP structure

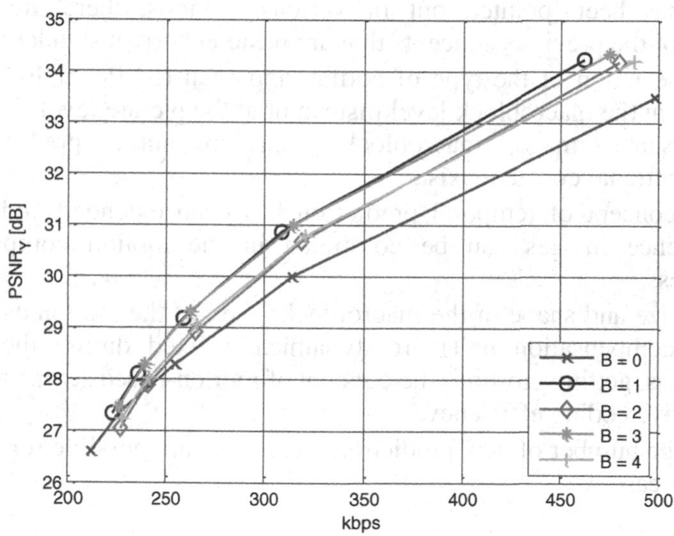

Fig. 9.45 Relation between mean PSNR and bit rate for 40 frames of the *Foreman* sequence for different number of B frames in the GOP structure

Table 9.2 Average number of bits needed to represent the 40 frames and mean PSNR for all 40 frames in the *Foreman* sequence

	B=0	B=1	B=2	B=3	B=4
kbits	16.61	15.39	15.97	15.81	16.24
PSNR (Y)	33.48	34.22	34.16	34.33	34.19

9.3 Going further

In this chapter, we have presented the basic concepts of video coding systems and illustrated using a few examples from two sequences. With the tools that have been provided, several other experiments could be conducted to further analyze these concepts. For example, at the end of *Tennis Table* sequence, there is a shot cut. The usefulness of correctly placing the beginning of a GOP at the beginning of the shot cut (i.e., use an I-Picture for an image that has no previous reference) can be studied.

Current standards, although relying on the basic concepts presented in this chapter, are extremely more complex than what has been presented here. As has been pointed out in various sections, there are several extensions of the previous concepts that are basic in current standards:

- The selection of the type of coding approach (I-, P-, or B-Mode) is made at the macroblock level instead of at the picture level. That is, in the same image, macroblocks that are intra, predictive, or bidirectional coded coexist.
- The concept of temporal prediction has been extended and several reference images can be combined in the motion compensation process.
- The size and shape of the macroblocks (i.e., of the motion estimation and compensation unit) are dynamically fixed during the coding process, getting closer to the concept of motion-based segmentation to improve coding efficiency.
- A large number of new prediction techniques are possible for the intra mode case.

Finally, let us comment that, nowadays, video coding systems aim at different functionalities rather than only compression. This way, concepts such as scalability, multiple description coding, error resilience, and transcoding capabilities are very important when currently designing a coding system. An excellent summary of the current state of the art in these concepts can be found in Zhu et al. (2005).

9.4 Conclusion

This chapter has concentrated on the basic concepts behind all current video coding standards. Specially, we have focused on the interaction between the various components of the system and, mainly, on analyzing the so-called motion estimation and compensation blocks, which are the most relevant ones in the whole chain.

For that purpose, the Block Matching algorithm has been presented and extensively studied, although in a restricted scenario: that of using the same type of motion prediction for all macroblocks in a given image.

References

ISO/IEC 11172-2 (1993) Information technology – Coding of moving pictures and associated audio for digital storage media at up to about 1,5 Mbit/s – Part 2: Video

Rec. ITU-R BT.601 (1983) Studio encoding parameters of digital television for standard 4:3 and wide screen 16:9 aspect ratios

Clarke RJ (1995) Digital Compression of Still Images and Video. Academic Press: Orlando, FL, USA

Cover T, Thomas A (1991) Elements of Information Theory. New York: John Wiley & Sons, Inc.

Fleet DJ, Wiess Y (2006) "Optical Flow Estimation", in Paragios et al.: Handbook of Computer Vision. Berlin: Springer

Gersho A, Gray RM (1993) Vector Quantization and Signal Compression. Boston: Kluwer Academic Publishers

Ghanbari M (2003) Standard Codecs: Image compression to advanced video coding. London: IEE Press

Noll NS, Jayant P (1984) Digital Coding of Waveforms: Principles and Applications to Speech and Video. Englewood Cliffs, NJ: Prentice Hall

Sullivan G, Wiegand T (2005) "Video Compression: From Concepts to the H.264/AVC Standard", Proceedings IEEE, Special Issue on Advances in Video Coding and Delivery, vol. 93, n.1, pp. 18–31.

Tekalp M (1995) Digital video processing. Prentice Hall Signal Processing Series

Zhu W, Sun MT, Chen LG, Sikora T (2005) Proceedings IEEE, Special Issue on Advances in Video Coding and Delivery

Chapter 10

How does digital cinema compress images?

A. Descampe(•), C. De Vleeschouwer(°), L. Jacques(°+), F. Marqués(*)

(•) intoPIX S.A., Belgium
(°) Université catholique de Louvain, Belgium
(+) Ecole Polytechnique Fédérale de Lausanne, Switzerland
(*) Universitat Politècnica de Catalunya, Spain

In spite of the strong temporal redundancy of video, in the Digital Cinema industry each image from a movie is compressed separately[1]

The development of digital technologies has drastically modified the requirements and constraints that a good image representation format should meet. Originally, the requirements were to achieve good compression efficiency while keeping the computational complexity low. This has led in 1992 to the standardization of the JPEG format, which is still widely used today (see Chapter 8). Over the years however, many things have evolved: more computing power is available and the development of Internet has required image representation formats to be more flexible and network-oriented, to enable efficient access to images through heterogeneous devices.

[1] This work has been partially supported by the Belgian National Funds for Scientific Research (FRS-FNRS).

T. Dutoit, F. Marqués (eds.), *Applied Signal Processing*,
DOI 10.1007/978-0-387-74535-0_10,

In this context, the JPEG committee worked on an advanced and versatile image compression algorithm, called *JPEG2000* (Rabbani and Joshi 2002, Skodras et al. 2001, Taubman and Marcellin 2002). It became an International Standard from the ISO[2] in 2001. Since then, it has been adopted on a growing number of professional markets that require both high-quality images and intrinsic *scalability*, i.e. intrinsic ability to seamlessly adapt to the user needs and available resources.[3] Among those markets, we should cite digital cinema (DC),[4] which has adopted JPEG2000 as its official standard (DCI 2005) or the *contribution*[5] segment of broadcast solutions (Symes 2006). Other potential applications are medical imaging (Tzannes and Ebrahimi 2003, Foos et al. 2003), remote sensing (Zhang and Wang 2005) and audio-visual archives (Janosky and Witthus 2004).

Figure 10.1 illustrates the high compression efficiency (Santa-Cruz et al. 2002) by comparing a JPEG and a JPEG2000 image at two different compression ratios. However, the feature that makes JPEG2000 really unique is its scalability. From a functional point of view, image scaling can be done at four different levels (see Fig. 10.2):

1. The *resolution*: The wavelet transform (further described below) reorganizes the information in the so-called *resolution levels*, each of them incrementally refining the spatial resolution of the image. Starting from the original full resolution, each successive level transforms its input image into a four times smaller[6] image plus detail coefficients. Independent encoding of low-resolution image and detail coefficients enables access at multiple resolutions.
2. The *bit depth*: Data is entropy-encoded on a "per bit-plane" basis. This means that most significant bits of all wavelet coefficients are encoded before less significant ones. By grouping encoded bits of equal significance, we obtain *quality layers*. The first quality layer gives a

---

[2] International Organization for Standardization.

[3] In a more general sense, scalability is defined as "the ability of a computer application or product (hardware or software) to continue to function well when it (or its context) is changed in size or volume in order to meet a user need" (from http://www.whatis.com).

[4] It might sound surprising that a still image codec is used instead of an MPEG video standard (see Chapter 9). However, extensive tests have revealed that the advantage of exploiting temporal redundancy between successive frames is significantly reduced for the high resolution and quality levels required by digital cinema (Smith and Villasenor 2004, Fossel et al. 2003, Marpe et al. 2003).

[5] The contribution, in broadcast, is the transfer of high-quality versions of the distributed media between different broadcast providers.

[6] Each level divides the image width and height by 2.

coarse version of the image (only the most significant bits of each pixel are used), which is further refined by subsequent quality layers.

3. The *spatial location*: Any spatial area inside an image can easily be extracted from a JPEG2000 codestream without having to process other parts of this image.
4. The *colour component*: When an image is made of several components (like in colour images or, more generally, multi-modal images), each component is coded independently and can therefore be extracted separately.

Fig. 10.1 Subjective comparison between JPEG (*left*) and JPEG2000 (*right*) compression efficiency. *First row*: Original image. *Second row*: Compression ratio of 170. *Third row*: Compression ratio of 65

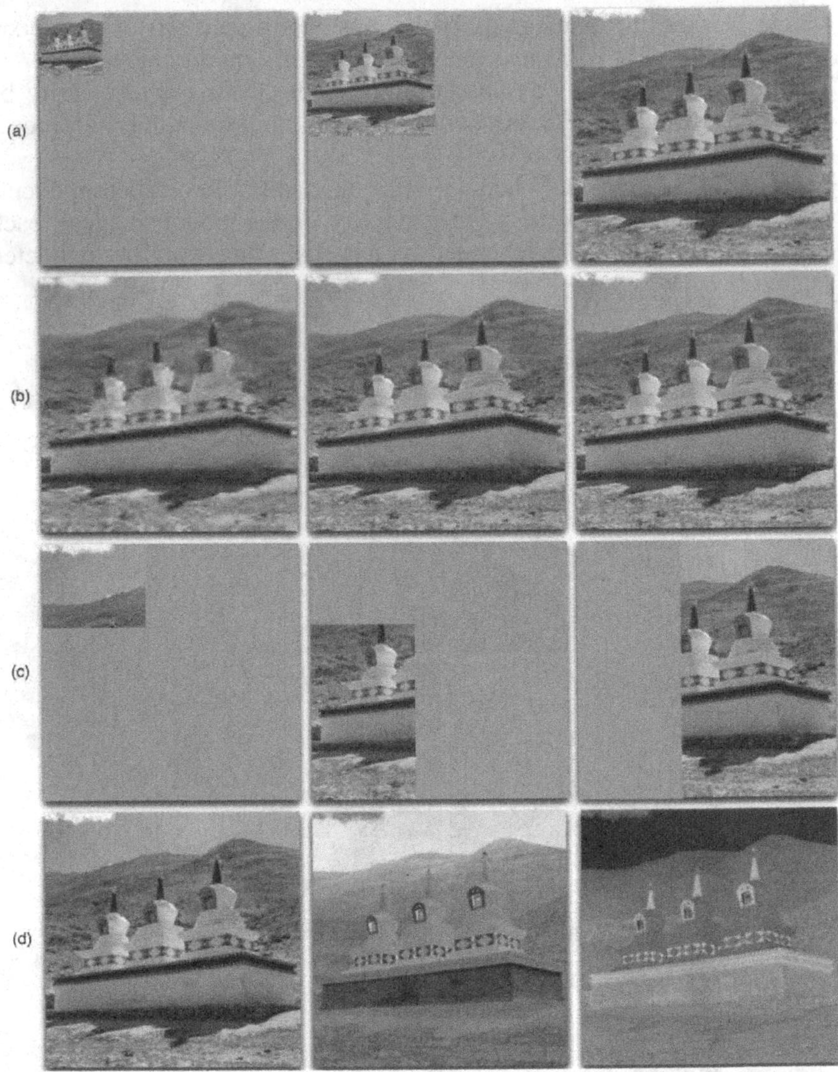

Fig. 10.2 Scalability in JPEG2000: Starting from a JPEG2000 codestream, image information can be extracted in several different ways: (**a**) by resolution, (**b**) by quality layer, (**c**) by spatial location and (**d**) by component (Y, Cb and Cr components, represented with a greyscale color map)

Based on the above image representation mechanisms, a JPEG2000 codestream organizes the image data in a hierarchical set of packets, each one containing the information related to a given quality layer from a given resolution, in a given spatial location of one of the image components.

Thereby one can easily extract the exact portion of information. that corresponds to his/her needs (in terms of image area, resolution, etc.) and available resources (bandwidth, display, etc.).

The rest of the chapter further explains and demonstrates the fundamental mechanisms that support such a versatile scalability. Specifically, the wavelet transform, the bit-plane encoder, and the grouping of bit-planes into quality layers are respectively considered in Section 10.1, Section 10.2 and Section 10.3. MATLAB proof-of-concept experiments are proposed in Section 10.4. For completeness, Section 10.5 concludes with a survey of the complete pipeline implemented to generate a JPEG2000 codestream, based on the concepts introduced along the chapter.

10.1 Background – Introduction to wavelet and multi-resolution transforms

This section introduces some basic concepts about wavelet analysis of 1D and 2D signals. Essentially, wavelets aim at changing the representation of a signal (i.e. the association of a time or a position with a certain value) so as to reorganize the information contained in the signal and reveal some properties that did not appear clearly in the initial representation.

Before the birth of wavelet analysis, most of the signal processing was performed using global tools such as global signal statistics, Fourier transform/series and global discrete cosine transform (DCT). These decompositions are global in the sense that they do not provide any information about the local signal structure. The DCT of a signal, for instance, points us "how many" frequencies are present inside a temporal sequence, but we do not know when each one was produced: *there is no way to produce a music partition with the Fourier reading of an orchestral symphony, you can just count the number of particular notes produced during the whole concert.*

To address this weakness, some early attempts were made to "artificially" localize these techniques by computing them within several limited time intervals or support areas. However, this solution has its drawbacks. The image block artefact illustrated in Fig. 10.1 for the JPEG compression is, for instance, due to an image representation that is split across a set of independent and non-overlapping, block-based DCT.

In contrast, this windowing process is incorporated naturally within a time–frequency 1D signal representation known as the *wavelet transform* (WT). As explained above, the WT gives birth to *multi-resolution descriptions* of any signal and can easily be generalized to image representation.

10.1.1 Think globally, act locally

Let us start by dealing with 1D continuous signal and see later how to practically manipulate signals digitally recorded as a discrete sequence of values. Our signal is a continuous function $s(t)$ representing the recording of some physical process at every time $t \in \mathbb{R}$. For convenience, s is not displaying weird behaviours and has, for instance, a finite energy, that is:

$$s \in L^2(\mathbb{R}) = \left\{ u : \|u\|^2 \triangleq \int_{\mathbb{R}} |u(t)|^2 \, dt < \infty \right\} \tag{10.1}$$

For the purpose of analysing the content of s, we may first compute its *approximation* $A_0 s$ in the set V_0 of signals that are constant on each interval $I_{0,n} = [n, n+1) \in \mathbb{R}$ for $n \in \mathbb{Z}$. To compute such an approximation, it is convenient to define

$$\varphi(t) = \begin{cases} 1 & \text{if } 0 \le t < 1 \\ 0 & \text{elsewhere} \end{cases} \tag{10.2}$$

The function φ, named the Haar *scaling function*, is fundamental for describing elements of V_0. The set $\Phi_0 = \{\varphi_{0,n}(t) = \varphi(t-n) : n \in \mathbb{Z}\}$, made of translated copies of φ, is a basis for V_0, i.e. any function of V_0 can be described as a linear combination of the φ_0 elements. This basis is actually orthonormal according to the usual scalar product $\langle u, v \rangle = \int_{\mathbb{R}} u(t) v(t) dt$, i.e. $\langle \varphi_{0,n}, \varphi_{0,m} \rangle = \delta_{nm}$ with δ_{nm} the Kronecker symbol, which is equal to 1 if $n=m$ and to *0* otherwise.

Thanks to φ, the average of the signal s in interval $I_{0,n} = [n, n+1)$ is simply computed by

$$a_0(n) \triangleq \langle \varphi_{0,n}, s \rangle = \int_n^{n+1} s(t) dt \tag{10.3}$$

Thus, $a_0(n)\varphi_{0,n}(t)$ is nothing but the approximation of s in $I_{0,n}$ by a constant function of height $a_0(n)$. For the whole timeline, the approximation reads

$$A_0 s \triangleq \sum_{n\in\mathbb{Z}} a_0(n)\varphi_{0,n}(t) \tag{10.4}$$

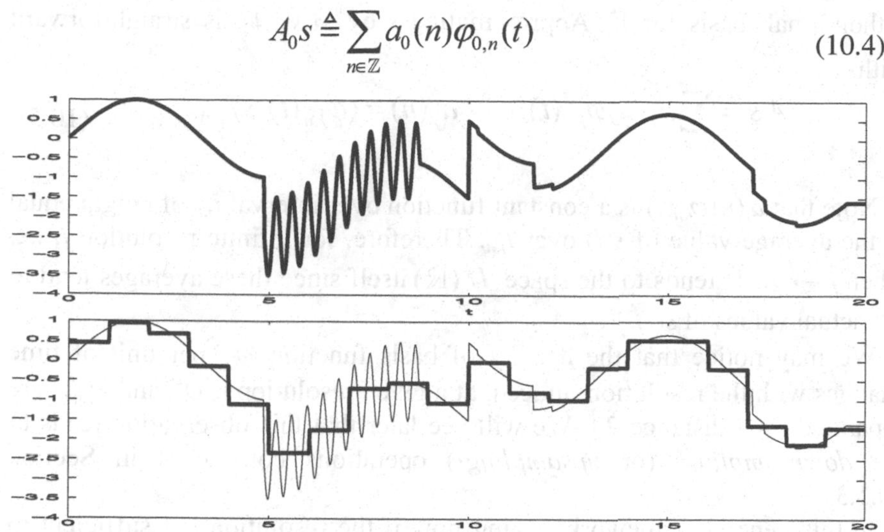

Fig. 10.3 First row: Original signal. Second row: Approximation $A_0 s$ of this signal in V_0

A simple approximation illustration is drawn in Fig. 10.3 for a toy signal s made of smooth, transient and oscillatory parts on the time interval [0,20]. As expected, $A_0 s \in V_0$ approximates the initial s but we can notice that smoothed parts of s are better rendered than the transient ones. The very oscillating part between t=5 and t=9 is, for instance, completely smoothed, while its sloppy trend is preserved.

10.1.2 Approximate ... but details matter

The numerical example of the last section suggests that we could tune the level of resolution of the piecewise constant function set on which s is projected to approximate some parts of a given signal with better success.

In other words, we want to work now with the general space V_j of functions constant at intervals $I_{j,n} = \left[2^{-j}n, 2^{-j}(n+1)\right)$. Indices j and n are named the *resolution* and the *position* parameters, respectively. The higher the j, the better the approximation of s in V_j.

These spaces V_j are hierarchically organized, i.e. each V_j is included into V_{j+1}. Moreover, $\Phi_j = \left\{\varphi_{j,n}(t) = 2^{j/2}\varphi(2^j t - n) : n \in \mathbb{Z}\right\}$ is the new orthonormal[7] basis for V_j. Approximating s by $A_j s \in V_j$ is straightforward with

$$A_j s \triangleq \sum_{n\in\mathbb{Z}} a_j(n)\varphi_{j,n}(t), \qquad a_j(n) = \langle \varphi_{j,n}(t), s \rangle \tag{10.5}$$

Note that $a_j(n)\varphi_{j,n}(t)$ is a constant function over interval $I_{j,n}$ of height equal to the average value of $s(t)$ over $I_{j,n}$. Therefore, for infinite resolution j, i.e. when $j \to \infty$, V_j tends to the space $L^2(\mathbb{R})$ itself since these averages tend to the actual values of $s(t)$.

We may notice that the density of basis function $\varphi_{j,n}$ per unit of time changes with the resolution. Indeed, at a given resolution j, $\varphi_{j,n}$ and $\varphi_{j,n+1}$ are separated by a distance 2^{-j}. We will see later that this observation leads to the *downsampling*[8] (or *upsampling*[9]) operations considered in Section 10.1.3.

In this general framework, asking now if the resolution j is sufficient to represent with high quality a local signal behaviour is related to determining the information that is lost when switching from one resolution j+1 to the coarser one j.

Therefore, we would like to compute the *details* that have to be added to an approximation $A_j s$ to get the approximation at level (j+1). Trivially, we have

$$A_{j+1}s = A_j s + \left(A_{j+1}s - A_j s\right) = A_j s + D_j s \tag{10.6}$$

where $D_j\, s$ are those details. We may remark that $D_j s \in V_{j+1}$ but the V_{j+1} space is actually too big to describe well any possible $D_j\, s$. In fact $D_j\, s$ belongs to a new space of functions W_j, which is the *detail space* of resolution j. It is also referred to as *subband* of resolution j in the literature and can be generated by the orthogonal basis $\Psi_j = \left\{\psi_{j,n}(t) = 2^{j/2}\psi(2^j t - n) : n \in \mathbb{Z}\right\}$ based on the Haar *wavelet*:

$$\psi(t) = \begin{cases} -1 & \text{if } 0 \le t < 1/2 \\ 1 & \text{if } 1/2 \le t < 1 \\ 0 & \text{elsewhere} \end{cases} \tag{10.7}$$

---

[7] The multiplicative constant $2^{j/2}$ in the definition of $\varphi_{j,n}$ guarantees the unit normalization at all resolutions, i.e. $\|\varphi_{j,n}\| = 1$.

[8] That is deleting every other position n.

[9] That is inserting zeros between every position n.

The wavelet function ψ has a vanishing average, i.e. $\int_{\mathbb{R}} \psi(t)dt = \langle \psi, 1 \rangle = 0$, and is also orthogonal to φ, i.e. $\langle \psi, \varphi \rangle = 0$. Both functions φ and ψ are represented in Fig. 10.4a and Fig. 10.4b respectively.

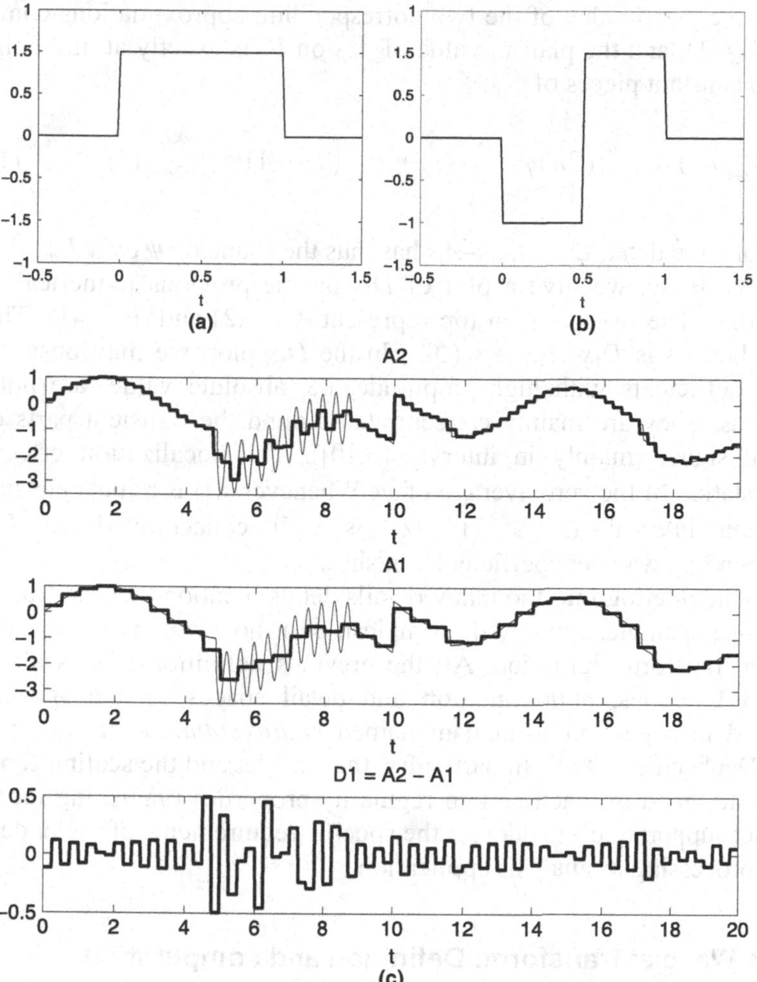

Fig. 10.4 (**a**) Haar scaling function. (**b**) Haar wavelet. (**c**) Example of detail coefficients from two resolution levels. The signal analysed is the same as in Fig. 10.3, which is drawn in grey on the two top figures

The piecewise constant shape of ψ is implicitly determined by Equation (10.7) and by the definition of φ. In short, for $t \in I_{j,n}$, we know that

$A_j s(t) = a_j(n)\varphi_{j,n}(t)$. The height of this constant function, i.e. $2^{j/2}a_j(n)$, is also the average of $s(t)$ over interval $I_{j,n}$. By construction, this interval merges two adjacent intervals from the higher level of approximation, i.e. $I_{j,n} = I_{j+1,2n} \cup I_{j+1,2n+1}$. As a consequence, an approximation at level j is exactly the mean value of the two corresponding approximations computed at level (j+1), and the plateau value of $A_j\, s$ on $I_{j,n}$ is exactly at mid-height of the two constant pieces of

$$A_{j+1}s(t) = a_{j+1}(2n)\varphi_{j+1,2n}(t) + a_{j+1}(2n+1)\varphi_{j+1,2n+1}(t) \qquad (10.8)$$

Therefore, the detail $D_j s = A_{j+1}s - A_j s$ has thus the shape of ψ over $I_{j,n}$.

In Fig. 10.4c, we give a plot of $D_1 s$ on the previous numerical signal (Fig. 10.3). The two plots on top represent $A_2 s$ (A2) and $A_1 s$ (A1). The last on the bottom is $D_1 s = A_2 s - A_1 s$ (D2). In the $D_1 s$ plot, we may observe that detail coefficients with high amplitude, i.e. absolute value, are not very numerous. They are mainly concentrated around the transient parts of the original signal (mainly in interval [5,10]). This localization effect is a manifestation of the zero average of ψ. Whenever $s(t)$ is almost constant on a certain interval $U \in \mathbb{R}$, if $\psi_{j,n}$ is well concentrated on U, the corresponding wavelet coefficient vanishes.

Without entering into too many details, let us mention that the Haar basis is just one example, actually the simplest, of orthonormal basis leading to a wavelet transform definition. All the previous definitions, i.e. scaling and wavelet functions, approximation and detail spaces V_j and W_j, can be extended in a general formalism named *multiresolution analysis* (Mallat 1999, Daubechies 1992). In particular, the wavelet and the scaling functions can be designed to reach certain regularity properties (vanishing moments, compact support, etc.) to address the specific requirements of the underlying signal processing or analysis application.

10.1.3 Wavelet transform: Definition and computation

Now that we know how to switch from one resolution to another, we can iterate the decomposition of Equation (10.6) from resolution level J as follows:[10]

---

[10] As it will be shown in Section 10.1.4, typically the highest resolution level for discrete signals is the original signal itself.

$$s(t) = \lim_{j\to\infty} A_j s(t) = A_J s(t) + \sum_{j=J}^{\infty} D_j s(t)$$
$$= \sum_n a_J(n)\varphi_{J,n}(t) + \sum_{j=J}^{\infty} d_j(n)\psi_{j,n}(t) \tag{10.9}$$

The (*Haar*) *wavelet transform* of a signal s is the collection of all the detail (or wavelet) coefficients $d_j(n) = \langle \psi_{j,n}, s \rangle$ plus the approximation coefficients $a_J(n)$. Equation (10.9) is referred to as the inverse WT, taking coefficients and rebuilding the original signal $s(t)$.

The WT implies a new representation of the signal: a temporal or spatial description of s has been replaced by a 2D resolution/position description $\{d_j(n), a_J(n)\}$. Moreover, the locality of the functions φ and ψ involves that these coefficients are actually a local measurement of s. As a mathematical microscope, coefficients $d_j(n)$ are thus probing locally the content of the signal s with a lens ψ of magnification 2^j, i.e. with a size 2^{-j}, and at position $2^{-j}n$.

About the practical computation of the WT, up to now it was suggested that scalar products, i.e. complete integrations on $\mathbb{R}$ (or Riemann sums for numeric simulations), between s and basis elements $\psi_{j,n}$ or $\varphi_{J,n}$ must be computed. There is, however, a recursive technique to follow based on the idea that φ and ψ respect each *scaling equation*, i.e. there exist filters h and g such that

$$\varphi_{j-1,0}(t) = \sum\nolimits_n h(n)\varphi_{j,n}(t) \tag{10.10}$$

$$\psi_{j-1,0}(t) = \sum\nolimits_n g(n)\psi_{j,n}(t) \tag{10.11}$$

The sequence $h(n) = \langle \varphi_{j,n}, \varphi_{j-1,0} \rangle$ and $g(n) = \langle \psi_{j,n}, \psi_{j-1,0} \rangle$ are named the *conjugate mirror filters* (or CMF) of φ and ψ. Interestingly, it may be noted that the values of h and g are actually independent of the resolution j. In other words, the link existing between the scaling functions and the wavelets between two consecutive resolutions are always the same. It can also be shown that h and g remain the same if we translate in time the functions on the left of Equations (10.10) and (10.11), i.e. if we develop $\varphi_{j-1,m}$ and $\psi_{j-1,m}$ for a given integer m in the same way. In consequence, as shown below, the knowledge of the sequences h and g dramatically simplifies the computation of the WT.

For the Haar basis, they are quite simple to calculate; the only non-zero elements are $h(0) = h(1) = 1/\sqrt{2}$ and $g(0) = -g(1) = 1/\sqrt{2}$. Note that for general (φ, ψ) i.e. outside of the Haar system, we have always $g(n) = (-1)^{1-n} h(1-n)$ (Daubechies 1992).

Thanks to these scaling relations, it is easy to prove (Mallat 1999) that

$$a_{j-1}(p) = \sum_n h(n-2p) a_j(n) = \left(a_j * \bar{h}\right)(2p) \tag{10.12}$$

$$d_{j-1}(p) = \sum_n g(n-2p) a_j(n) = \left(a_j * \bar{g}\right)(2p) \tag{10.13}$$

with $\bar{u}(n) = u(-n)$ and * stands for the discrete *convolution* between two sequences. Note the presence of a factor 2 in the argument of the convolution. As formally described in Section 10.1.2, this actually represents a *downsampling* of the density of positions by a factor 2 when passing from resolution j to resolution $j-1$.

Therefore, from an "engineer" point of view, a wavelet transform can be seen as the recursive application of a filter bank, i.e. the combination of the discrete filters h and g, thereby resulting in a low-resolution version of the signal plus a set of detail subbands.

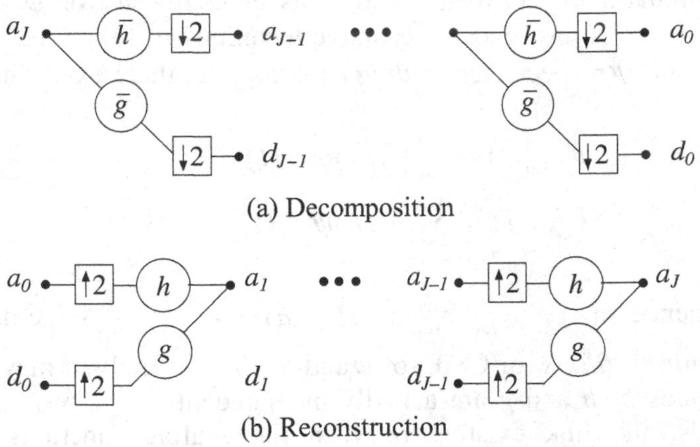

Fig. 10.5 (**a**) Decomposition and (**b**) reconstruction scheme for approximation and detail coefficients computations. Symbols ↑2 and ↓2 represent upsampling and downsampling operations, respectively. Circles mean convolution with filter name inside

In the converse sense, approximation coefficients can be rebuilt from approximation and detail coefficients at a coarser level with

$$a_{j+1}(p) = \left(\breve{a}_j * h\right)(p) + \left(\breve{d}_j * g\right)(p) \tag{10.14}$$

where, for any sequence u, $\breve{u}(2n) = u(n)$ and $\breve{u}(2n+1) = 0$, i.e. the operator inserts zero between sequence elements (*upsampling* operator).

In short, a_j and d_j are computed from a_{j+1}, and a_{j+1} can be rebuilt from a_j and d_j, in each case knowing the CMF filters h and g. This hierarchic computing (summarized schematically in Fig. 10.5) drastically simplifies the computation of the WT compared to an approach in which each coefficient should be computed based on the convolution of the original signal with rescaled filters.

Biorthogonality

Let us include a remark on the concept of *biorthogonality*. The mentioned orthogonality between φ and ψ may be lifted leading, for instance, to the definition of *biorthogonal* systems. This requires the addition of two *dual* functions $\tilde{\varphi}$ and $\tilde{\psi}$. The set $\{\varphi, \psi, \tilde{\varphi}, \tilde{\psi}\}$ is more flexible and easier to design than an orthogonal system $\{\varphi, \psi\}$.

In the induced biorthogonal WT, the direct functions φ and ψ are used to compute the approximation and the detail coefficients a_j and d_j, while the dual functions are related to the reconstruction process.

Table 10.1 Examples of CMF filters. Filters g and $\tilde{g}$ are not represented for Daubechies filters since $g(n) = (-1)^{1-n}\tilde{h}(1-n)$ and $\tilde{g}(n) = (-1)^{1-n}h(1-n)$.

	Haar	
n	$\tilde{h}[n]$	$h[n]$
0	$2^{-1/2}$	$-2^{-1/2}$
1	$2^{-1/2}$	$2^{-1/2}$

	Daubechies 7/9	
n	$\tilde{h}[n]$	$h[n]$
0	0.78848561640637	0.85269867900889
±1	0.41809227322204	0.37740285561283
±2	−0.04068941760920	−0.11062440441844
±3	−0.06453888262876	−0.02384946501956
±4		−0.03782845554969

On the filter side, we work also with four filters $\{h, g, \tilde{h}, \tilde{g}\}$. The first two, the direct filters h and g, are used in the decomposition process, while the *dual filters* $\tilde{h}$ and $\tilde{g}$ are involved in the reconstruction. For instance, JPEG2000 uses either the Daubechies 9/7 filters, i.e. h and $\tilde{h}$ have 9 and 7 non-zero elements, respectively (see Table 10.1), or the Daubechies 5/3 filters (named also LeGall 5/3).

10.1.4 WT and discrete signals: DWT

In the previous sections, we have seen that, thanks to a very simple basis, a continuous signal can be decomposed into several resolutions of details. In practice, signals are provided as a discrete (and finite) sequence of numbers, corresponding to a *sampling* of the original signal s, i.e. with values $s_d(n)$.

To extend the above theory from continuous to discrete signals, we simply assume that s_d corresponds to the approximation coefficients[11] $a_J(n) = \langle \varphi_{J,n}, s \rangle$ of some hypothetical continuous s at resolution J. Thus, pyramidal rules of Section 10.1.3 can be directly applied onto $s_d(n) = a_J(n)$. This is what defines the discrete WT or DWT.

Taking into account the downsampling arising from the computations of Equation (10.12), the whole DWT of a discrete sequence s_d of $N=2^J$ elements[12] at a resolution $0 \leq J_0 < J$ is thus composed of $2^{J-1} + 2^{J-2} + \ldots + 2^{J_0} = 2^J - 2^{J_0}$ wavelet coefficients d_j ($J_0 \leq j < J$) and 2^{J_0} approximation coefficients a_{J_0}. Therefore, the DWT provides exactly the same number of coefficients, i.e. $N = 2^J$, as the number of samples in s_d. No *redundancy* has been introduced in the transform.

For the inverse DWT (iDWT), the rule is thus simply to apply Equation (10.13) to recover $a_J(n)=s_d(n)$ from the DWT coefficients. About the computation complexity of the DWT and the iDWT, thanks to both the downsampling and the pyramidal computations, all can be performed in not more than $O(N)$ operations, i.e. the total number of multiplications–additions involved increases linearly with the number N of signal samples.

---

[11] See Mallat (1999) for further details on this subject.

[12] To be exact, assuming s_d limited to N values induces particular boundary treatments in the DWT like making the sequence periodic or using different wavelets to compute WT coefficients involving boundaries. The interested reader may refer to Daubechies (1992) or Mallat (1999).

10.1.5 WT and DWT for Images: 1+1=2

In previous sections, we were concerned only about wavelet transforms of 1D signal. How can we extend it to the manipulations of images, i.e. 2D functions? The recipe that we are going to describe below is quite simple: 2D DWT can be obtained by applying 1D DWT successively to each dimension of the image.

As previously commented, a wavelet transform can be seen as the recursive application of a filter bank, resulting in a low-resolution version of the signal plus a set of detail subbands. When dealing with images, the computation of the wavelet transform can be performed by applying the filter bank successively in the vertical and horizontal directions, resulting in the computation process shown in Fig. 10.6.

Here, we follow the JPEG2000 terminology and refer to *L* to represent a low-pass convolution by *h* (computing a signal approximation) and to *H* to denote a high-pass convolution by *g* (computing signal details). Step by step, in the notations of Section 10.1.3, the initial image *I*=`LL0` is first decomposed into two vertically downsampled convolutions

$$\mathtt{L1}(\mathbf{n}) = \left(\mathtt{LL0} *_y \bar{h}\right)(n_x, 2n_y) \text{ and } \mathtt{H1}(\mathbf{n}) = \left(\mathtt{LL0} *_y \bar{g}\right)(n_x, 2n_y) \quad (10.15)$$

where $\mathbf{n}=(n_x, n_y)$ is the 2D index, $*_u$ stands for the 1D convolution (as in Equation (10.12) performed in direction u∈{“x”, “y”}, i.e. horizontal or vertical).

By downsampling effect, `L1` and `H1` are thus two rectangular images as depicted in the second stage in Fig. 10.6. Then, the first level of a WT decomposition is eventually reached by applying horizontal convolutions to the two previous outputs so that, for instance, $\mathtt{LL1}(\mathbf{n}) = \left(\mathtt{L1} *_x \bar{h}\right)(2n_x, n_y)$ (third stage in Fig. 10.6). Finally, any other level of decomposition *n*+1 is obtained iteratively by working on the last approximation `LLn` (fourth stage in Fig. 10.6). Note that when *n* increases, resolution *j* decreases.

As subsampling operations arise in both horizontal and vertical directions, each subband *n* contains four times fewer coefficients than at resolution *n*+1, and the *N*×*N* pixel values are transformed into *N*×*N* coefficients.

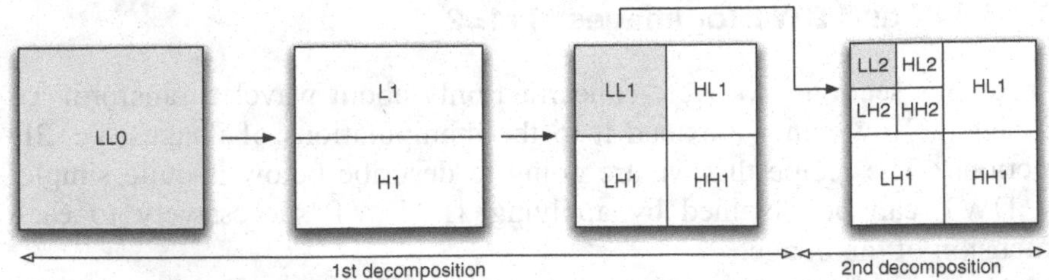

Fig. 10.6 Two-level discrete wavelet transform: continuing with the JPEG2000 terminology, note that the index of the different bands in the JPEG2000 context increases when the resolution decreases. This is highlighted here since JPEG2000 indexing is inverting the notation used in the previous sections

It is important to understand here that no compression has been done up to this point. The image information has just been reorganized and decorrelated (see Chapter 8) so as to concentrate the image energy in the upper left corner. By doing so, the image has been "prepared" for compression; most high-frequency subband coefficients are indeed close to zero (see grey points in Fig. 10.7) and can therefore be efficiently compressed.

10.2 Background – Context-based modelling of wavelet coefficient bit planes

This Section defines how the wavelet coefficients corresponding to the multi-resolution representation of an image, as presented in Section 1, can be entropy-coded, both efficiently and in a way that supports random spatial access and progressive decoding capabilities.

10.2.1 Spatial and bit-depth scalability

First, let us see how spatial and quality scalability is obtained. In short, each subband is split into several rectangular entities, named *codeblocks*, which are compressed independently to preserve random spatial access. To offer the capability to decode the image at multiple quality levels, the coefficients in a codeblock are bit-plane-encoded, which means that a coefficient is primarily defined by its most significant bit and progressively refined by adding bits in decreasing order of significance. The decomposition of the image into planes of bits is illustrated in Fig. 10.7.

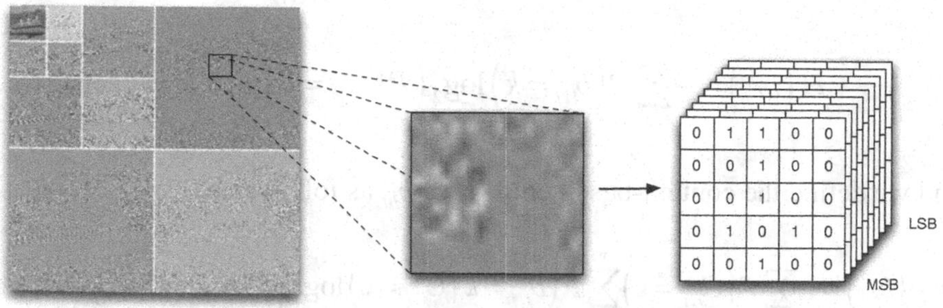

Fig. 10.7 Encoding of a codeblock on a per bit-plane basis. Codeblocks are usually 32×32 or 64×64 blocks of wavelet coefficients (here, only 25 coefficients have been represented for visual convenience)

10.2.2 Efficient entropy coding

Let us now describe how coefficient bit planes can be efficiently entropy-coded. Entropy coding has been introduced in Chapter 8, and consists in compacting the sequence of messages generated by a random source, based on the knowledge of the statistics of the source. When dealing with bit planes, efficient compression thus relies on accurate estimation of the probability distribution associated with the sequence of binary symbols encountered while scanning the bit planes, typically in a most-to-least significant and raster scan order (from left to right and from top to bottom). For improved estimation, a *context-based image modeller* has been introduced by JPEG2000 designers. Context-based means that the probability of a binary symbol is estimated based on its neighbourhood, also named its *context*.

In a sense, the context-based modeller partitions the aggregated image source into a set of sources characterized by distinct contexts. As we know that the aggregation of two sources (with known probability distributions) into a single source (modelled by an aggregated distribution) leads to an entropy increase, we conclude that the context-based approach increases coding efficiency, as long as the statistics associated with each context can be accurately estimated. The benefit is significant only when the statistics of the distinct sources are sufficiently different.

Formally, let $b_{i,j}$ denote the binary random variable associated with the *jth* bit of the *ith* bit plane and define $C_{i,j}$ as the random state variable associated with the context of $b_{i,j}$, with possible realizations $\mathbf{c}$ of $C_{i,j}$ belonging to Γ. Using $P(.)$ to denote the probability distribution of a random variable, the entropy of $b_{i,j}$ is

$$H\left(b_{i,j}\right) = -\sum_{k=0}^{1} P\left(b_{i,j} = k\right) \log_2 \left(P\left(b_{i,j} = k\right)\right) \tag{10.16}$$

and we define the context-based entropy of $b_{i,j}$ as follows:

$$H_C\left(b_{i,j}\right) = -\sum_{\mathbf{c}=\Gamma} P\left(C_{i,j} = \mathbf{c}\right) \sum_{k=0}^{1} P\left(b_{i,j} = k \mid C_{i,j} = \mathbf{c}\right) \log_2 \left(P\left(b_{i,j} = k \mid C_{i,j} = \mathbf{c}\right)\right) \tag{10.17}$$

In practice, the probability distributions $P(b_{i,j})$ and $P(b_{i,j} \mid C_{i,j}=\mathbf{c})$ are estimated based on histogram computations, i.e. probabilities are approximated by frequencies of occurrence. Those estimated values can be computed either based on the signal to encode or based on a predefined and representative set of images. In the first case, the frequencies of occurrence have to be transmitted as side information, while in the latter case, they are a priori known by both the coder and the decoder.

In JPEG2000, the context is computed based on state variables related to surrounding coefficients and to the processed coefficient itself. The most important state variable is the *significance status* of a coefficient. Initially, all coefficients are labelled as insignificant and bit planes are processed from the most to the least significant one. A coefficient is said to switch from insignificant to significant at the most significant bit plane for which a bit equal to "1" is found for this coefficient. Once significant, the coefficient keeps this status for all the remaining less significant bit planes. Other variables affecting the context are the kind of subband (LL, HL, LH or HH), the sign of the coefficient and its *first refinement status*.[13]

Intuitively, it is easy to understand that such a context improves the predictability of encoded binary values. Indeed, in a given bit plane, if a non-significant coefficient is surrounded with significant ones, it is more likely to become significant (i.e., get a "1" bit) than if it is surrounded with non-significant coefficients. We will therefore use a higher probability of getting a "1" when encoding bits that belong to coefficients that are in this situation.

Based on the above arguments, we understand that context-based modelling is likely to significantly decrease the entropy of the binary source associated with bit-plane scanning. To turn such entropy gain into actual bit-

[13] This variable is always equal to "0", except at the bit plane immediately following the bit plane where the coefficient became significant, where it is set to "1"

budget reduction, it is important to implement an efficient entropy coder. In JPEG2000, this is done by a *MQ-coder*, which is a derivative of the arithmetic Q-coder (Mitchell and Pennebaker 1988). According to the provided context, the coder chooses a probability for the bit to encode among predetermined probability values supplied by the JPEG2000 Standard and stored in a look-up table.[14] Using this probability, it encodes the bit and progressively generates code words.

10.3 Background – Rate–distortion optimal bit allocation across wavelet codeblocks

This section describes how the single and complete codestream generated by the entropy coder can be adapted to meet a given bit-budget constraint, while preserving image quality. Such adaptation is typically required when storage or transmission resources get scarce.

In Section 10.2, we have explained that the coefficients of an image codeblock are encoded on a bit plane by bit-plane basis. Hence, bit-budget reduction can simply be obtained by dropping the bitstream segments associated with the least significant codeblock bit planes. Converse to entropy coding, which does not cause any loss of information, such dropping mechanisms obviously affect image quality. Hence, a fundamental problem consists in deciding for each codeblock about the number of bit planes to drop, so as to minimize the distortion of the reconstructed image while meeting the given bit-budget (storage or rate) constraint.

The problem of *rate–distortion* (RD) *optimal allocation* of a bit budget across a set of image blocks characterized by a discrete set of RD trade-offs[15] has been extensively studied in the literature (Shoham and Gersho 1988, Ortega et al. 1994, Ortega 1996). Under strict bit-budget constraints, the problem is hard, and its resolution relies on heuristic methods or dynamic programming approaches (Ortega et al. 1994). In contrast, when some relaxation of the rate constraint is allowed, Lagrangian optimization and convex-hull approximation can be considered to split the global optimization problem in a set of simple block-based, local decision problems (Shoham and Gersho 1988, Ortega et al. 1994, Ortega 1996). This

[14] For improved coding efficiency, JPEG2000 dynamically updates the probability distribution associated with each context along the coding process. In this way, the context-based modeller adapts to the image content and to the evolution of the probability distribution across the bit planes.

[15] In the JPEG2000 context, for each codeblock.

approach is described in detail in the rest of this section. In short, the convex-hull approximation consists in restricting the eligible transmission options for each block to the RD points sustaining the lower convex hull of the available RD trade-offs of the block.[16] Global optimization at the image level is then obtained by allocating the available bit budget among the individual codeblock convex hulls in decreasing order of distortion reduction per unit of rate.

10.3.1 Problem definition

We assume that N known input codeblocks have to be encoded using a given set Q of M admissible quantizers such that the choice of the quantizer j for a codeblock i induces a distortion d_{ij} for a cost in bits equal to b_{ij}.

The objective is then to find the allocation $\mathbf{x} \in Q^N$, which assigns a quantizer $x(i)$ to codeblock i, such that the total distortion is minimized for a given rate constraint.

In our case, the index j of the quantizer refers to the number of encoded bit planes, $0 \le j \le M$. We also assume an additive distortion metric for which the contribution provided by multiple codeblocks to the entire image distortion is equal to the sum of the distortion computed for each individual codeblock.

In practice, the distortion metrics are computed based on the square error (SE) of wavelet coefficients, so as to approximate the reconstructed image square error (Taubman 2000). Formally, let $c_b(n)$ and $\hat{c}_b(n)$ denote the 2D sequences of original and approximated subband samples, respectively, in codeblock b. The distortion d_{ij} associated with the approximation of the *ith* codeblock by its j first bit planes is then defined by

$$\mathrm{d}_{ij} = \sum_{n \in i} w_{sb}^2 \left(\hat{c}_i^j(n) - c_i(n) \right)^2 \tag{10.18}$$

where $c_i(n)$ and $\hat{c}_i^j(n)$ denote the original and the quantized *nth* coefficient of codeblock i, respectively, while w_{sb} denotes the L2-norm of the wavelet basis functions for the subband sb to which codeblock b belongs (Taubman 2000).

Formally, the rate–distortion optimal bit allocation problem is then formulated as follows:

---

[16] The convex hull or convex envelope of a set of points X in a vector space is the (boundary of the) minimal convex set containing X.

Optimal rate-constrained bit allocation: For a given target bit budget B_T, find $\mathbf{x}^*$ such that

$$\mathbf{x}^* = \arg\min_{\mathbf{x}} \sum_{i=1}^{N} \mathrm{d}_{ix(i)} \tag{10.19}$$

subject to

$$\sum_{i=1}^{N} \mathrm{b}_{ix(i)} < B_T \tag{10.20}$$

10.3.2 Lagrangian formulation and approximated solution

Strictly speaking, the above formulation is known in the literature as a Knapsack problem, which can be solved at high computational cost using dynamic programming (Kellerer 2004, Wolsey 1998). Hopefully, in most communication applications, the bit-budget constraint is somewhat elastic. Buffers absorb momentary rate fluctuations so that the bits that are saved (overspent) on the current image just slightly increment (decrement) the budget allocated to subsequent images, without really impairing the global performance of the communication.

Hence, we are interested in finding a solution to Equation (10.19) subject to a constraint B' that is reasonably close to B_T. This slight difference dramatically simplifies the RD optimal bit allocation problem, because it allows the application of the Lagrange multiplier method. We now state the main and fundamental theorem associated with Lagrangian optimization, because it sustains our subsequent developments.

Theorem: For any $\lambda \geq 0$, the solution $\mathbf{x}_\lambda{}^*$ to the unconstrained problem

$$\mathbf{x}_\lambda^* = \arg\min_{\mathbf{x}} \left(\sum_{i=1}^{N} \mathrm{d}_{ix(i)} + \lambda \sum_{i=1}^{N} \mathrm{b}_{ix(i)} \right) \tag{10.21}$$

is also the solution to the constrained problem of Equation (10.17) when the constraint

$$B_T = \sum_{i=1}^{N} \mathrm{b}_{ix_\lambda^*(i)} \tag{10.22}$$

Proof: Let us define $D(\mathbf{x})=\sum_{i=1}^{N} \mathrm{d}_{ix(i)}$ and $B(\mathbf{x})=\sum_{i=1}^{N} \mathrm{b}_{ix(i)}$. By definition of $\mathbf{x}_\lambda{}^*$, we have $D(\mathbf{x}_\lambda{}^*)+\lambda B(\mathbf{x}_\lambda{}^*)\leq D(\mathbf{x})+\lambda\, B(\mathbf{x})$, for all $\mathbf{x}\in Q^N$. Equivalently, we have $D(\mathbf{x}_\lambda{}^*)-D(\mathbf{x})\leq\lambda\, B(\mathbf{x})-\lambda B(\mathbf{x}_\lambda{}^*)$, for all $\mathbf{x}\in Q^N$. Hence, because $\lambda\geq 0$, for all $\mathbf{x}\in Q^N$ such that $B(\mathbf{x})\leq B(\mathbf{x}_\lambda{}^*)$, we have $D(\mathbf{x}_\lambda{}^*)-D(\mathbf{x})\leq 0$. That is, $\mathbf{x}_\lambda{}^*$ is also the solution to the constrained problem when $B_T=B(\mathbf{x}_\lambda{}^*)$. □

This theorem says that to every non-negative λ, there is a corresponding constrained problem whose solution is identical to that of the unconstrained problem. As we sweep λ from zero to infinity, sets of solutions $\mathbf{x}_\lambda{}^*$ and constraints $B(\mathbf{x}_\lambda{}^*)$ are created. Our purpose is then to find the solution that corresponds to the constraint that is close to the target bit budget B_T.

We now explain how to solve the unconstrained problem. For a given λ, the solution to Equation (10.21) is obtained by minimizing each term of the sum separately. Hence, for each codeblock i

$$x(i)_\lambda^* = \arg\min_j \left(\mathrm{d}_{ij}+\lambda \mathrm{b}_{ij}\right) \tag{10.23}$$

Minimizing Equation (10.23) intuitively corresponds to finding the operating point of the *ith* codeblock that is “first hit” by a line of absolute slope λ (see the examples in Fig. 10.8). The convex-hull RD points are defined as the $(\mathrm{d}_{ij}, \mathrm{b}_{ij})$ pairs that sustain the lower convex hull of the discrete set of operating points of the *ith* codeblock.

For simplicity, we relabel the $M_H(i)\leq M$ convex-hull points and denote $(\mathrm{d}_{ik}{}^H, \mathrm{b}_{ik}{}^H)$, $k\leq M_H(i)$ to be their rate–distortion coordinates. When sweeping the λ value from infinity to zero, the solution to Equation (10.21) goes through the convex-hull points from left to right. Specifically, if we define $S_i(k)=\left(\mathrm{d}_{ik}^H-\mathrm{d}_{i(k+1)}^H\right)/\left(\mathrm{b}_{i(k+1)}^H-\mathrm{b}_{ik}^H\right)$ to be the slope of the convex hull after the *kth* point, the *kth* point is optimal when $S_i(k-1)>\lambda>S_i(k)$, i.e. as long as the parameter λ lies between the slopes of the convex hull on both sides of the *kth* point.

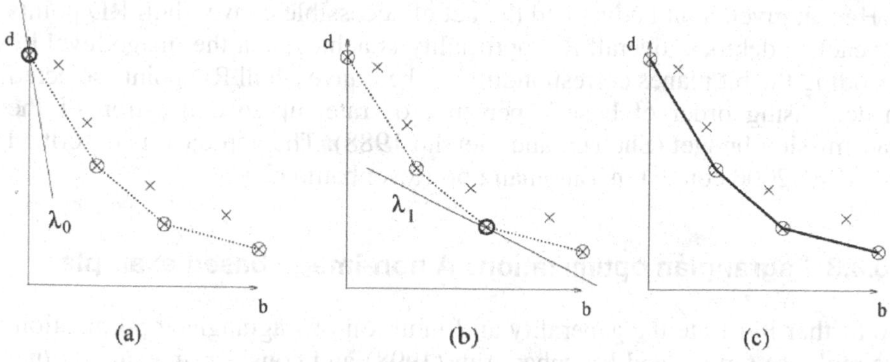

Fig. 10.8 Examples of Lagrangian-based bit allocation. In all graphs, the crosses depict possible operating points for a given codeblock. Circled crosses correspond to RD convex-hull points, which provide the set of solutions to the unconstrained bit allocation problem. (**a**) and (**b**) depict the "first-hit" solution for two distinct values of λ, while (**c**) plots the lower convex hull

At the image level, RD optimality is achieved by ensuring that each codeblock selects its operating point based on the same rate–distortion trade-off, as determined by the λ parameter. Since the λ slope is the same for every block, this algorithm is also referred to as a *constant slope optimization*. As illustrated in Section 10.3.3 and further described in Ortega and Ramchandran (1998), the intuitive explanation of the algorithm is simple. By considering operating points at constant slope, all blocks are made to operate at the same marginal return for an extra bit in rate–distortion trade-off. By marginal return, we mean the incremental reduction of distortion obtained in return for an extra bit. Hence, the same marginal return for all blocks means that the distortion reduction resulting from one extra bit for any given block is equal to the distortion increase incurred in using one less bit for another block (to maintain the same overall budget). For this reason, there is no allocation that is more efficient for that particular budget.

Now that we have solved the unconstrained problem, we explain how to find the solution whose constraint is close to the target budget B_T. While reducing the value of λ, the optimal solution to Equation (10.21) progressively moves along the convex hull for each codeblock (e.g. going from λ_0 to λ_1 in Fig. 10.8), ending up in decoding more and more bit planes. The process naturally covers the entire set of solutions to the unconstrained problem in increasing order of bit consumption and image reconstruction quality. Under a budget constraint B_T, we are interested in the solution that maximizes the quality while keeping the bit budget below the constraint.

Hence, given a bit budget and the set of accessible convex-hull RD points for each codeblock, overall RD optimality is achieved at the image level by decoding the bit planes corresponding to the convex-hull RD points selected in decreasing order of benefit per unit of rate, up to exhaustion of the transmission budget (Shoham and Gersho 1988). The approach is described in a JPEG2000 context in Taubman and Rosenbaum (2003).

10.3.3 Lagrangian optimization: A non-image based example

To further illustrate the generality and intuition of Lagrangian optimization, we rephrase Ortega and Ramchandran (1998) and consider an example that is outside the scope of image coding. This should hopefully highlight the general applicability of those solutions to resource allocation problems. The example is described as follows.

Nora is a student dealing with two questions during a 2-hour exam. Both questions worth 50% of the grade for the course, and we assume Nora is able to project her expected performance on each question based on how much time she devotes to them, as depicted in Fig. 10.9.

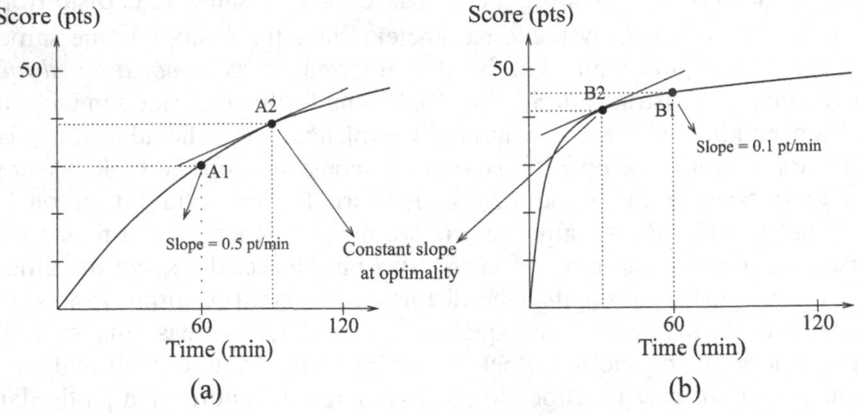

Fig. 10.9 Illustration of Lagrangian optimization. Graphs (**a**) and (**b**) depict the score–time trade-offs corresponding to first and second questions of Nora's exam, respectively. Optimal overall score is obtained when the allocation of an additional time unit provides the same score increment for both questions

Since the examination time is limited, Nora has to budget her time carefully. One option could be to devote half of the time to each question. This would amount to operating points A_1 and B_1 in Fig. 10.9, which results in an expected score of 75 (30+45). Can Nora make better use of her time? The answer lies in the slopes of the (Time, Score) trade-off curves that characterize both questions. Operating point A_1 has a slope of about 0.5 point/minute, while operating point B_1 has a slope of only 0.1 point/minute. Clearly, Nora could improve her score by diverting some time from the second to the first question. Actually, she should keep on stealing time from the second question as long as it provides a larger return for the first question than the corresponding loss incurred on the second question. At the optimum, the same marginal return (i.e., score increment) would be obtained from any additional time spent on any question. This is exactly the operating points A_2 and B_2 in Fig. 10.9, which live on the same slope of the trade-off characteristics, and correspond to a complete allocation of the 2-hour budget for an optimal score equal to 80 (38+42).

Here, it is worth emphasizing that the above reasoning relies on the convex nature of the time/score characteristics (see Section 10.3.2). We now consider a slightly different example to better capture the intuition lying behind the convex-hull constraint. In this example, one of the two questions is composed of a hierarchy of four dependent sub-questions. By dependent, we mean that a sub-question can be answered only if the answers to previous sub-questions are known. Figure 10.10 depicts Nora's expected performance to such a hierarchical question, as a function of the time devoted to it. Since the points resulting from a correct sub-answer might not be in line with the amount of time required to solve the corresponding sub-problem, the time/score trade-off might be non-convex. In our example, answers to sub-question A, B, C and D, respectively, worth 10, 10, 15 and 15 points but are expected to require 10, 30, 20 and 60 minutes to be answered correctly and completely by Nora.

Solving the unconstrained time-allocation problem $t^*=\text{argmax}_t(s+\lambda t)$ when the slope parameter λ sweeps from infinity to zero ends up in scanning the convex hull (depicted as dots in Fig. 10.10) of the time/score characteristic. Hence, the Lagrangian framework will prevent considering non-convex hull operating points as eligible solutions to the resource-allocation problem. In Fig. 10.10, all operating points lying on the time/score characteristic between A and C become ineligible solutions to the unconstrained optimization problem, whatever the slope parameter is. Intuitively, those non-convex points correspond to cases for which time has been spent with relatively little return compared to what will be obtained by a longer effort. Those non-convex operating points should thus also be omitted by the iterative solution described in Section 10.3.2. This is the reason why it is crucial to first compute the trade-off characteristic convex hull before running the iterative solution proposed in Section 10.3.2.

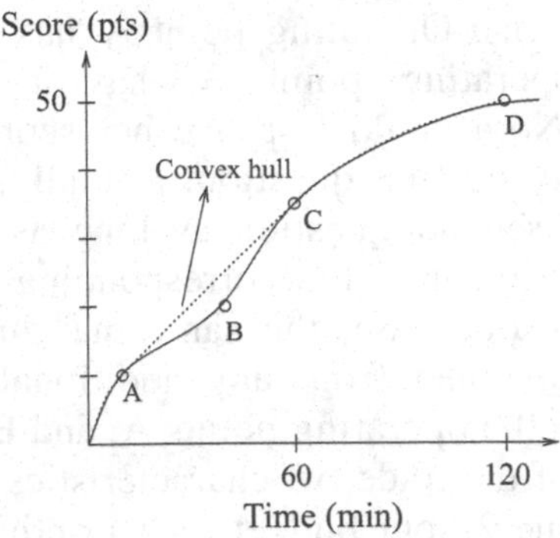

Fig. 10.10 Illustration of convex-hull approximation. Operating points lying between A and C on the time/score characteristic do not belong to the time/score convex hull and are thus ineligible solutions for the unconstrained allocation problem

10.4 MATLAB proof of concept

This section demonstrates the theoretical concepts previously introduced in this chapter. A number of experiments are first presented to illustrate the multi-resolution and decorrelative properties of the wavelet transform. In the second part, an image is selected and processed through a simplified JPEG2000 scheme.

All along this section, compressed image quality is estimated based on mean-squared error (MSE) or peak signal-to-noise ratio (PSNR) metrics. The MSE between an $N \times M$ image I and its approximation $\tilde{I}$ is equal to

$$MSE(I,\tilde{I}) = \frac{1}{MN} \sum_{m=1}^{M} \sum_{n=1}^{N} \left(I(m,n) - \tilde{I}(m,n) \right)^2 \tag{10.24}$$

In turn, the PSNR is measured in decibels (dB) and is defined by

$$PSNR(I,\tilde{I}) \triangleq 10 \log_{10} \left(\frac{Dynamic}{MSE(I,\tilde{I})} \right) \tag{10.25}$$

where *Dynamic* denotes the dynamic range of the image and is equal to 255 for an 8 bits/pixel image.

10.4.1 Experiments with the wavelet transform

Four numerical experiments are presented to make the reader more familiar with the practical aspects of the 1D and 2D DWT implementations in MATLAB. The third and fourth experiments go slightly further this exploration by studying a very simple image compression procedure in the wavelet domain.

Experiment 1: Computing the Haar DWT

Let us present an example of DWT computation using the MATLAB Wavelet Toolbox. The following sequence of MATLAB commands performs the Haar DWT of our favourite 1D signal (see Fig. 10.3) from resolution J=10 to resolution J_0=5. Detail and approximation coefficients are represented with the `waveshow()` command and are depicted in Fig. 10.11. Localization of important coefficients close to transient signal parts is now obvious.

```
%Loading the signal, N=1024=2^10
load 1d-sig;
% Performing the DWT,
J = 10; J0 = 5;
[W,L] = wavedec(sig, J-J0, 'haar');
% Showing it (waveshow.m is part of the CDROM)
figure; waveshow(sig,W,L)
```

Using the detail and approximation coefficients, we can also create the hierarchy of different approximation signals from resolution J0=5 to J=10, which is the original signal itself. These signals are represented with the `appshow ()` command and are depicted in Fig. 10.12.

```
%% Showing the ladder of approximations
%% (appshow.m is part of the CDROM)
figure; appshow(sig,W,L, 'haar')
```

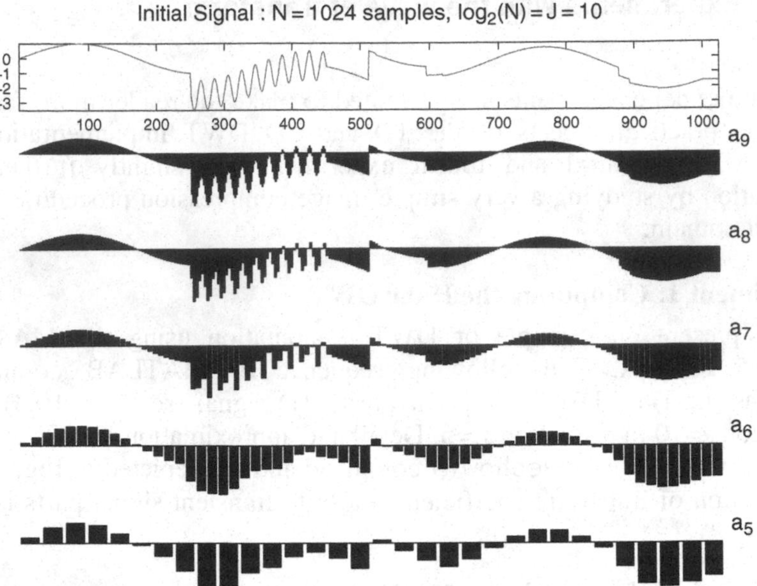

Fig. 10.11 Haar wavelet transform of the 1D signal presented in Fig. 10.3

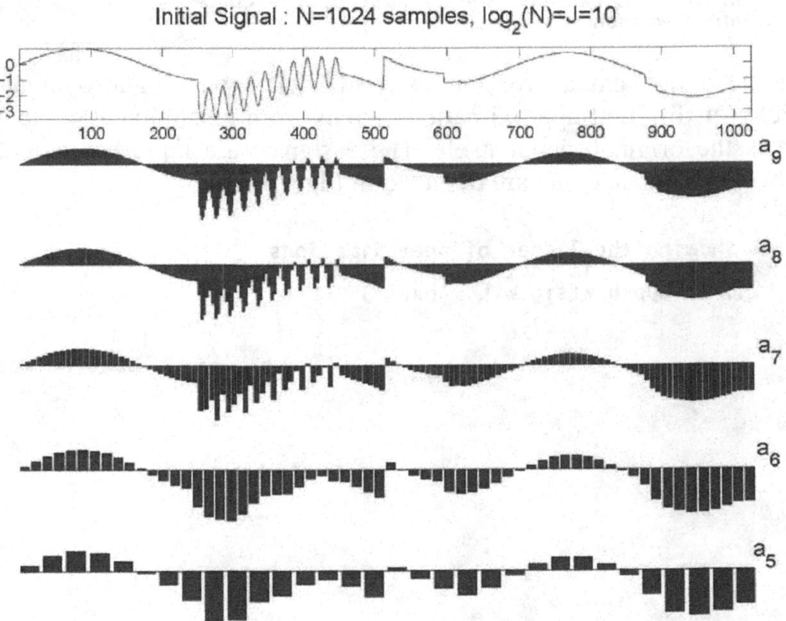

Fig. 10.12 Hierarchy of approximations obtained with the Haar wavelet transform of the 1D signal presented in Fig. 10.3

Experiment 2: Performing 2D DWT

The following routine computes the Daubechies 9/7 2-D DWT of the input images (see Fig. 10.13) and displays the wavelet coefficients in the same way as explained in Fig. 10.6. In `waveshow2()`, horizontal, vertical and diagonal coefficients are shifted around 128 and normalized by their maximal absolute value at each resolution. The concept is illustrated in Fig. 10.14. Note that for visualization purposes, the pixel values of the various subbands have been normalized (i.e. coefficients with zero value are represented with a 128 value). We observe that the most significant coefficients appear only in the transient parts, i.e. close to the edges or in the textures of the image (e.g. the grass in the bottom of Fig. 10.13a).

```
% Loading the image : The Cameraman picture
% img is a 256x256 size array
im = double(imread('cameraman.tif'));
figure; imagesc(im); colormap(gray); axis equal tight;
set(gca,'clim',[7 253])
% 2D DWT Computations with Daubechies 9/7 (== 'bior4.4')
% W contains the DWT coefficients in a column shape.
J = log2(256); J0 = 2; wname = 'bior4.4';
[W,S] = wavedec2(im, J-J0, wname);
% The DWT array
figure, waveshow2(W,S,'haar');
```

Experiment 3: On the compression road

We now focus on the "sparseness" of the wavelet coefficients for the representation of *natural* images. The relevance of the WT regarding compression holds in the following concept: *very few coefficients concentrate the essential of the image information*. This can already be perceived in Fig. 10.14. Converse to initial pixel values, detail coefficients with high amplitude are not very numerous and are well localized on image edges.

The following code is the continuation of Experiment 2. Here we compress images by keeping only the 10% highest energy wavelet coefficients. This is achieved by sorting the wavelet values by decreasing order of magnitude and recording the amplitude of the *Kth* largest element (with K the closest integer to $N^2/10$). Then, all other wavelet coefficients with non-zero magnitude are set to 0, i.e. their information is lost (see Fig. 10.15).

```
% Sorting wavelet coefficient amplitude
sW = sort(abs(W(:)), 'descend');

% Number of elements to keep
% Compression of 90% !
K = round(256*256/10);
T = sW(K);
```

```
% Thresholding values of W lesser than T
% i.e. we keep the K strongest
nW = W;
nW(abs(nW) < T) = 0;

% Rebuilding the compressed image
Timg = waverec2(nW, S, wname);
figure; imagesc(Timg); colormap(gray); axis equal tight;
set(gca,'clim',[7 253])
```

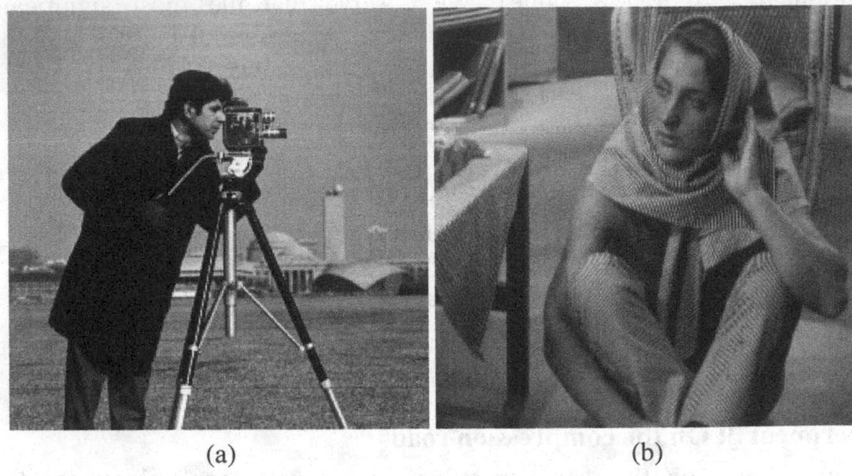

(a) (b)

Fig. 10.13 (**a**) Original *Cameraman* image (256×256 pixels) and (**b**) original *Barbara* image (512×512 pixels)

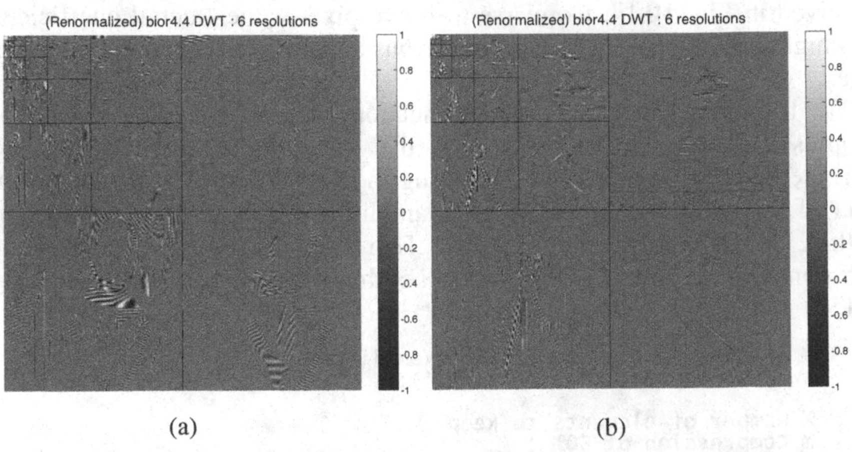

(a) (b)

Fig. 10.14 2D DWT of (**a**) *Cameraman* and (**b**) *Barbara* using Daubechies 9/7

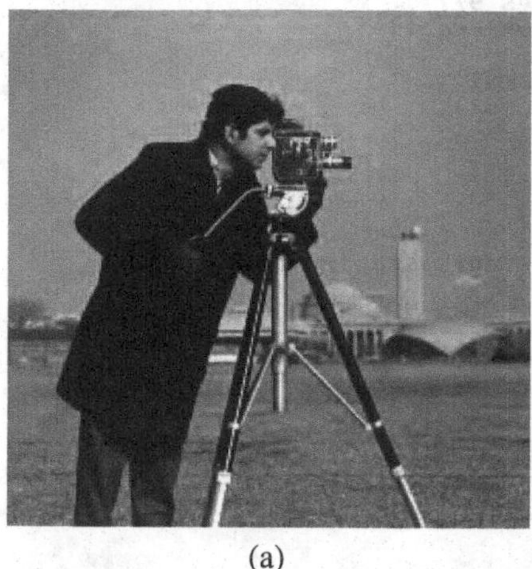

(a)

(b)

Fig. 10.15 Reconstruction of the previous images using 10% of DWT coefficients with highest energy: (**a**) *Cameraman* (PSNR=32.87 dB) and (**b**) *Barbara* (PSNR=31.75 dB)

Experiment 4: Quantifying compression quality

Let us now quantify a bit further the quality reached by the compression scheme of Experiment 3 as a function of both the number of DWT coefficients kept during the thresholding and the type of wavelet filters used.

In Fig. 10.16, the quality curve obtained for different percentages of wavelet coefficients is drawn (from 5 to 30%) for the *Cameraman* image of Fig. 10.13. We can clearly see that the Daubechies 9/7 filter provides the best quality. However, quality is not the only criterion which makes a filter better than another. Daubechies (or Legall) 5/3 filters can be expressed by rational numbers and, therefore, are used for *lossless compression* in JPEG2000.

```
nbpix = 256*256;

% Fixing the percentage of pixels to keep in the compression
% between 5% and 30%
K = round(((5:5:30)/100) * nbpix);
nbK = length(K);

% Cameraman Image decomposed on J levels
im = double(imread('cameraman.tif'));
J = log2(256); J0 = 0;

% Quality metrics between two images : MSE and PSNR.
% Assuming an 8 bits original image
MSE = @(X,Y) norm(X(:) - Y(:), 2)^2 / nbpix;
PSNR = @(X,Y) 10*log10( (256-1)^2 / MSE(X,Y) );
```

```
wname1 = 'bior4.4'; %% Daubechies 9/7
wname2 = 'bior3.3'; %% Daubechies/Legall 5/3

[W1,S1] = wavedec2(im, J-J0, wname1);
[W2,S2] = wavedec2(im, J-J0, wname2);

sW1 = sort(abs(W1(:)), 'descend');
sW2 = sort(abs(W2(:)), 'descend');

for k = 1:nbK,
   % Setting all the DWT coefficients smaller than the Kth
magnitude to
   % zero.
   % For DB97
   T1 = sW1(K(k));
   nW1 = W1;
   nW1(abs(nW1) < T1) = 0;
   Timg1 = waverec2(nW1, S1, wname1);

   % For DB53
   T2 = sW2(K(k));
   nW2 = W2;
   nW2(abs(nW2) < T2) = 0;
   Timg2 = waverec2(nW2, S2, wname2);

   % Recording quality
   curve_97(k) = PSNR(im, Timg1);
   curve_53(k) = PSNR(im, Timg2);
end
```

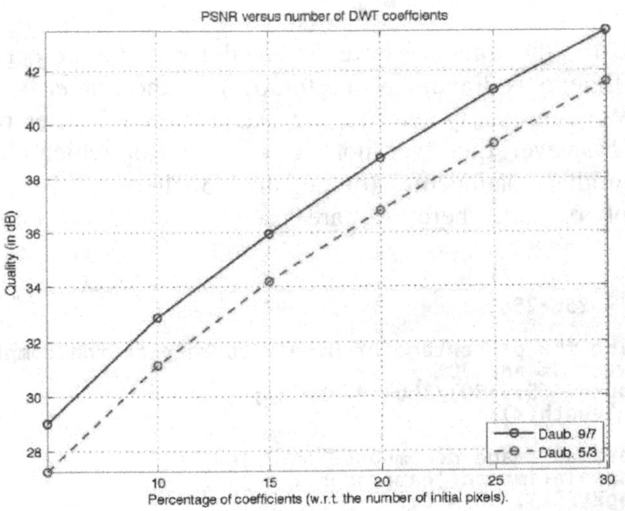

Fig. 10.16 Quality curve of compressed images (for the *Cameraman* image) for different percentages of DWT coefficients and for different filters

10.4.2 A simplified JPEG2000 scheme

In this Section, we provide a proof of concept of the JPEG2000 image compression standard. We start by transforming the input image using 2D DWT. We then consider context-based entropy coding of the wavelet coefficient bit planes and conclude by performing a rate–distortion optimal allocation.

All along this code, a structure `img` that will contain the required settings to process the image is defined. These settings include the path to the image to be processed (`img.path`), the number of wavelet decompositions to be applied (`img.nwdec`) and the kind of wavelet filters to be used (`img.wfilt`).

First of all, we load a greyscale image and shift the coefficients from an unsigned to a signed representation.

```
X = imread(img.path);
X = double(X);
img.bdepth = ceil(log2(max(X(:)+1)));
[img.h img.w] = size(X);
X = X - pow2(img.bdepth-1); % DC-level shifting
```

Discrete wavelet transform

The JPEG2000 standard defines two different filters, namely the 5/3 and the 9/7 transforms. The former is used in JPEG2000 when lossless coding is required. The latter, with a slightly higher decorrelating power, is used for lossy coding.

The norms of the synthesis filters[17] are denoted by `wnorm_53` and `wnorm_97`. They will be used to approximate square errors in the pixel domain based on the ones in the wavelet domain. According to the parameter `wfilt` defining the wavelet transform to be used, we apply the 5/3 or the 9/7 transform. The function `wavedec2` from the Wavelet Toolbox is used.

```
if img.wfilt==0
    [C,S] = wavedec2(X,img.nwdec,lo_53_D,hi_53_D);
elseif img.wfilt==1
    [C,S] = wavedec2(X,img.nwdec,lo_97_D,hi_97_D);
else
    error('wavelet filter not recognized');
end
```

[17] The norm of a filter corresponds to the sum of the squared coefficients of the filter. It can be seen as the average amplitude change that will occur when filtering the signal.

Context-based modelling of coefficient bit planes

After the DWT, the JPEG2000 algorithm performs bit-plane context-based entropy coding of each subband. This Section computes the incremental bit budget and distortion reduction resulting from the addition of bit planes to refine DWT coefficients. Thereby, it provides the inputs required by the rate–distortion optimal bit allocation mechanisms envisioned in the next Section. Moreover, this Section illustrates the advantage of context-based modelling by comparing two measures of the entropy associated with the binary representation of the wavelet coefficients. In the first case, the binary symbols are assumed to be generated by an i.i.d. sequence of random variables and the probability distribution of each binary random variable is estimated based on the frequency of occurrence of 1's and 0's symbols. In the second case, we use the conditional probabilitics associated with each binary symbol, knowing its context. The reduction of entropy between the first and the second case corresponds to the benefit obtained from the chosen context model.

It should be noted that to avoid a too long section, we do not actually implement the entropy coder. Only the performance of such coder is evaluated through estimation of the source entropy. This estimation enables to compute the output rate that would be obtained for a given distortion level.

Before being entropy-coded, wavelet coefficients are quantized and mapped on a certain amount of bits. To do so, we first separate the sign and magnitude of the wavelet coefficients. They will indeed be encoded separately.

```
Csign = sign(C);
Cmagn = abs(C);
```

Then, coefficients are quantized; this quantization is different from the one that will implicitly occur later by dropping some of the least significant bit planes, done by dividing them by a pre-defined quantization stepsize. In our case, the quantization stepsize chosen follows the rule used in the OpenJPEG library[18] and depends on the kind of subband and on the norm of the synthesis filter. The following code corresponds to the quantization stepsize for the LL subband.

```
if img.wfilt==0
    img.res(1).sb(1).qstep = 1;
elseif img.wfilt==1
    img.res(1).sb(1).qstep = 1/wnorm_97(1,img.nwdec+1);
end
```

[18] Open-source JPEG2000 codec from the TELE lab, UCL, Belgium: http://www.openjpeg.org.

Eventually, quantized coefficients are mapped onto a certain amount of bits (fixed-point representation). In this simple experiment, we simply keep the integer part of the coefficients that we represent using 16 bits. For the LL subband, this is done through the following code (`Am` corresponds to the magnitude of the wavelet coefficients from the LL subband):

```
img.res(1).sb(1).coeff =
uint16(floor(Am/img.res(1).sb(1).qstep));
```

The code for the other subbands is similar and is therefore not reproduced here. As we can see, all quantized coefficients are stored in the `IMG` structure, resolution per resolution and subband per subband.

In JPEG2000, wavelet coefficients are processed bit plane by bit plane and not coefficient by coefficient. Now that the coefficients are quantized and mapped onto a fixed number of bits, we can truly observe the scalability offered by such bit-plane representation. To illustrate this, we choose a subband, let us say the HL subband of last resolution, and display its k most significant bit planes for k=1,.., K, where K is the number of significant bit planes for this subband:

```
sbcoeff = img.res(end).sb(1).coeff;
sbsign = img.res(end).sb(1).sign;
K = ceil(log2(max(double(sbcoeff(:)+1))));
```

It should be noted that K is not necessarily equal to the maximum number of bit planes (16 in our experiment). Indeed, in most subbands and especially in those corresponding to the high frequencies, 16 bits will be far too much to represent the coefficient values. Consequently, many of the most significant bit planes will often remain at zero. In practice, when encoding a subband, rather than encoding these all-zero bit planes, we skip them until the first "1" bit is encountered. The number of skipped bit planes will then simply be stored in a header of the compressed bitstream.

The truncation of the coefficients is done by applying successively a "AND" mask and an "OR" mask to the wavelet coefficients. The first one sets the required number of least significant bits to "0". The second one moves the truncated coefficient value to the middle of the truncation step by setting the most significant truncated bit to "1".

```
mask_AND = bitshift(uint16(65535),nbp_discard);
mask_OR = bitshift(uint16(1),nbp_discard-1);
...
m_trunc = bitor(bitand(sbcoeff,mask_AND),mask_OR);
```

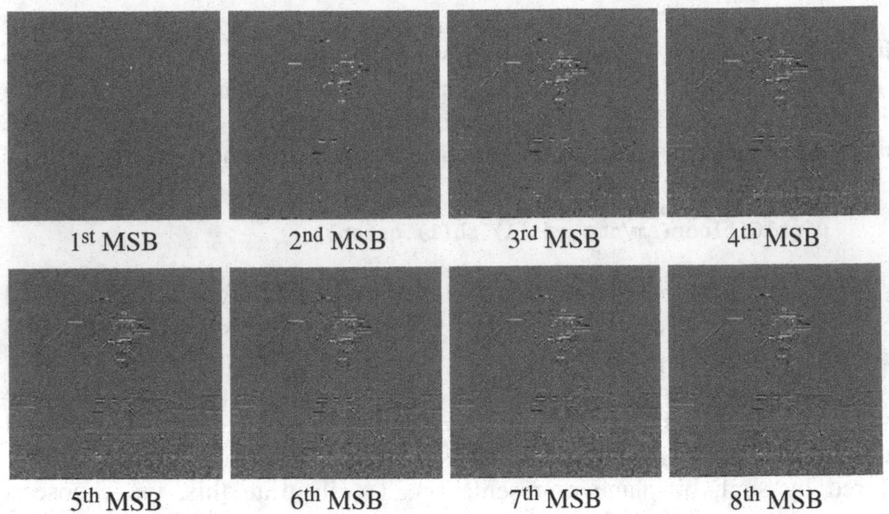

Fig. 10.17 Progressive refinement of wavelet coefficients through inclusion of an increasing number of bit planes (MSB: most significant bit)

As we see in Fig. 10.17, wavelet coefficients are progressively refined, as the bit planes (from the most to the least significant one) are included in the coefficient estimation.

We now compute the probability distribution of the binary representation of our image. This distribution will then be exploited to compute the related entropy.

As explained above, we will compare two kinds of distribution. The first one is based on the frequency of occurrence of "1" and "0" symbols over the whole image. In the second case, the conditional probabilities of binary random variables are estimated, knowing their context. The relevance of such an approach has been intuitively justified in Section 10.2.2.

Practically, the context of a bit corresponds to a set of state variables related to (i) the coefficient to whom the bit belongs and (ii) its neighbouring coefficients. In this experiment, two state variables were used.

1. The *significant status* of a coefficient. Let us remind that a coefficient is said to become significant in a given bit plane if a "1" bit is encountered for the first time for this coefficient (all other more significant bits were "0"s).
2. The *first refinement status* of a coefficient. Among already significant coefficients, we will distinguish those that became significant in the previous (more significant) bit plane.

In this proof of concept, we have considered 12 different contexts. They are presented in the preamble of function `get_context` and are actually a subset of the contexts used in the JPEG2000 standard (which uses 19 different contexts). Nine out of the 12 contexts are used to code not yet significant coefficients, while the last three ones are used for already significant coefficients. The sign of each coefficient is introduced "as is" in the codestream and is not entropy-coded.

Let us first initialize the number of contexts and the context distribution table (CDT). This vector stores the probability of having a "1" for each context. The last element of the vector is used to store the global probability of getting a "1" on the whole image.

```
global nctxt;
nctxt = 12;
CDT = zeros(nctxt+1,1);
```

As explained in Section 10.2.1, before being entropy-coded, subbands are divided into small entities called codeblocks. Each codeblock will then be entropy-coded separately, starting from the most significant bit plane to the least significant one.

In the code, several `for` loops are embedded so that each subband from each resolution level is processed and divided into such codeblocks.

```
for resno=1:numel(img.res)
    for sbno=1:numel(img.res(resno).sb)
    ...
        for y0=1:img.cbh:size(coeff,1)
            for x0=1:img.cbw:size(coeff,2)
```

Then, each codeblock is analysed. First, as explained above, the number of all-zero most significant bit planes is computed (`NBPS` field of each codeblock).

```
cb.nbps = ceil(log2(max(double(cb.coeff(:)+1))));
```

Now, each codeblock is analysed through the `analyze_cb` function.

```
[cb.ctxt,cb.disto] = analyze_cb(cb);
```

This function will process each bit plane starting from the first most significant non-zero bit plane and return two variables:

(1) a matrix `ctxt` that gives two values for each context and each bit plane: (i) the total number of bits and (ii) the number of "1" bits,

(2) a vector `disto` that computes, for each bit plane `BP`, the square error between the original codeblock and the codeblock whose least significant bit planes are truncated, starting from and including bit-plane BP.

This `disto` vector will be used to decide where to truncate the codeblock when performing the R–D allocation. However, it has to be adapted to reflect the square error in the pixel domain. To do so, the values of the `disto` vector (that correspond to the square errors between original and truncated wavelet coefficients) are multiplied by two factors. The first one is the squared stepsize, which was used for the quantization that took place after the wavelet transform. By doing so, we actually perform an inverse quantization to get the square errors back in the dynamic range obtained just after the wavelet transform. Then, these "dequantized" square errors are multiplied by the norm of the corresponding synthesis filter. As explained at the beginning of Section 10.4.2, this last operation gives the corresponding square errors in the pixel domain, which are the ones of interest as we want to precisely approximate the distortion reduction in this domain.

```
if img.wfilt==0
    cb.disto=cb.disto.*((img.res(resno).sb(sbno).qstep)^2*...
                (wnorm_53(cb.sbtype,img.nwdec+2-resno))^2);
elseif img.wfilt==1
    cb.disto=cb.disto.*((img.res(resno).sb(sbno).qstep)^2*...
                (wnorm_97(cb.sbtype,img.nwdec+2-resno))^2);
end
```

Once matrix `ctxt` has been computed for each codeblock, we can easily compute a context distribution table (`CDT`) which will be used for entropy coding. The `CDT` stores the probability for each of the 12 contexts to get a "1" bit according to the processed image. A 13th value stores the global probability to get a "1" bit, independently from the neighbourhood of the coefficient. These values are presented in Fig. 10.18.

The `CDT` is then saved so that it can be reused when encoding other images.

```
save(CDTpath,'CDT');
```

Once the `CDT` has been computed, we can use it to estimate the entropy of each codeblock. If `img.useownCDT=0`, we do not use the CDT computed on the processed image but the one specified in `img.CDT2use`. It should be noted that using a context distribution computed on an image whose content is very different from the one of the processed image does not change the performances drastically. This can be explained by the fact that the context mainly characterizes local image features, which reduces their dependency on the global appearance and statistics of the image. Moreover, in our proof-of-concept example, the conditional distributions are computed on all bit planes of all subbands, which further reduces the dependency of those distributions on the global image features. In JPEG2000, however, the distribution is reinitialized for each codeblock and is dynamically adapted while encoding the bit planes. In this case, the entropy coder uses a more adapted distribution and is therefore more efficient.

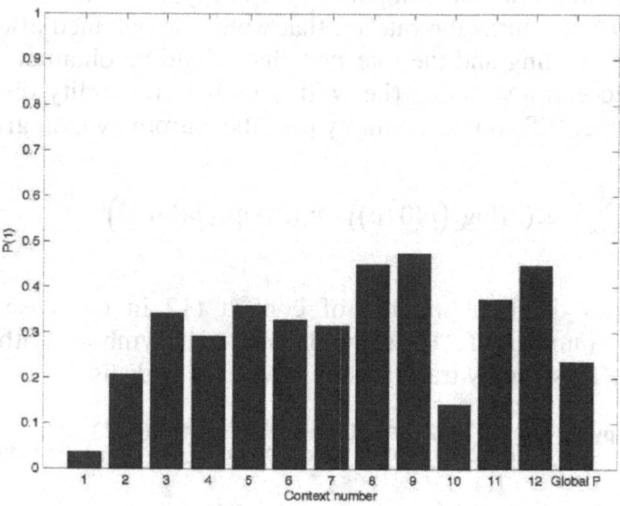

Fig. 10.18 Comparison of conditional probability of getting a "1", knowing the context of the bit, with the global probability, independent of the context

Practically, for each subband, we progressively fill in a matrix `RD`: `img.res(resno).sb(sbno).RD(cbno,bpno,i)`, where `i=1` for the rate values and `i=2` for the distortion values. Each pair of values (`i=1,2`) in the matrix RD therefore gives the amount of bits that are needed to encode a given bit plane from a given codeblock in the processed subband (R) and the distortion reduction that this bit plane brings when decoding the image (D). As all codeblocks do not have the same number of significant bit planes in a subband, dimension 2 of matrix RD (counting the bit planes) is taken equal to the maximum number of significant bit planes among the codeblocks from the subband. Note that `resno=1` for the smallest resolution and `bpno=1` for the most significant bit plane.

```
if img.useownCDT==0
    CDT=load(img.CDT2use);
    ...
end
```

Nested in loops on resolutions, subbands, codeblocks and bit planes, two simple functions are applied to fill in the RD matrix.

```
[rc rnc]=get_rate(ctxt,coeff,bpno,CDT);
RD(cbno,offset+bpno,1)=rc;
RD(cbno,offset+bpno,2)=get_disto(disto,bpno);
```

The `get_rate` function computes the entropy of each bit plane based on the given `CDT`. It returns the rate `rc` that would be obtained after a context-based entropy coding and the rate `rnc` that would be obtained after a non-context-based entropy coding (i.e. with a global probability distribution on the whole image). To do so, it simply uses the entropy, which gives

$$\sum_{c=1}^{nctxt}\left(-n_0(c)\log_2\left(p(0\,|\,c)\right)-n_1(c)\log_2\left(p(1\,|\,c)\right)\right) \tag{10.26}$$

where *nctxt* is the total number of context (12 in our case), $n_0(c)$ and $n_1(c)$ are the number of "0" symbols and "1" symbols with context *c*, respectively. This simply translates in MATLAB code as

```
entropy = sum(-(log2(CDT(1:end-1))).*ctxt(:,2) ...
          -(log2((1-CDT(1:end-1)))).*(ctxt(:,1)-ctxt(:,2)));
```

The `get_disto` function computes the difference between square errors obtained with successive bit planes.

```
if bpno<numel(disto)
    D = disto(bpno)-disto(bpno+1);
else
    D = disto(end);
end
```

The matrix `RD` of each subband will then be used in the next Section to find for each codeblock the best truncation point, i.e., the one that will minimize the distortion for a given global bit budget.

Here we simply use an intermediary matrix `R_res` to store and compare the global compression performance obtained with and without context. Figure 10.19 shows the compression ratio obtained for each resolution level. The total number of uncompressed bits taken into account to compute these values is the number of bits truly processed by the entropy coder, i.e. wavelet coefficients excluding the non-significant bit planes. Figure 10.20 computes the global compression ratio, comparing the original number of bits in the pixel domain, with the rate obtained after the entropy coding step. As expected, the context-based approach is more efficient than the other one, as it more accurately estimates the probability of getting a "1" or a "0" depending on the coefficient location. In particular, in Fig. 10.19, we see that the non-context-based entropy coding actually expands the LL subband rather than compressing it. This is because the global probability distribution used in this case is very different from the one really observed for this resolution level. On the contrary, when using contexts, and even if those contexts do not take the resolution level into account, the entropy coder still achieves compression on the low frequencies.

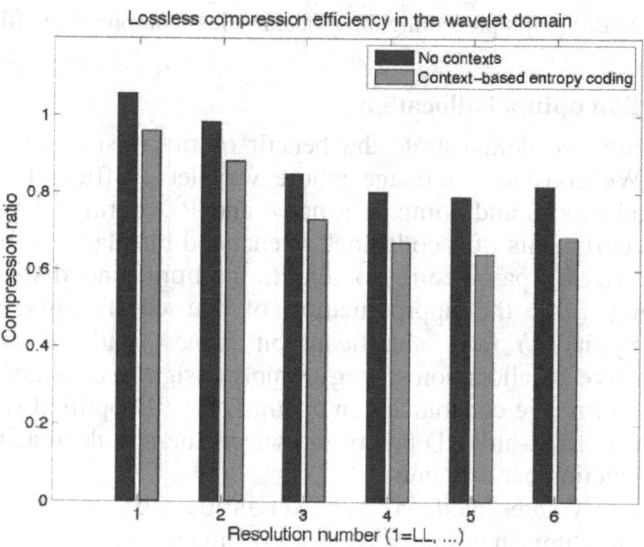

Fig. 10.19 Bar graph plotting the compression ratio obtained for each resolution level, with and without contexts

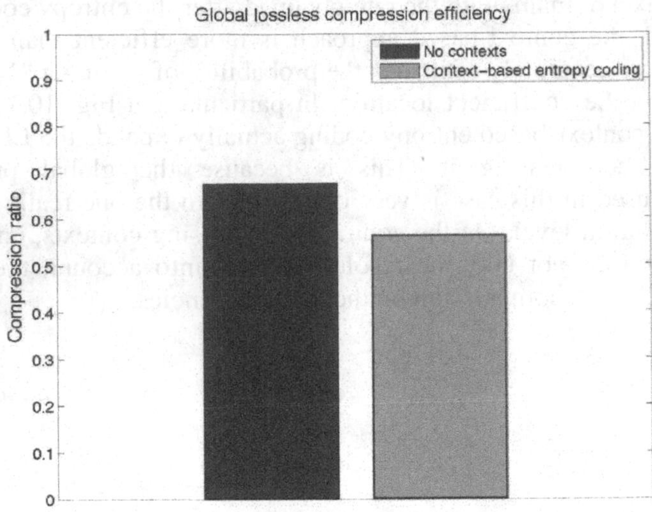

Fig. 10.20 Bar graph presenting the global lossless compression efficiency

Rate distortion optimal allocation

In this section, we demonstrate the benefit of rate–distortion optimal bit allocation. We consider an image whose wavelet coefficients have been split into codeblocks and compare a naive and RD optimal bit allocation. The set of coefficients of a codeblock is encoded bit plane by bit plane to define `bpno` (d_{ij}, b_{ij}) pairs, corresponding to the `bpno` rate–distortion trade-offs resulting from the approximation of the coefficients of the *ith* codeblock by its *jth* most significant bit planes, with $0 \leq j <$ `bpno`. The envisioned naive bit allocation strategy simply assigns a constant number of bit planes to all image codeblocks. In contrast, the RD optimal strategy first computes the convex-hull RD points and selects them in decreasing order of distortion reduction per cost unit.

As previously presented, `img.res(resno).sb(sbno).RD(i,j,k)` denotes a structure that conveys, for all image resolution and image subband, the incremental reduction of distortion and the increase of bits corresponding to the addition of the *jth* bit plane to the definition of codeblock *i*. Index *k* is used to differentiate the cost in bits (*k*=1) from the decrease in distortion (*k*=2).

To implement the RD optimal bit allocation method, the *hull structure* `HS` is defined to collect the convex-hull RD points of every codeblock in the image. For each convex-hull RD point, the `HS` structure records the decrease in distortion per bit unit provided by the bit planes corresponding to the convex-hull RD point. It also records the codeblock resolution, subband, index and number of bit planes corresponding to the convex-hull RD point.

```
% Loop on codeblocks of resolution i and subband j.
for k=1:size(img.res(i).sb(j).RD,1)
  % Number of bitplanes for the codeblock.
  nbbp_cb = size(img.res(i).sb(j).RD(k,:,:), 2);
  % Number of bitplanes corresponding to the last
  % operating point found on the convex-hull.
  lhbp=0;
  % To enter in the while loop.
  gain_max=1;
  while (lhbp<nbbp_cb && gain_max>0)
    nhbp=lhbp;
    gain_max=0;
    for bp=(lhbp+1):nbbp_cb
      gain=sum( img.res(i).sb(j).RD(k,(lhbp+1):bp,2) )
                 / sum( img.res(i).sb(j).RD(k,(lhbp+1):bp,1) );
      if gain > gain_max
        gain_max=gain;
        nhbp=bp;
      end
    end
    nbHS=nbHS+1;
    deltaR = sum( img.res(i).sb(j).RD(k,(lhbp+1):nhbp,1) );
    deltaD = sum( img.res(i).sb(j).RD(k,(lhbp+1):nhbp,2) );
    HS(nbHS,:)=[gain_max,i,j,k,nhbp, deltaR, deltaD];
    lhbp=nhbp;
  end % End while loop.
end % End for loop.
```

Once the `HS` structure has been defined, the RD optimal allocation strategy simply consists in selecting the convex-hull RD points in decreasing order of distortion reduction per cost unit. Hence, the `HS` structure is sorted to control the bit-plane allocation strategy for a target rate `R_T` as follows:

```
ascendingHS = sortrows(HS);
indHS = size(HS,1);

Rtmp = 0;
while (Rtmp < R_T && indHS>0)
        Rtmp = Rtmp + ascendingHS(indHS,6);
        Dtmp = Dtmp - ascendingHS(indHS,7);
        img.res(ascendingHS(indHS,2)).sb(ascendingHS(indHS,3))
                .nbrplanes(ascendingHS(indHS,4)) =
ascendingHS(indHS,5);
        indHS = indHS - 1;
end
```

In this MATLAB code sample, `Rtmp` and `Dtmp` define the global image bit budget and distortion, respectively, while the variable `img.res(i).sb(j).nbrplanes(k)` records the number of bit planes allocated to the *kth* codeblock of the *jth* subband of the *ith* resolution.

On the MATLAB CDrom, the global image RD trade-offs obtained when naively allocating a constant number of bit planes to each codeblock are compared with the RD points obtained when RD optimal allocation targets the same bit budgets as the ones obtained for naive allocation. In Fig. 10.21, we present a comparison of the RD points achieved using both allocation strategies.

MATLAB code is also provided to reconstruct the image based on the number of bit planes allocated to each codeblock. It permits to compare both allocation methods from a perceptual point of view. Reconstructed images using naive and optimum bit allocation strategies are presented in Figs. 10.22 and 10.23, respectively. It is clear that the visual quality of the optimum bit allocation version largely outperforms that of the naive allocation for both cases.[19]

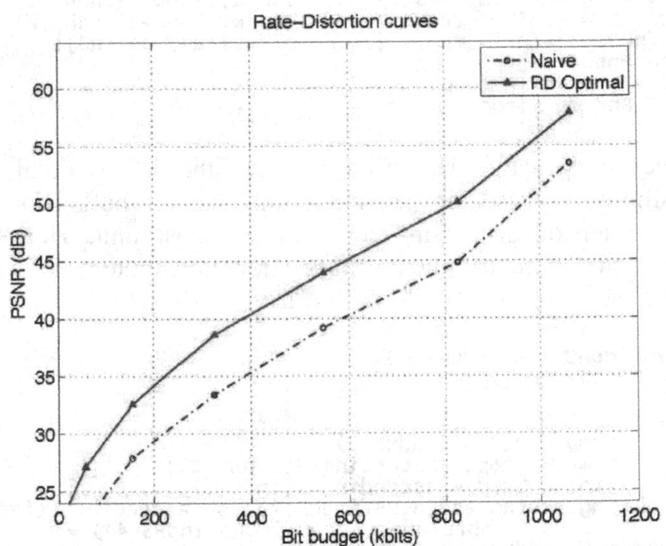

Fig. 10.21 Rate–distortion curves comparing the naïve and the optimal bit allocation strategies

---

[19] Recall that Cameraman size is 256×256, while Barbara size is 512×512; hence the so different target rates.

Reconstructed image of cameraman.tif at 7.67 kbits, Naive allocation.
Reconstructed image of images/Barbara512.pgm at 56.09 kbits, Naive allocation.
(a)
(b)

Fig. 10.22 Reconstructed version, using naïve bit allocation, of the image: (**a**) *Cameraman* at 9.4 kbits (PSNR=26.06 dB) and (**b**) *Barbara* at 65.56 kbits (PSNR=20.59 dB)

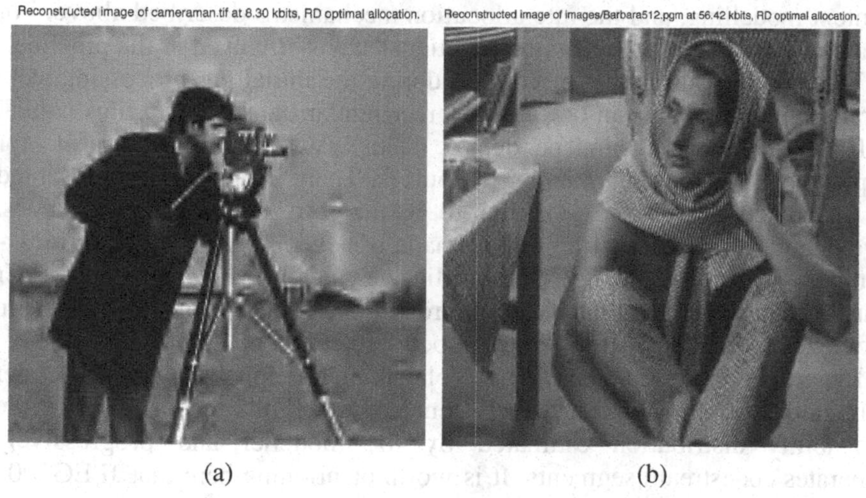

Reconstructed image of cameraman.tif at 8.30 kbits, RD optimal allocation.
Reconstructed image of images/Barbara512.pgm at 56.42 kbits, RD optimal allocation.
(a)
(b)

Fig. 10.23 Reconstructed version, using optimal bit allocation, of the image: (**a**) *Cameraman* at 9.4 kbits (PSNR=30.67 dB) and (**b**) *Barbara* at 66.00 kbits (PSNR=25.86)

10.5 Going further: From concepts to compliant JPEG2000 codestreams

To complete this chapter, Fig. 10.24 presents the entire pipeline to generate a JPEG2000-compliant codestream.

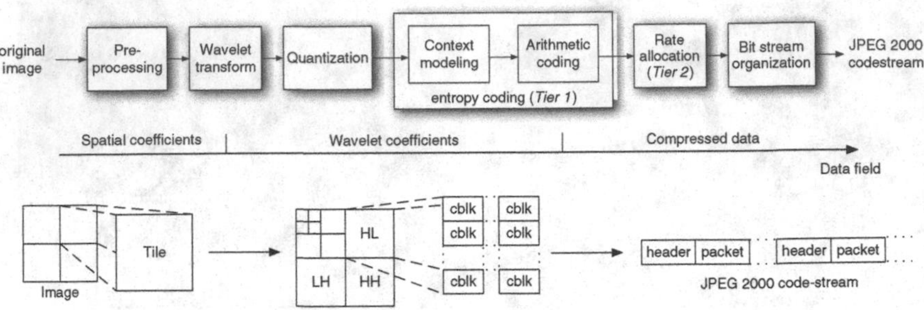

Fig. 10.24 The JPEG2000 coding steps. Pixels are first transformed into wavelet coefficients. Then, the various subbands from each resolution level are divided into codeblocks that are independently entropy-coded. Eventually, entropy-coded data are distributed in packets that compose the final JPEG2000 codestream

We recognize the wavelet transform, the bit-plane quantization, the context modelling, and the rate allocation mechanisms described above. For completeness, we now survey the additional stages involved in the pipeline.

Three operations can be performed during the initial pre-processing step. First of all, the image can be split into rectangular blocks called tiles, which will be compressed independently. This is particularly useful for applications with limited memory resources. Then, if pixels are represented by unsigned values, they are shifted to get the corresponding signed values. Eventually, in the case of an image made of several components, an inter-component decorrelation can be applied. Typically, the RGB to YCbCr transformation is exploited to increase the subsequent compression efficiency by reducing the correlation between components.

Regarding *arithmetic coding*, as explained in Section 10.2, the MQ-coder (Mitchell and Pennebaker 1988) encodes the bits according to the probability distribution estimated by the modeller and progressively generates codestream segments. It is worth mentioning here that JPEG2000 further refines the way bits are scanned within a bit plane, so as to first encode the bits for which a large benefit in quality is expected per unit of rate. This process is based on the observation of already encoded bit planes, and the algorithm used to select the bits that will be encoded first is named

Embedded Block Coding with Optimized Truncation (EBCOT) (Taubman 2000) by the JPEG2000 community.

Note that when decoding an image, the context modelling part only provides the context to the arithmetic coding part and waits for the decoded bit. This bit is computed by the MQ-decoder that progressively consumes the compressed bitstream.

The last step involves *packetization* and *bitstream organization*. Once all bit planes have been encoded, rate–distortion optimal allocation is considered for a set of increasing target bit budgets or equivalently for a decreasing sequence of λ parameters. As explained in Section 10.3, the sequence of corresponding RD optimal allocations progressively adds bit planes to the image codeblocks.

In JPEG2000, the incremental contributions added from one target rate to another (or from one Lagrangian parameter to another) are grouped in the so-called *quality layers*. Several quality layers are thus defined for an image, corresponding to distinct rate–distortion trade-offs.

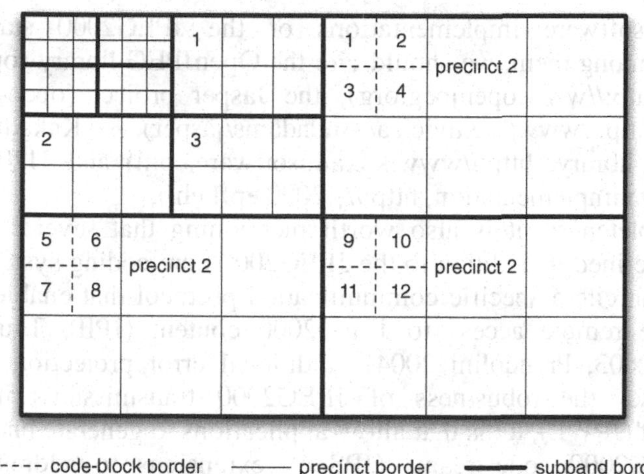

Fig. 10.25 Precinct and codeblock subdivisions. A precinct includes all codeblocks belonging to a given resolution level and covering a given spatial area. For example, in the highest resolution, the first precinct is made of 12 codeblocks, four from each subband. The size (i.e. the number of pixels) of the area covered by a precinct can vary from one resolution level to another. This is the case in the figure where this area size has been adapted in each resolution level so as to have the same number of precincts (6) in each level

Once the rate allocation step has distributed the incremental contributions from all codeblocks over the specified quality layers, the compressed data are divided into packets. A packet contains the information related to a certain quality layer from a certain resolution, in a certain spatial location of one of the image components. A spatial location is called a *precinct* and corresponds to the set of codeblocks from all subbands of a given resolution and covering a given spatial area (see Fig. 10.25). Packets, along with additional headers, form the final JPEG2000 codestream.

10.6 Conclusion

Although quite elaborated in its functional capabilities, the JPEG2000 image compression standard relies on the combination of rather simple basic principles: discrete wavelet transform, context-based entropy coding and rate–distortion optimal allocation mechanisms. This combination leads to a flexible and powerful image compression standard that is mainly used in professional markets requiring a high-quality level and a solid intrinsic scalability.

Various software implementations of the JPEG2000 standard are available. Among them, we should cite the OpenJPEG library (open-source C-library, http://www.openjpeg.org), the Jasper project (open-source C-library, http://www.ece.uvic.ca/~mdadams/jasper), Kakadu (C++ commercial library, http://www.kakadusoftware.com) and JJ2000 (freely available java implementation, http://jj2000.epfl.ch/).

For completeness, it is also worth mentioning that several extensions have been defined in addition to the JPEG2000 core coding system. Among them, we can cite a specific communication protocol that enables efficient and flexible remote access to JPEG2000 content (JPIP, Taubman and Prandolini 2003, Prandolini 2004), additional error protection techniques that increases the robustness of JPEG2000 transmissions in wireless environment (JPWL), tools that allow applications to generate and exchange secure JPEG2000 codestreams (JPSec), extensions to address 3D and floating point data (JP3D), etc. We invite the interested reader to visit www.jpeg.org/jpeg2000 for more information on this topic.

References

Daubechies I., Ten lectures on wavelets. Society for Industrial and Applied Mathematics, 1992, ISBN 0-89871-274-2.

DCI. Digital cinema system specifications. Digital Cinema Initiatives (DCI), March 2005, URL http://www.dcimovies.com/.

Foos D.H., Muka E., Slone R.M., Erickson B.J., Flynn M.J., Clunie D.A., Hildebrand L., Kohm K.S., Young S.S., JPEG2000 compression of medical imagery. Proceedings of SPIE, 3980:85, 2003.

Fossel S., Fottinger G., Mohr J., Motion JPEG2000 for high quality video systems. IEEE Transactions on Consumer Electronics, 49(4):787–791, 2003.

Janosky J., Witthus R.W., Using JPEG2000 for enhanced preservation and web access of digital archives. In IS&T's 2004 Archiving Conference, pages 145–149, April 2004.

Kellerer H., Pferschy U., Pisinger D. Knapsack Problems. Springer Verlag, Berlin, 2004, ISBN 3-540-40286-1.

Mallat S., A Wavelet Tour of Signal Processing. Academic Press, San Diego, CA, 2nd edition, 1999.

Marpe D., George V., Cycon H.L., Barthel K.U., Performance evaluation of Motion-JPEG2000 in comparison with H. 264/AVC operated in pure intracoding mode. Proceedings of SPIE, 5266:129–137, 2003.

Mitchell J.L., Pennebaker W.B., Software implementations of the Qcoder. IBM Journal of research and Development, 32(6):753–774, November 1988.

Ortega A., Optimal bit allocation under multiple rate constraints. In Data Compression Conference, pages 349–358, Snowbird, UT, April 1996.

Ortega A., Ramchandran K., Rate–distortion methods for image and video compression. IEEE Signal Processing Magazine, 15(6):23–50, November 1998.

Ortega A., Ramchandran K., Vetterli M., Optimal trellis-based buffered compression and fast approximation. IEEE Transactions on Image Processing, 3(1):26–40, January 1994.

Prandolini R., 15444-9:2004 JPEG2000 image coding system – Part 9: Interactivity tools, apis and protocols. Technical Report, ISO/IEC JTC1/SC29 WG1, March 2004.

Rabbani M., Joshi R., An overview of the JPEG2000 still image compression standard. Signal Processing: Image Communication, 17(1):3–48, January 2002.

Santa-Cruz D., Grosbois R., Ebrahimi T., JPEG2000 performance evaluation and assessment. Signal Processing: Image Communication, 17(1):113–130, January 2002.

Shoham Y., Gersho A., Efficient bit allocation for an arbitrary set of quantizers. IEEE Transactions on Signal Processing, 36(9):1445–1453, September 1988.

Skodras A., Christopoulos C., Ebrahimi T., The JPEG2000 still image compression standard. Signal Processing Magazine, IEEE, 18(5):36–58, 2001.

Smith M., Villasenor J., Intra-frame JPEG-2000 vs. Inter-frame compression comparison: The benefits and trade-offs for very high quality, high resolution sequences. SMPTE Technical Conference, Pasadena, CA, October 2004.

Symes P., JPEG2000, the Professional compression scheme. Content Technology Magazine, 3(3), June 2006, URL http://svc126.wic003tv.server-web.com/CT-pdf/CT-May-June-2006.pdf.

Taubman D., High performance scalable image compression with ebcot. IEEE Transactions on Image Processing, 9(7):1158–1170, July 2000.

Taubman D., Marcellin M.W., JPEG2000: Image Compression Fundamentals, Standards and Practice. Kluwer Academic, Boston, MA, USA, 2002.

Taubman D., Prandolini R., Architecture, philosophy and performance of JPIP: Internet protocol standard for JPEG2000. In International Symposium on Visual Communications and Image Processing (VCIP), Lugano, Switzerland, July 2003.

Taubman D., Rosenbaum R., Rate–distortion optimized interactive browsing of JPEG2000 images. In IEEE International Conference on Image Processing (ICIP), September 2003.

Tzannes A., Ebrahimi T., Motion JPEG2000 for medical imaging. ISO/IEC wg1n2883, Medical Imaging Ad Hoc Group, 2003.

Wolsey L., Integer Programming. Wiley, New York, 1998.

Zhang D.R., Wang X.X., The manipulation approach of JPEG2000 compressed remote sensing images. Proceedings of SPIE, 6044:315–324, 2005.

Chapter 11

How can physicians quantify brain degeneration?

M. Bach Cuadra(°), J.-Ph. Thiran(°), F. Marqués(*)

(°) Ecole Polytechnique Fédérale de Lausanne, Switzerland
(*) Universitat Politècnica de Catalunya, Spain

"If the human brain were so simple that we could understand it, we would be so simple that we couldn't" E.M. Pugh

Life expectation is increasing every year. Along with this aging population, the risk of neurological diseases (e.g., dementia)[1] is considerably increasing[2] as well. Such disorders of the human nervous system affect the patient from both a physical and a social point of view, as in the case of Alzheimer's disease, one of the most known brain disorders (Mazziotta et al. 2000).

To pose their diagnosis, in addition to the usual clinical exam, physicians increasingly rely on the information obtained by means of 2D or 3D imaging systems: the so-called *image modalities*. Nowadays there exists a wide range of 3D medical image modalities that allow neuroscientists to see inside a living human brain. 3D imaging is of precious help since it allows, for instance, the physician to better localize specific areas inside the body

[1] Dementia is a progressive degenerative brain syndrome, which affects memory, thinking, behavior, and emotion.

[2] In 2006, 24.3 million people in the world had dementia, with 4.6 million new cases annually. By 2040, the number would have risen to 81.1 million. Source: http://www.alz.co.uk/.

T. Dutoit, F. Marqués (eds.), *Applied Signal Processing*,
DOI 10.1007/978-0-387-74535-0_11,

and to understand the relationships between them. Among all medical image modalities, magnetic resonance imaging (MRI) has been increasingly used since their invention in 1973 by P. Lauterbur and Sir P. Mansfield.[3] As a matter of fact, MR imaging has useful properties; it provides an anatomical view of the tissues and deep structures of the body (see Fig. 11.1), thanks to the magnetic properties of water, without the use of ionic radiation (Haacke et al. 1999).

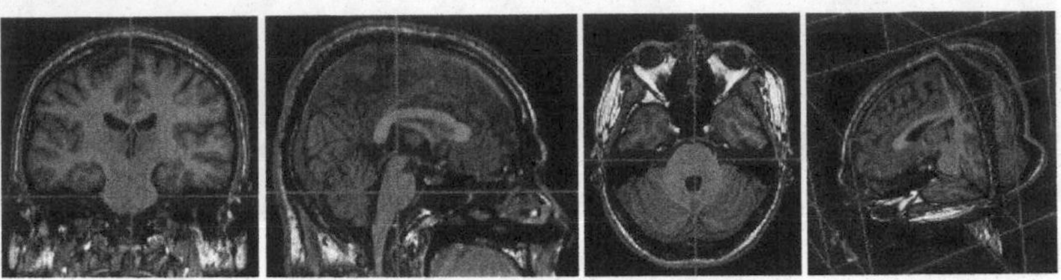

Fig. 11.1 3D MR image of a human brain: coronal, sagittal, axial, and 3D views. The red (dark gray) cross indicates the same spatial point in three views

For a long time, MRI has been used qualitatively, for visualization purpose only, to help physicians in their diagnosis or treatment planning. For instance, when physicians look at MR images to evaluate lesions or tumors, they first segment the suspicious growth mentally and then use their training and experience to assess its properties. However, due to several factors such as the ever-increasing amount of clinical data generated in clinical environments and their increasingly complex nature, it is nowadays necessary to rely on *computerized segmentation techniques* (Suri et al. 2005). By segmentation we mean dividing an image into homogeneous regions for a set of relevant properties such as color, intensity, and texture, which allow a quantitative analysis of medical images (automatic estimation of quantitative measures and regions of interests).

In the specific case of brain disorders, accurate and robust brain tissue segmentation from MR brain images is a key issue for quantitative studies (Bach Cuadra et al. 2005). Studying MR brain images of such patients is a key step, first, to differentiate from a normal tissue loss due to normal aging; second, to better understand and characterize the anatomy of such patients; and third, to better assess drug treatments. Thus, there is a need to quantify MR image changes to determine *where* the tissue loss is and its *amount*.

---

[3] They were awarded the 2003 Nobel Prize in Medicine or Physiology for their discoveries concerning MRI.

In this chapter, we will address the problem of quantifying brain degeneration in a patient suffering from focal brain degeneration, causing a progressive aphasia. To do so, we will do a longitudinal study, that is, we will compare two MR images from the same patient acquired in a year interval (I1 in 2000 and I2 in 2001). First, the image classification problem must be solved for every 3D image acquisition (see Fig. 11.2). Second, to detect and quantify the tissue loss, we need to compare the I1 segmentation with the I2 segmentation in a point to point basis. Therefore, an *image registration* problem arises; a point-to-point correspondence is not ensured between both MR scans since the patient has not been lying in the same position during the two acquisition process. This can be solved by a rigid registration process that looks for the spatial transformation that will bring one image to a point-to-point correspondence with the other one. In this application, only translation and rotation are needed to compensate for a different position of the patient in the two scans. Even in such a simplified framework (only translation and rotation), image registration is a very complex problem (Hajnal et al. 2001), whose complete discussion is out of the scope of this chapter. For the sake of completeness, the registration problem is briefly addressed in Section 11.3.

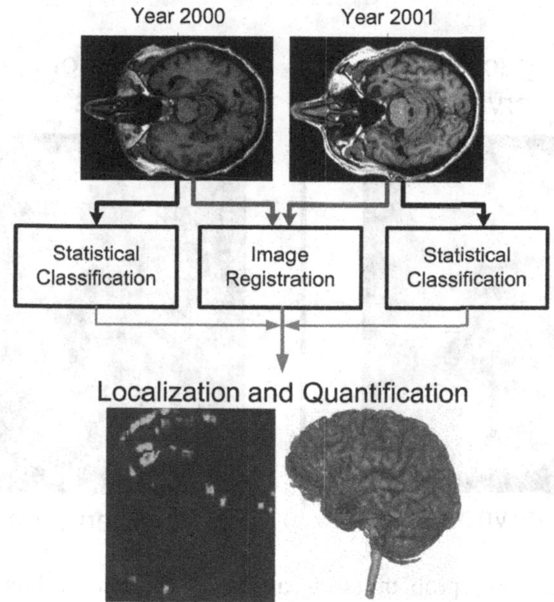

Fig. 11.2 Image processing tools presented in this chapter

The chapter is organized as follows. In Section 11.1, we present the theoretical framework to solve the image classification problem. In Section 11.2, a MATLAB proof of concept for the image classification is developed. As previously commented, the problem of image registration is briefly discussed in Section 11.3. Finally, in Section 11.4, a discussion concludes the chapter.

11.1 Background – Statistical pattern recognition for image classification

Numerous approaches have been proposed for image classification. They can be divided into two main groups: (i) supervised classification, which explicitly needs user interaction and (ii) unsupervised classification, which is completely automatic. In this chapter, we describe an unsupervised statistical classification approach. In fact, this can be seen as an estimation problem: estimate the underlying class (*hidden data*) of every *voxel* (3D extension of the pixel concept) from the 3D intensity image (*observed data*). This concept is illustrated in Fig. 11.3.

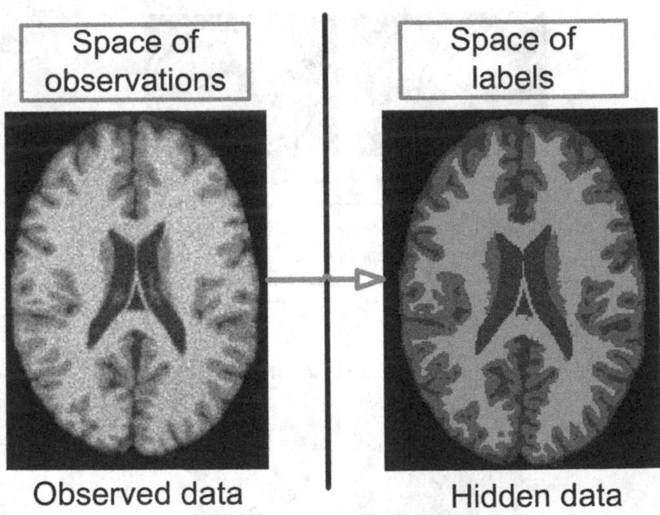

Fig. 11.3 Classification problem seen as an estimation problem. MR image intensities (on the *left*) are the observed data. The labeled image (on the *right* side) is what we are looking for: *blue* (*dark gray*) voxels are the cerebrospinal fluid (CSF) class, *red* (*medium gray*) voxels are Gray matter (GM), and *green* (*light gray*) voxels are white matter (WM)

11.1.1 Statistical framework

Formally, let us consider N data points (typically voxels) and the observed data features $y_i \in \mathbb{R}$, with $i \in S = \{1,2,\ldots,N\}$ [4]. In the MR image case, Y_i is the random variable associated with the intensity of voxel i, with the set of possible outcomes D, and y_i represents the observed intensity of voxel i. The ensemble of all random variables defining the MR image random process is denoted by Y and any simultaneous observation of the random Y_i (i.e., any observed MR image) is denoted by

$$\boldsymbol{y}=\{y_1,y_2,\cdots,y_N\}\subset D^N \subset \mathbb{R}^N \tag{11.1}$$

The classification process aims at classifying the data into one of the hidden underlying classes labeled by one of the symbols L={*CSF, GM, WM*}. CSF stands for cerebrospinal fluid (blue or dark gray voxels in Fig. 11.3), GM for gray matter (red or medium gray voxels in Fig. 11.3), and WM for white matter (green or light gray voxels in Fig. 11.3). X_i is the random variable associated with the class of voxel i, with the set of possible outcomes L, and x_i represents the class assigned to voxel i. The ensemble of all random variables defining the hidden underlying random process is denoted by X and any simultaneous configuration of the random variables X_i (i.e., any label map) is given by

$$\boldsymbol{x}=\{x_1,x_2,\cdots,x_N\}\subset L^N \tag{11.2}$$

In this chapter, we are looking for the *most probable* label map $\boldsymbol{x}^*$, given the image observation **y**. As already introduced in Chapter 4, this can be expressed mathematically by the so-called *Bayesian*, or *maximum a posteriori* (MAP), decision rule as: [5]

---

[4] S represents the grid in which voxels are defined and, therefore, i is a specific site in this grid.

[5] As in Chapter 4, here we will often use a shortcut notation for probabilities, when this does not bring confusion. The probability $P(A=a|B=b)$ that a discrete random variable A is equal to a given the fact that random variable B is equal to b will simply be written $P(a|b)$. Moreover, we will use the same notation when A is a *continuous* random variable for referring to *probability density* $p_{A|B=b}(a)$.

$$\boldsymbol{x}^* = \arg\max_{x \in X} P(\boldsymbol{x} \mid \boldsymbol{y}, \Theta) \tag{11.3}$$

where $\boldsymbol{x}$ represents *a possible* label map and the conditional probability is evaluated over all possible label maps X, and Θ represents the set of parameters used to estimate the probability distribution. For instance, in the Gaussian case, these parameters are the mean vector and the covariance matrix of the process.

Bayes' rule can then be applied to Equation (11.3), yielding

$$P(\boldsymbol{x} \mid \boldsymbol{y}, \Theta) = \frac{P(\boldsymbol{y} \mid \boldsymbol{x}, \Theta) P(\boldsymbol{x})}{P(\boldsymbol{y} \mid \Theta)} \tag{11.4}$$

where $P(\boldsymbol{y} \mid \boldsymbol{x}, \Theta)$ represents the contribution of the so-called *intensity image model* (i.e., the likelihood that a specific model $\boldsymbol{x}$ has produced the intensity image observation $\boldsymbol{y}$), which models the image intensity formation process for each tissue type; $P(\boldsymbol{x})$ represents the contribution of the so-called *spatial distribution model* (i.e., the a priori probability of the corresponding label map) that models the spatial coherence of the images; and $P(\boldsymbol{y} \mid \Theta)$ represents the a priori probability of the observed image.

Based on the above, we thus have to address the following *classification* problem. Given an image intensity $\boldsymbol{y}$, find the most probable label map $\boldsymbol{x}^*$ such that

$$\boldsymbol{x}^* = \arg\max_{x \in X} \frac{P(\boldsymbol{y} \mid \boldsymbol{x}, \Theta) P(\boldsymbol{x})}{P(\boldsymbol{y} \mid \Theta)} \tag{11.5}$$

The conditional probability $P(\boldsymbol{y} \mid \Theta)$ is *often assumed to be constant for all hypotheses of label maps and, therefore, it* can be ignored in the maximization process so that Equation (11.5) simplifies to

$$\boldsymbol{x}^* = \arg\max_{x \in X} P(\boldsymbol{y} \mid \boldsymbol{x}, \Theta) P(\boldsymbol{x}) \tag{11.6}$$

which itself implies two steps:

- *Image intensity modeling*: Image modeling refers to the estimation of $P(\boldsymbol{y} \mid \boldsymbol{x}, \Theta)$, and this is typically carried out using Gaussian mixture models (GMM; see Section 11.1.2). As we will see, assuming data independency, classification can be performed relying only on this model. However, this assumption is too strong and the resulting classifiers lead to low-quality classification results (see Section 11.2.3).

- *Spatial distribution modeling*: Spatial prior probabilities of classified images $P(\boldsymbol{x})$ are modeled using Markov random fields (MRF; see Section 11.1.4).

Actually, the global model to be used comes from the combination of the two previous models. This way, in Section 11.1.5, we will introduce the concept of hidden Markov random fields (HMRF), which allows including dependency among random variables. Finally, in Section 11.1.6, Gaussian hidden Markov random fields (GHMRF) are presented as a specific case of the previous HMRF model.

Note that there are in fact many aspects of the theory, which we will present here that are shared with Chapter 4. The same MAP method is used, thus the mathematical formalism is almost the same. As we shall see, however, some differences exist: first, here we will deal with 3D data; and second, the prior probability $P(\boldsymbol{x})$ will be modeled by hidden Markov random fields, where emission probabilities will not be estimated as in Chapter 4 but modeled by an exponential function, thanks to the Hammersley-Clifford theorem.

11.1.2 Gaussian mixture models (GMM)

We define in this section the initial mathematical model of the intensity distribution of an MR image volume. Let us suppose, for now, that all the random variables representing the image intensity, Y_i (with $i \in S$), are identically and independently distributed. This is a very strong assumption of the data, which will be later refined using the notion of conditional independence in the context of hidden Markov random fields (see Section 11.1.5). Nevertheless, under the previous assumptions, the *probability density function* of the intensity voxel at *ith location* can be defined by

$$P(Y_i) = \sum_{\forall l \in L} P(X_i = l) P(Y_i \mid X_i = l), \tag{11.7}$$

where $P(X_i = l)$ is the probability of voxel i being of tissue class l, and $P(Y_i \mid X_i = l)$ is the probability density function of Y_i, given the tissue class l. A density function of this form is called *finite mixture* (FM) density. The conditional densities $P(Y_i \mid X_i = l)$ are called *component densities* and, in this case, they encode the intensity information. The a priori probabilities, $P(X_i = l)$, are the *mixing parameters* and they encode the spatial information (see Section 11.1.5).

The simplest component density considers that the probability density function for the observed intensity Y_i, given the pure tissue class $X_i = l$, is given by the Gaussian function

$$P(Y_i \mid X_i = l, \theta_l) = \frac{1}{\sigma_l \sqrt{2\pi}} e^{\frac{-(Y_i - \mu_l)^2}{2\sigma_l^2}} = f(Y_i \mid \theta_l) \tag{11.8}$$

where the model parameters $\theta_l = \{\mu_l, \sigma_l\}$ are the mean and the variance of the Gaussian function *f*, respectively. Let us note that this is a good approximation since the noise present, in an MRI follows a Rician distribution (Gudbjartsson and Patz 1995, Haacke et al. 1999), which, at high signal-to-noise ratio, can be approximated by a Gaussian distribution.

In the particular case where the random variables X_i (with $i \in S$) are also assumed to be independent of each other[6] means that

$$P(X_i = l) = \omega_l \tag{11.9}$$

Then, the probability density function for the observed intensities Y_i involved in the image intensity model term of Equation (11.6) can be written as

$$P(Y_i \mid \Theta) = \sum_{l \in L} \omega_l f(Y_i \mid \theta_l) \tag{11.10}$$

Note that the mixing parameters ω_l are not known in general and must therefore also be included among the unknown parameters. Thus, the mixture density parameter estimation tries to estimate the parameters $\Theta = \{\theta_l, \omega_l\}$ such that

$$\sum_{l \in L} \omega_l = 1. \tag{11.11}$$

[6] As it will be shown in Section 11.1.4, this implies that no Markov random field model is considered.

11.1.3 The Expectation–Maximization algorithm (EM)

A possible approach to solve the parameter estimation problem in Equation (11.10) is to find the value of Θ using either the *maximum likelihood* (ML) or the *maximum log likelihood criterion*[7] (Bilmes 1998):

$$\Theta^* = \arg\max_{\Theta} P(Y \mid X_i, \Theta) = \arg\max_{\Theta} \log(P(Y \mid X_i, \Theta)) \tag{11.12}$$

One of the most used methods to solve the maximization problem is the EM algorithm (see Chapter 4). For the particular case of Gaussian distributions, the EM algorithm leads to the following equations:

- *Initialization step*. Choose the best initialization for $\Theta^{(0)}$.
- *Expectation step.* Calculate the a posteriori probabilities:

$$\hat{P}^{(k)}(X_i = l \mid y_i, \hat{\Theta}_l^{(k-1)}) = \frac{P\left(y_i \mid \hat{\Theta}_l^{(k-1)}\right)\hat{P}^{(k-1)}(X_i = l)}{\sum_{l \in L} P\left(y_i \mid X_i = l, \hat{\Theta}_l^{(k-1)}\right)\hat{P}^{(k-1)}(X_i = l)} \tag{11.13}$$

- *Maximization step*:

$$\hat{\omega}_l^{(k)} = \hat{P}^{(k)}(X = l) = \frac{1}{N}\sum_{i \in S} \hat{P}^{(k)}(X_i = l \mid y_i, \hat{\Theta}_l^{(k-1)}) \tag{11.14}$$

$$\hat{\mu}_l^{(k)} = \frac{\sum_{i \in S} y_i \hat{P}^{(k)}\left(X_i = l \mid y_i, \hat{\Theta}_l^{(k-1)}\right)}{\sum_{i \in S} \hat{P}^{(k)}\left(X_i = l \mid y_i, \hat{\Theta}_l^{(k-1)}\right)} \tag{11.15}$$

$$(\hat{\sigma}_l^{(k)})^2 = \frac{\sum_{i \in S} (y_i - \hat{\mu}_l^{(k)})^2 \hat{P}^{(k)}\left(X_i = l \mid y_i, \hat{\Theta}_l^{(k-1)}\right)}{\sum_{i \in S} \hat{P}^{(k)}\left(X_i = l \mid y_i, \hat{\Theta}_l^{(k-1)}\right)} \tag{11.16}$$

[7] Although both criteria are equivalent, it is usually more convenient to work with the sum of *log* likelihoods (see Chapter 4).

11.1.4 Markov random fields (MRF)

In Section 11.1.2, we have made the assumption that all random variables X_i (with $i \in S$) were independent of each other (see Equation (11.9)). If we want to introduce some relationship between neighbor variables in the random field (which seems a very natural constrain), we have to refine our random field model.

We have seen in Chapter 4 that a Markov model (or *Markov chain)* can be used for estimating the probability of a random sequence, based on the assumption that the value taken by *ith* observation in the sequence only depends on its immediate neighborhood, i.e., on the $(i-1)th$ observation. Similarly, *a Markov random field* (MRF) can be used to estimate the probability of a random image,[8] based on the assumption that the dependency of the value of a given pixel (or voxel) on the values taken by all other pixels in an image, $P(X_i)$, can be reduced to the information contained in a local neighborhood of this pixel.

More formally, let us relate all the pixel (or voxel) locations (*sites*) in a grid S with a *neighborhood system* $N = N_i \ (i \in S)$, where N_i is the set of sites neighboring *i*, with $i \notin N_i$ and $i \in N_j \Leftrightarrow j \in N_i$. A *random field X* is then said to be an MRF on S with respect to a neighborhood system *N* if and only if, for all *i* in S

$$P\left(x_i \mid x_{S-\{i\}}\right) = P\left(x_i \mid x_{N_i}\right) \tag{11.17}$$

where x_i denotes the actual value of X_i, i.e., of the random field *X* at location *i*; $x_{S-\{i\}}$ denotes the values of the random field *X* at all the locations in *S* except *i*; and x_{N_i}, denotes the values of the random field *X* at all locations within the neighborhood of location *i*. This property of an MRF is known as *Markovianity.*

According to the *Hammersley–Clifford theorem*, an MRF can be equivalently characterized by a *Gibbs distribution*,

$$P(\boldsymbol{x}) = \frac{e^{-U(\boldsymbol{x},\beta)}}{Z} \tag{11.18}$$

---

[8] The definitions here proposed can be related to 2D or 3D images. In the case of 2D images, the single elements are pixels, whereas in the case of 3D images, they are voxels.

where $\boldsymbol{x}$ represents a simultaneous configuration of the random variables, X_i, that is, a given label map (see Section 11.1.1).

The expression in Equation (11.18) has several *free* parameters to be determined: the *normalization factor* Z, the *spatial* parameter β, and the *energy function U*. Let us briefly discuss how these parameters can be determined in the particular framework of image segmentation. We refer the interested reader to Li (2001) for further details.

The energy function *U*

Given the previously commented Hammersley–Clifford theorem, we can include a specific definition of the spatial distribution model in Equation (11.6). This way, the maximization proposed in Equation (11.6) can be formulated in terms of the energy function:

$$\log\left[P(\boldsymbol{y}\,|\,\boldsymbol{x},\Theta)P(\boldsymbol{x})\right] = \log P(\boldsymbol{y}\,|\,\boldsymbol{x},\Theta) - U(\boldsymbol{x},\beta) + \text{const} \qquad (11.19)$$

The definition of the energy function is arbitrary and several choices for $U(x)$ can be found in the literature related to image segmentation. A general expression for the energy function can be denoted by the following equation:

$$U(\boldsymbol{x},\beta) = \sum_{\forall i\in S}\left(V_i(x_i) + \frac{\beta}{2}\sum_{j\in N_i} V_{ij}(x_i,x_j)\right) \qquad (11.20)$$

This is known as the *Potts model* with an *external field*, $V_i(x_i)$, that weights the relative importance of the various classes present in the image (see, for instance, Van Leemput et al. (2003)). However, the use of an external field includes additional parameter estimation, thus this model is less used in image segmentation. Instead, a simplified Potts model with no external energy, $V_i(x_i) = 0$, is used. Then, only the local spatial transitions $V_{ij}(x_i,x_j)$ are taken into account and all the classes in the label image are considered equally probable. A common way to define them is, for instance, as

$$V_{ij}(x_i, x_j) = -\delta(x_i, x_j) = \begin{cases} -1, \text{ if } x_i = x_j \\ 0, \text{ otherwise} \end{cases} \quad (11.21)$$

Intuitively, the above equation encourages one voxel to be classified as the tissue to which most of its neighbors belong. That is, it favors configurations where homogenous classes are present in the image.

The spatial parameter β

Given the energy function defined in Equations (11.20) and (11.21), the maximization proposed in Equation (11.6) can be formulated as a minimization problem:

$$\boldsymbol{x}^* = \arg\min_{x \in X} [-\log P(\boldsymbol{y} \mid \boldsymbol{x}, \Theta) + \beta U(\boldsymbol{x})] \quad (11.22)$$

As it can be observed, the *spatial* value of controls the influence of the energy factor due to the spatial prior, $U(\boldsymbol{x})$, over the factor due to the intensity, $-\log P(\boldsymbol{y} \mid \boldsymbol{x}, \Theta)$. Note that its influence on the final segmentation is important. For instance, β=0 corresponds to a uniform distribution over the possible states, i.e., the maximization is done only on the conditional distribution of the observed data $P(\boldsymbol{y} \mid \boldsymbol{x}, \Theta)$. On the contrary, if the spatial information is dominant over the intensity information, i.e., if $\beta \to \infty$, MAP tends to classify all voxels into a single class.

The value of β can be estimated by ML estimation. However, many problems arise due to the complexity of MRF models and alternative approximations have to be done. Commonly, this is done by simulated annealing, such as Monte-Carlo simulations (Geman and Geman 1984) or by maximum pseudo-likelihood approaches (Besag 1974). The β parameter can be also determined *arbitrarily* by gradually increasing its value over the algorithm iterations. Here, the value of β will be fixed empirically and we will study its influence on the final segmentation in Section 11.2.5.

The normalization factor Z

The normalization factor of Gibbs distribution is theoretically defined as

$$Z(U) = \sum_{x \in X} e^{-U(\boldsymbol{x}, \beta)} \quad (11.23)$$

but this definition implies a high computational cost or it is even intractable since the sum, among all the possible configurations, of X is usually not known (Geman and Geman 1984). Note also its dependence on the definition of the energy function U. Nevertheless, in optimization problems, the solution does not depend on the normalization factor since it is a constant value shared by all possible solutions. Therefore, commonly, the exact computation of the normalization factor is circumvented.

11.1.5 Hidden Markov random fields (HMRF)

In Sections 11.1.2 and 11.1.3, we have examined how GMMs can be used for estimating the image intensity model, while assuming that Y_i are independent and identically distributed. In this section, we will introduce some form of dependency among variables Y_i through the use of an underlying, hidden MRF for spatial distribution modeling.

The theory of hidden Markov random field (HMRF) models is derived from hidden Markov models (HMMs), as seen in Chapter 4. HMMs are defined as stochastic processes generated by a Markov chain whose state sequence X cannot be observed directly but only through a sequence of observations Y. As a result, HMMs model a random sequence of observations Y with two sets of probabilities: the emission and the transition probabilities, which respectively account for the local and the sequential randomness of the observations. Here we consider a special case since, instead of a Markov chain, an MRF is used as the underlying stochastic process, i.e., the sequential correlation between observations will be replaced by the spatial correlation between voxels.

The HMM concept can then be adapted to MRFs, leading to the notion of hidden MRF (HMRF). Like HMMs, HMRFs are defined with respect to a pair of random variable families (X,Y), where X accounts for spatial randomness, while Y accounts for local randomness. In summary, a HMRF model is characterized by the following:

- **Hidden random field**: $X = \{X_i, i \in S\}$ is an underlying MRF assuming values in a finite state space L with probability distribution as defined in the Gibbs equation (11.18).
- **Observable random field**: $Y = \{Y_i, i \in S\}$ is a random field with a finite state space D. Given any particular configuration, $\boldsymbol{x} \in L^{\mathbb{N}}$, every Y_i follows a known conditional probability distribution $P(Y_i \mid X=\boldsymbol{x})$ of the same functional form $f(Y_i \mid \theta_x)$, where θ_x are the involved parameters.

This distribution is called *emission probability function* and Y is also referred to as the *emitted random field.*

- **Conditional independence**: For any $\boldsymbol{x} \in L^{\mathbb{N}}$, the random variables Y_i are supposed to be independent, which means that

$$P(\boldsymbol{y} \mid \boldsymbol{x}) = \prod_{i \in S} P(y_i \mid x_i) \tag{11.24}$$

So now, as in Section 11.1.2, it is possible to compute the probability density function for the observed intensities Y_i involved in the image intensity model term of Equation (11.6). In the HMRF case, the probability density function of Y_i dependent on the parameter set Θ and the values in the associated neighborhood of the hidden random field, x_{N_i}, is

$$\begin{aligned} P(Y_i \mid X_{N_i}, \Theta) &= \sum_{l \in L} P(Y_i, X_i = l \mid x_{N_i}, \Theta_l) \\ &= \sum_{l \in L} P(X_i = l \mid x_{N_i}) P(Y_i \mid \theta_l). \end{aligned} \tag{11.25}$$

where $\Theta = \{\theta_{x_i}, x_i \in \ell\}$. This is again a *finite mixture* (FM) density as in Equation (11.10), where, now, the a priori probabilities $P(X_i = l \mid x_{N_i})$ are the *mixing parameters.*

11.1.6 Gaussian hidden Markov random field model (GHMRF)

In the Gaussian case we can assume

$$P(Y_i \mid \theta_l) = \frac{1}{\sigma_l \sqrt{2\pi}} e^{\frac{-(Y_i - \mu_l)^2}{2\sigma_l^2}} = f(Y_i \mid \theta_l), \tag{11.26}$$

where, $f(Y_i \mid \theta_{X_i})$ is a Gaussian distribution defined by $\theta_l = \{\mu_l, \sigma_l\}$. The probability density function of the intensity voxels can now be written as function of the parameter set Θ and of the voxel neighborhood X_{N_i} as

$$P(Y_i \mid X_{N_i}, \Theta) = \sum_{l \in L} P(X_i = l \mid x_{N_i}) f(Y_i \mid \theta_l). \tag{11.27}$$

where $P(X_i = l \mid x_{N_i})$ represents the locally dependent probability of the tissue class l. Note that in Section 11.1.2 we simplified the analogous probability: $P(X_i = l) = P(X = l) = \omega_l$. Here, the independence assumption is not valid any more.

To solve the parameter estimation problem, an adapted version of the EM algorithm presented above, called the HMRF-EM, is used (Zhang et al. 2001). The update equations for the $\theta_l = \{\mu_l, \sigma_l\}$ parameters are actually the same update equations as for the *finite Gaussian mixture model (FGMM)* (Equations 11.15 and 11.16) except that

$$\hat{P}^{(k)}(X_i = l \mid y_i, \hat{\Theta}_l^{(k-1)}) = \frac{P\left(y_i \mid \hat{\Theta}_l^{(k-1)}\right) \hat{P}^{(k-1)}\left(X_i = l \mid x_{N_i}\right)}{\sum_{l \in L} P\left(y_i \mid X_i = l, \hat{\Theta}_l^{(k-1)}\right) \hat{P}^{(k-1)}\left(X_i = l \mid x_{N_i}\right)} \tag{11.28}$$

Note that the calculation of $\hat{P}^{(k-1)}\left(X_i = l \mid x_{N_i}\right)$ involves a previous estimation of the class labels, X_i, i.e., the classification step. Typically, an initial classification is computed using the GMM approach, thus without the HMRF model (see Section 11.2.4).

11.2 MATLAB proof of concept

11.2.1 3D data visualization

As previously commented, we will do a *longitudinal study*, i.e., we will compare two MR images from the same patient acquired in a year interval (I1 in 2000 and I2 in 2001, see Fig. 11.2). Studies on patterns of brain atrophy in medical images differ, however, from the proof of concept presented here. Usually, such studies are performed statistically, in comparison to a probabilistic pattern that represents the normal anatomy, i.e., a *transversal study*. Nevertheless, the statistical classification framework presented here remains valid.

The image data used here are 3D MR images of the human brain. They are T1-weighted[9] images of 186 × 186 × 121 voxels of dimension 1 mm × 1mm × 1.3 mm. Voxel type is *unsigned short* integers coded in 16 bits, i.e., voxel values can vary from 0 to 65,535.

MATLAB provides several techniques for visualizing 3D scalar volume data.[10] One of them is the `contourslice` function:

```
load ImageVolume.mat
phandles=contourslice(D,[],[45,64,78,90,100,110,120,127,135,142
,150,160,175,190,200,210,220,228],[60,65,70,75,80,85],7);
view(-169,-54); colormap(gray)
```

[9] Different image contrast can be generated in the MR imaging process; the most often used is T1-weighted but also T2 and proton density (Haacke et al. 1999).

[10] Search for "Techniques for Visualizing Scalar Volume Data" in the Help menu of MATLAB.

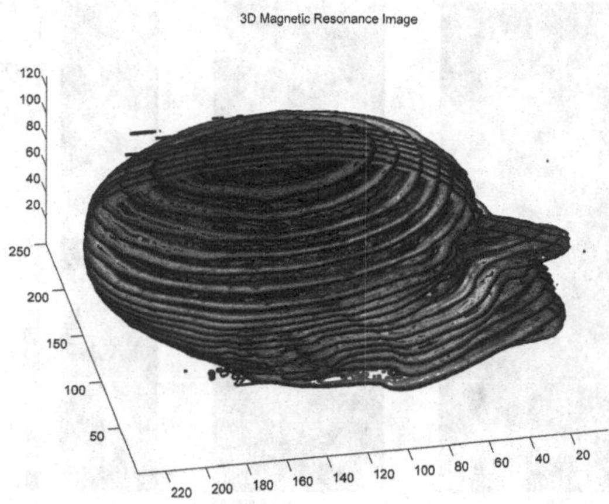

Fig. 11.4 Visualization of 3D MRI data

The resulting volume is shown in Fig. 11.4. The *view angle* can be changed by using *Tools* and *Rotate 3D* in the figure menu. However, medical doctors usually visualize such data using three main views, namely the *axial, coronal*, and *sagittal* views (Fig. 11.5). After loading the volume into the MATLAB *Workspace*, an `Axial` matrix that contains the image intensities appears. Its dimensions are

```
load Image1_voi.mat
size(Axial)
ans=    186    186    121
```

Let us have a look at the three views of the `Axial` matrix:

```
subplot(1,3,1); imagesc(squeeze(Axial(90,:,:)));
subplot(1,3,2); imagesc(squeeze(Axial(:,65,:)));
subplot(1,3,3); imagesc(squeeze(Axial(:,:,60))');
```

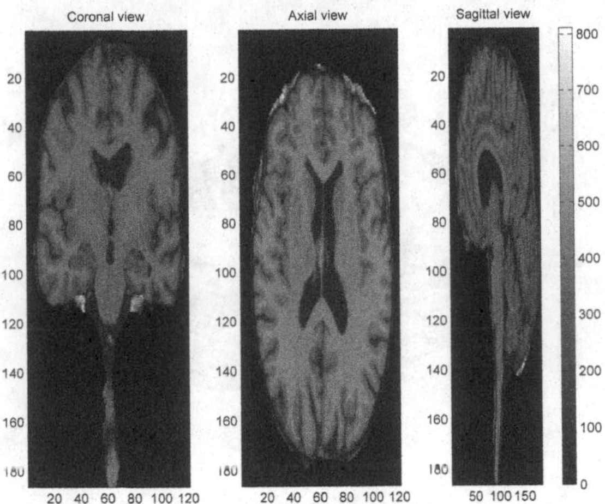

Fig. 11.5 Coronal, axial, and sagittal views of a 3D MR brain image at time 1

Thus the same volume as in Fig. 11.4 is seen here from three different 2D views: from the front of the head (coronal), from the top of the head (axial), and from a side of the head (sagittal). Note that only the brain is presented, since the skull has been removed.[11] The color map is as follows: dark gray is cerebrospinal fluid or CSF; middle gray is gray matter or GM; and light gray is white matter or WM. Note that few patches of the skull are still present.

Now we can run the same code for the image at time 2 (see Fig. 11.6).

```
load(Image2_voi.mat);
```

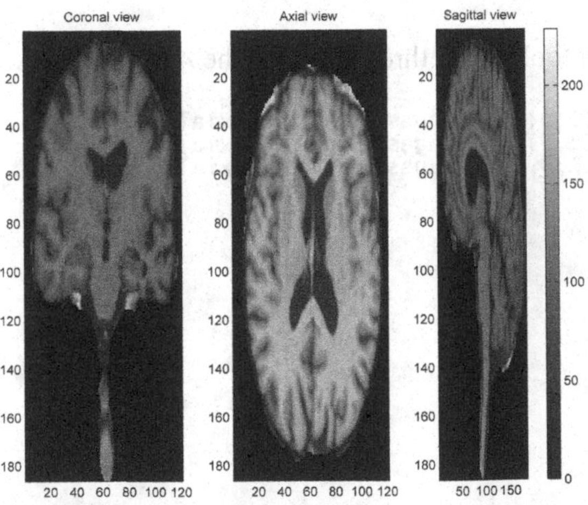

Fig. 11.6 Coronal, axial, and sagittal views of a 3D MR brain image at time 2

[11] Several techniques can be used to remove skull and other nonbrain tissues from brain. Here morphological operations (dilations and closings) have been applied.

Note that both images look very similar; they show the same brain, both have the same size, and there is a spatial correspondence (they have been already rigidly matched), i.e., they have a point-to-point correspondence (see Section 11.3). However, as the *colorbar* on the right side shows, there is a difference in the intensity level range. Thus we propose hereafter to study the image histogram.

11.2.2 Image histogram

The function involved in the computation of the image histogram is

```
[H,X,NbrBin,BinSize]=histogram(Axial,loffset,roffset);
```

The outputs are the values of the image histogram `H`, the bins vector `X`, the total number of bins, `NbrBin`, and the bin size is `BinSize`. `loffset` and `roffset` correspond respectively to the left and right offsets that define the start and end of the image intensities considered in the computation. Note that images in Figs. 11.5 and 11.6 present large black areas (0 gray level), which do not provide any information. To remove these areas from the statistical analysis, we set the left offset to 1. In turn, the right offset is set to 0. Results are shown in Fig. 11.7.

```
loffset=1;
roffset=0;

load Image1_voi.mat
[H,X,NbrBin,BinSize]=histogram(Axial,loffset,roffset);

load Image2_voi.mat
[H,X,NbrBin,BinSize]=histogram(Axial,loffset,roffset);
```

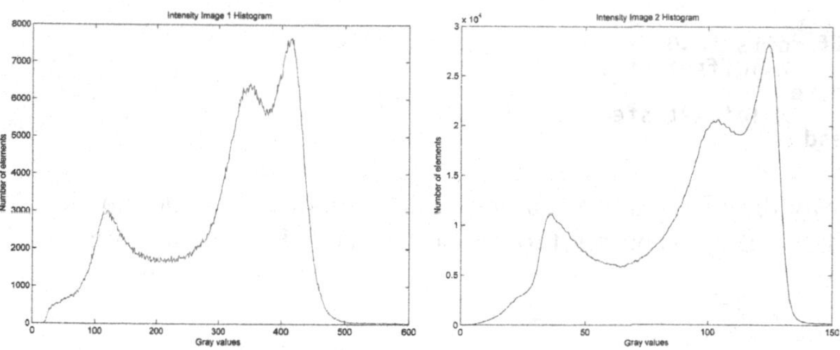

Fig. 11.7 Image histogram: *left* is image 1 (year 2000) and *right* is image 2 (year 2001)

As expected, intensity range values are different (as already seen with the *colorbar* in Fig. 11.5 and 11.6). Three peaks can be observed: the first one is the darkest value and corresponds to CSF, the second one represents GM, and the third one corresponds to WM. The current number of bins used in the representation is

```
NbrBin

ans = 1179; % For Image 1
ans = 301; % For Image 2
```

Our `histogram` function calls the `histc` MATLAB function:

```
H=histc(Axial(:),X);
```

This function counts the number of values in `Axial` that fall between the elements in the `X` vector of bins (which must contain monotonically nondecreasing values). Then, `H` is a vector of length `length(X)` containing these counts.

We can change the range of bins `X` interactively[12] by choosing a different left and right offset:

```
lims=getpts;
disp(' ')
loffset=floor(lims(1))
roffset=floor(NbrBin-lims(2))

loffset = 23
roffset = 696
```

The number of bins can be modified by not only varying the extremes of vector `X` but also changing the bin size. This is implemented by changing the `step` value in the `ASP_brain_segmentation_histogram.m` function:

```
if roffset==0
    X=loffset:step:n;
else
    X=loffset:step:(n-roffset);
end
```

This way the histogram resolution can be modified as seen in Fig. 11.8, where `step` value has been set to 0.1 (left plot) and 40 (right plot):

[12] The `getpts` function allows you to select points from the current figure with the mouse (left button) and select last point with the right button.

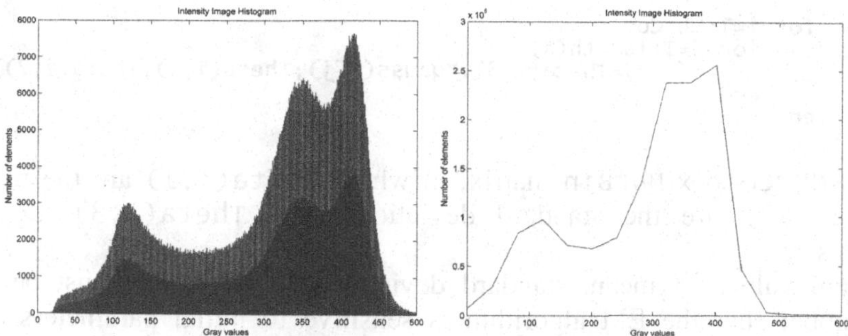

Fig. 11.8 Image histograms using step=0.1 (*left*) and step=40 (*right*)

The selection of the number and the size of the bins in the previous example is only a matter of visualization. However, these parameters can have an influence in the computation time and precision of some optimization problems where image histograms are computed very often. We will show in the next section the influence of the bin size on the Gaussian-fitting optimization problem.

In this work, we will approximate the probability density function (PDF) of the intensity values by the normalized image histogram. The normalization of the image histogram nH is computed as follows:

```
NbrSamples=sum(sum(H));
nH=H*1/NbrSamples;
```

11.2.3 Gaussian mixture model (GMM)

As presented in Section 11.1.3, to solve the parameter estimation of the Gaussian mixture model, we use the expectation–maximization (EM) algorithm. Such a model assumes that the probability density function of the image intensity of every class follows a Gaussian distribution. Before discussing the estimation problem, let us have a look at the MATLAB functions involved in the generation of the Gaussian probability densities:

```
y = exp(-0.5*(x-mu).*(x-mu)/(sigma*sigma))...
/(sqrt(2*pi)*sigma);
```

This is packed into the ASP_brain_segmentation_gauss.m function and corresponds to Equation (11.8). In this equation, x is the image intensity values and mu and sigma must be estimated. In fact, we are looking for the parameters of a weighted sum of Gaussian distributions that better fit the image intensity histogram (likelihood estimation). NbrComp is the number of mixtures used for the fitting, which, in our case, is equal to three since three classes are used: CSF, GM, and WM.

```
for i=1:NbrComp
    for j=1:length(X)
        G(i,j)=Theta(i,3).*gauss(X(j),Theta(i,1),Theta(i,2));
    end
end
```

`G` is a `NbrComp x NbrBin` matrix, in which `Theta(:,1)` are the mean, `Theta(:,2)` are the standard deviations, and `Theta(:,3)` are the weights.

Initial values of mean, standard deviations, and weights must be set. Remember that the EM algorithm is sensitive to initial parameters and, particularly, to initial mean values (see Chapter 4 for an example). We can modify the initial parameters by manually selecting the mean values and, this way, observe the convergence behavior. Let us first set three mean values on the left side of the image histogram:

```
mus=[130  220  330];

s=size(mus);
NbrComp=s(1);
```

Note that mean values can be also selected by clicking on the image histogram; the corresponding MATLAB code is provided in the main script, `ASP_brain_segmentation.m`.

Then, the standard deviation and the initial weight are usually set as

```
for i=1:NbrComp
    sig(i,1)=5.0;
    pw(i,1)=1/NbrComp;
end
```

That is, standard deviation can be set to a small but nonzero value and weights are all set equally. As a matter of fact, the EM algorithm is not as sensitive to these parameters as to the mean values. Finally, initial parameters are saved in `Theta`:

```
% compose Theta
Theta=cat(2,mus,sig,pw);

Theta =

   130      5.0000      0.3333
   220      5.0000      0.3333
   330      5.0000      0.3333
```

Now, the EM algorithm can be applied:

```
[S,e,Theta,fitper,G]=segmentem(Axial, loffset, roffset,
Theta);
```

This function computes the following output parameters: the new estimated `Theta` values, the corresponding Gaussian distributions `G`, the squared error `fitper` between the image histogram and the fit with the Gaussian mixture, the threshold boundaries `e` between each mixture, and the estimated classification `S` obtained by applying a threshold to the intensity values with `e`. The result is plotted in Fig. 11.9. In it, three different types of functions can be observed. The original histogram is presented in dot line, whereas the approximated histogram is in solid line. This approximated histogram is the sum of the three Gaussian histograms that have been estimated for each class, which are presented in dashed lines. The vertical lines represent the two different thresholds that have been finally obtained and applied to perform the classification.

```
fitper = 84.6362
Estimated means and variance:
u1 =
   126.3722
   225.6552
   368.8622
s1 =
    42.2332
    35.0064
    55.6226

Estimated thresholds for Bayesian decision:
e = 0    190    256
```

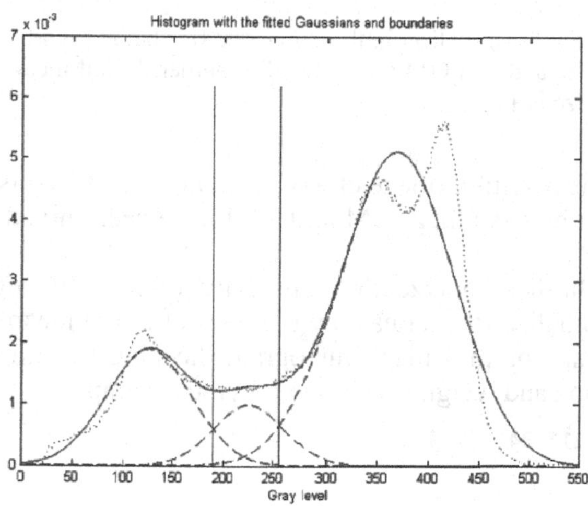

Fig. 11.9 Fitting of the image histogram: three Gaussians are used. Initial means are 130, 220, and 330 and bin size is 1

Let us now visualize the image segmentation `S` after thresholding the input image using the `e` values:

```
figure; clf;
subplot(1,3,1); imagesc(squeeze(S(90,:,:)));
iptsetpref('ImshowBorder', 'tight');
title('Coronal view'); hold on;
subplot(1,3,2); imagesc(squeeze(S(:,65,:)));
iptsetpref('ImshowBorder', 'tight');
title('Axial view'); hold on;
subplot(1,3,3); imagesc((squeeze(S(:,:,60)))');
axis fill;
iptsetpref('ImshowBorder', 'tight');
colorbar; title('Sagittal view');
```

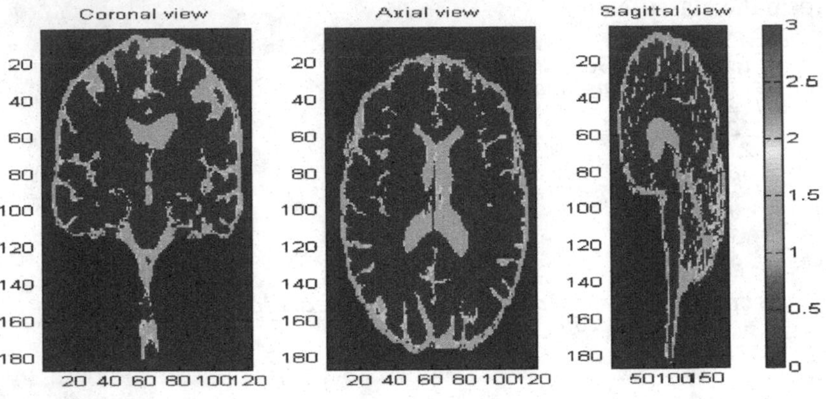

Fig. 11.10 Bayesian classification in three classes: 0 is background, 1 is CSF, 2 is a thin transition class, and 3 is GM and WM all together. Initial means are 130, 220, and 330 and bin size is 1

Obviously, the resulting image classification of Fig. 11.10 is wrong since the Gaussian fitting has failed. GM and WM have been mixed together into a single class.

In order to further analyze the sensitivity of the EM algorithm with respect to the initialization, let us now use a set of initial mean values closer to the three peaks of the image histogram, keeping the initial values of standard deviation and weights as in the previous example:

```
mus = [115 345 410]';
```

```
Theta =

   115     5.0000     0.3333
   345     5.0000     0.3333
   410     5.0000     0.3333
```

This time, the result of `ASP_brain_segmentation_segmentem.m` is

```
fitper =  87.3078

Estimated mean and variance:
u1 =
   138.6997
   347.4169
   414.2959
s1 =
    49.6081
    63.1638
    17.1897

Estimated thresholds for Bayesian decision:
e =
     0   219   401
```

This result is presented in Fig. 11.11 following the same layout that has been used in Fig. 11.9.

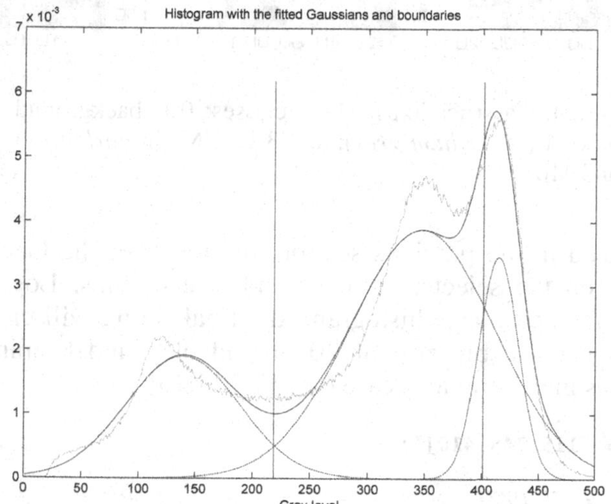

Fig. 11.11 Fitting of the image histogram: three Gaussians are used. Initial means are 115, 345, and 410 and bin size is 1

The fitting and classification results are now more satisfactory: classes converged to the underlying anatomical tissues. The resulting image classification in Fig. 11.12 shows a good convergence of the algorithm. Unfortunately it is quite noisy; note the spurious points that are misclassified as gray matter in the interior of white matter. This is due to the Bayesian decision, which is taken based on the intensity gray level information only. Better results can be obtained by including local spatial information about the neighboring voxels. This can be done by applying the

MRF theory as seen previously in the theory in Section 11.1.4 and presented hereafter.

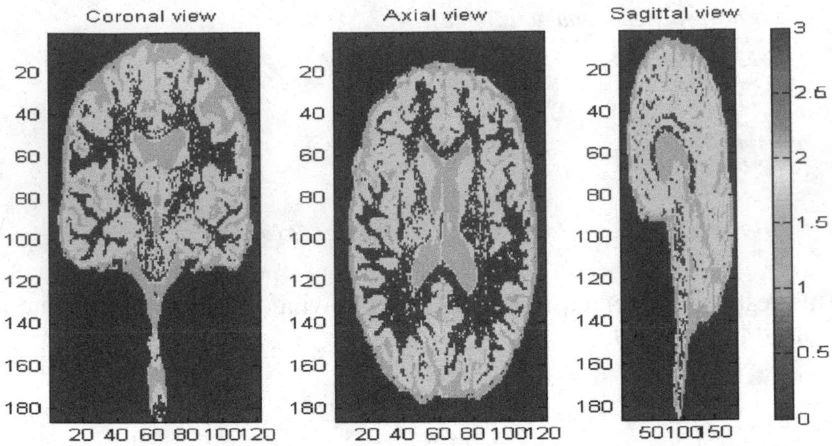

Fig. 11.12 Bayesian classification in three classes: 0 is background, 1 is CSF (in *light grey*), 2 is GM (in *medium grey*), and 3 is WM (in *dark grey*). Initial means are 115, 345, and 410.

As mentioned in the previous section, the result of the Gaussian fitting also depends on the selected number and size of bins. Logically, if we change the reference image histogram, the final fitting will change. Let us illustrate this using a bin size of 20 instead of 1 and keeping the same initialization as in the previous case (see Fig. 11.13):

```
mus = [115 345 410]';
```

```
fitper = 75.0771

Estimated means and variance:
ul =
  130.5112
  291.7118
  390.3842
sl =
   38.9105
   98.6497
   42.1912

Estimated thresholds for Bayesian decision:
e =
     0   161   321
```

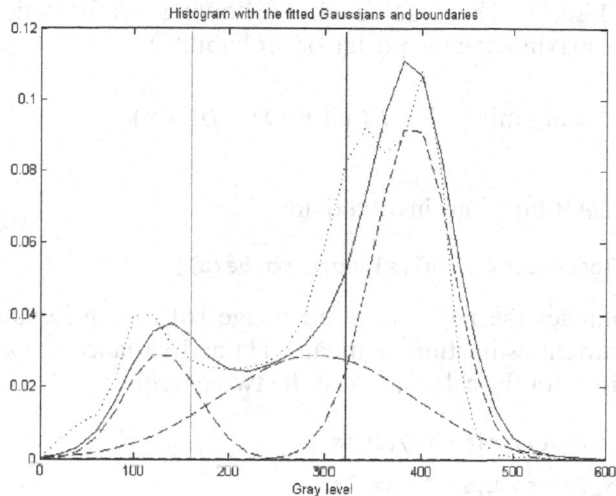

Fig. 11.13 Fitting of the image histogram using a different bin size to compute the image histogram. Initial means are 115, 345, and 410 and bin size is 20

In terms of computational time, the smaller the number of bins, the faster the fitting. However, as illustrated in this example, there is a compromise since the use of too few bins leads to an inaccurate histogram estimate. As it can be seen, the selected thresholds are not as accurate as in the previous case (see Fig. 11.12).

Moreover, in that particular case, the use of a too small number of bins presents another effect. Note that the threshold obtained between class 1 (CSF) and class 2 (GM) is not coincident with the intersection of the Gaussian functions representing their individual density functions. This is due to the rough quantification into a too small number of bins: the theoretical value (which is the intersection point between both density functions) is approximated by the center of the closest bin.

11.2.4 Hidden Gaussian mixture model

The strategy underlying the HMRF-EM algorithm consists in applying iteratively the following two steps:

1. Estimate the image labeling $\boldsymbol{x}$, given the current model θ and then use it to form the complete data set $\boldsymbol{x}$, $\boldsymbol{y}$ (y being the intensity image observation). The labeled image is initially estimated only using GMM

(Section 11.2.3). Then, solve the following minimization problem (instead of maximizing the posterior probability):

$$x^* = \arg\min_{x \in X} \left[-\log P(y \mid x, \Theta) + \beta U(x) \right] \tag{11.29}$$

The MATLAB functions involved are

```
nL=daloop(nL,Axial,ul,sl,NbrComp,beta);
```

which computes the new classified image (`nL`) as in Equation (11.22) with the current estimation of mean (`ul`) and variance (`sl`). It includes two functions for the calculation of the two energies:

```
calcuyx(mu(l),sigma(l),Image);
calcux_3class(L,Sx,Sy,Sz,l);
```

where `l` is the current label, `Sx`, `Sy`, and `Sz` are the voxel dimensions (1 mm, 1 mm, and 1.3 mm, respectively), and `L` is the current classified image.

2. Estimate a new θ by maximizing the expectation of the complete_data log likelihood, $E[\log P(x, y \mid \theta)]$. That is, re-estimate the mean and variance vectors as described in Section 11.1.6 and update the equations for mean and variance vectors in the EM algorithm.

   The MATLAB function involved is

   ```
   [ul,sl]=calculsl(Axial, nL, nH, X, ul, sl,NbrComp);
   ```

   The two parameters to set in the HMRF–EM algorithm are the number of iterations and the value of β (see Equation (11.22)) that weights the energy $U(x)$ with respect to $-\log P(y \mid x, \Theta)$. The loop is

   ```
   nit=5;
   beta=0.6;
   for ii=1:nit
       nL=daloop(nL,Axial,ul,sl,NbrComp,beta);
       [ul,sl]=calculsl(Axial, nL, nH, X, ul, sl,NbrComp);
   end
   ```

   The final label map after five MRF iterations is shown in Fig. 11.14. Note that it is much less noisy than the label map shown in Fig. 11.12.

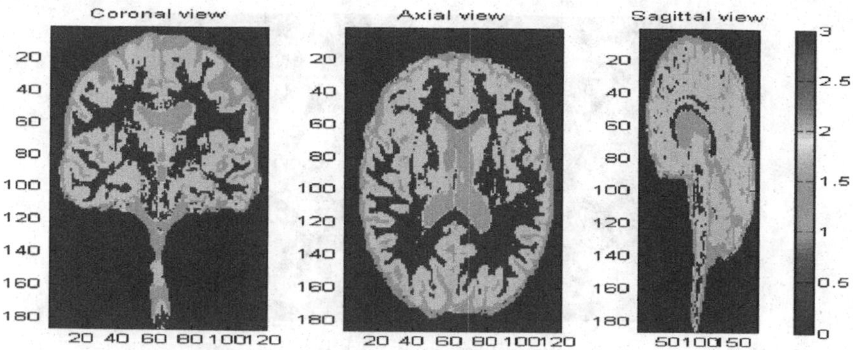

Fig. 11.14 HMRF classification in three classes after five iterations and β set to 0.6

11.2.5 Influence of the spatial parameter

As mentioned in Section 11.1.4, the β parameter is set here experimentally. In this Section we will analyze its influence on the final segmentation. Let us first observe the HMRF classification with β=*0.1*, *1.2*, and *10* after 60 iterations (see Fig. 11.15).

Let us now study the HMRF convergence. For this purpose, we compute the percentage of voxels (related to the total number of voxels of the image, without considering the background) that changed of assigned label, between two consecutive iterations, during 20 iterations (see Fig. 11.16). This will give us an idea about the *speed* of convergence to a local optimal solution for the MAP decision.

The corresponding MATLAB code is included in the main loop of HRMF:

```
nit=60;
beta=10;

nLold=nL;
Change=zeros(nit,1);

for ii=1:nit
    nL=daloop(nL,Axial,ul,sl,NbrComp,beta);

    Diff=(nLold-nL);
    Change(ii)=length(find(Diff))./length(find(nL))*100;
    nLold=nL;

    [ul,sl]=calculsl(Axial, nL, nH, X, ul, sl,NbrComp);
end
```

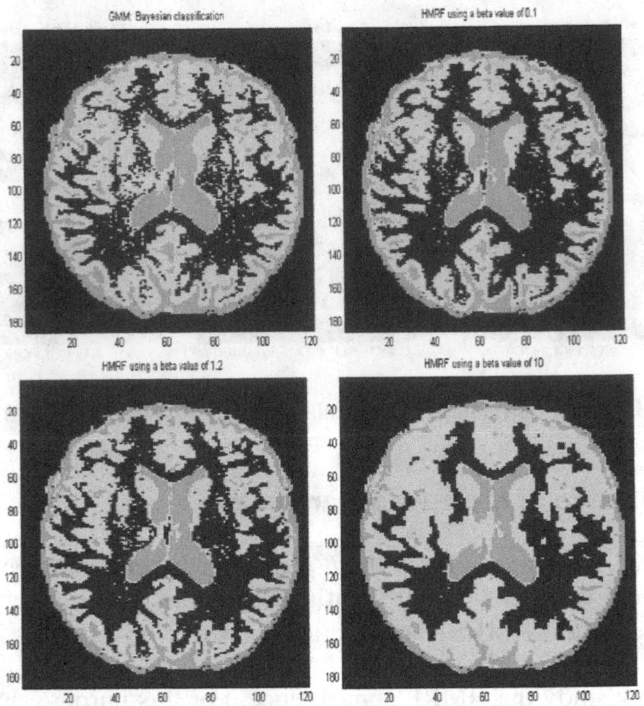

Fig. 11.15 Segmentation after 60 HMRF iterations. *Top left*: Bayesian classification without HMRF; *top right*: HMRF classification with β=0.1. *Bottom left*: HMRF classification with β=1.2; *bottom right*: same with β=10

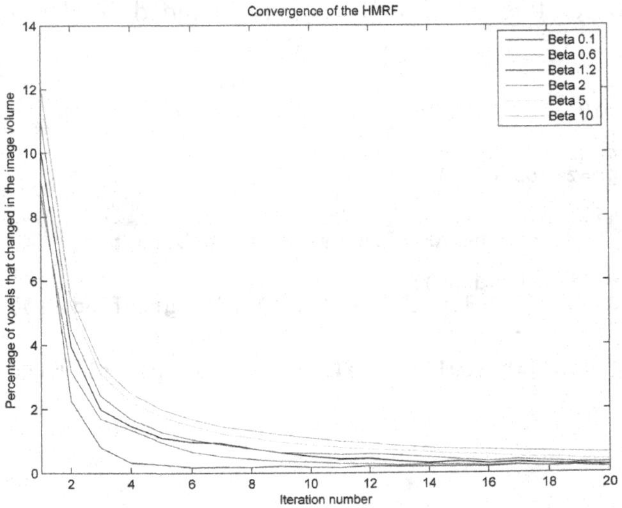

Fig. 11.16 HMRF convergence study

As seen in Fig. 11.16, the larger the β, the slower the convergence to a MAP solution. There is thus a compromise between speed (to reach a solution) and quality of the final segmentation. In this work, we will keep the values of $\beta = 0.6$ and `nit` = 5 iterations. This choice is still arbitrary.

11.2.6 Localization and quantification of brain degeneration

Our goal here is to localize and quantify the GM degeneration in the previous example. Globally, studies on patterns of brain atrophy in medical images are done mainly by two different approaches. The first consists of the manual detection and classification of a region of interest (ROI), leading to a time-consuming and subjective procedure (Galton et al. 2001). The second, and most used approach, is the *voxel-based morphometry* (VBM), usually having the following steps (Ashburner and Friston 2000):

- Normalization of all the subjects into the same space
- Gray matter extraction and smoothing of the normalized segmented images
- Statistical analysis of the differences between a priori reference data and the subjects

The method proposed here is in fact based on the VBM theory. In our study, the probabilistic reference is not available so the evolving degeneration is only studied from the sequence images, i.e., a so-called *longitudinal* study is performed.

Our interest is not in the final tissue classification (label map) but in the tissue probability maps, i.e., functions (images in our case) in which each point (pixel) has associated the likelihood of belonging to a given tissue. Particularly, as GM degeneration is studied, cerebrospinal fluid (CSF) and GM posterior probability maps are retained from the classification step. Working with probability maps will allow us to be more precise in the computation of average concentration maps.

The MATLAB functions involved are

```
ProbMaps(Axial, nL, ul, sl, NbrComp,filenameMRI(1:end-4));
```

This function outputs the posterior probability maps into .raw files for every tissue class:

```
%Write output files
fname=strcat(FileName,'_Pcsf.raw');
fid = fopen(fname,'wb');
fwrite(fid,A*255,'float');
fclose(fid)
```

All probability maps are rescaled to gray levels in [0,255] before being stored in *.raw* files. These probability maps are in fact computed as

```
ptly(1,:,:,:)=calcptly(Image, L, 1, nH, X, mu(1), sigma(1),
NbrComp);
```

which implements the equation described in Equation (11.28).

Then, a smoothing of the tissue probability maps by a large Gaussian kernel is applied. This way, each voxel of the tissue probability map represents the tissue concentration within a given region defined by the standard deviation of the Gaussian filter being used.

In order to determine the standard deviation, morphometry theory, as described in Ashburner and Friston (2000), suggests a value similar to the size of the *changes* that are expected. For the particular case of the image sequence under study, the value of $\sigma = 11$ mm is used. The resulting concentration maps of CSF and GM at time 1 are shown in Figs. 11.17 and 11.18, respectively.

```
Ccsf_1=smooth_3DGauss('Image1_voi_Pcsf.raw',s(1),s(2),s(3),
'float',7,11);
Cgm_1=smooth_3DGauss('Image1_voi_Pgm.raw',s(1),s(2),s(3),
'float',7,11);
Ccsf_2=smooth_3DGauss('Image2_voi_Pcsf.raw',s(1),s(2),s(3),
'float',7,11);
Cgm_2=smooth_3DGauss('Image2_voi_Pgm.raw',s(1),s(2),s(3),
'float',7,11);
```

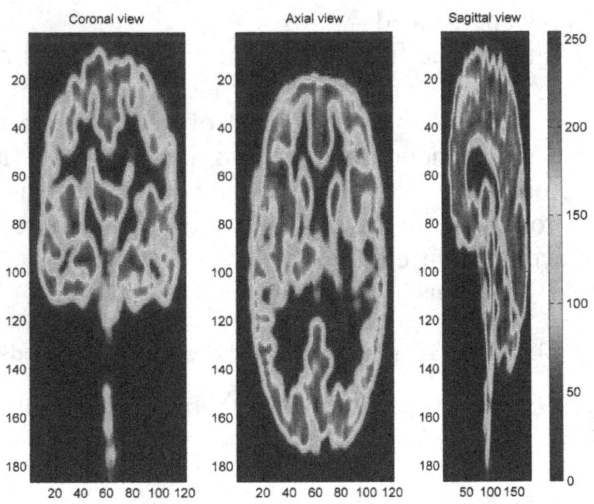

Fig. 11.17 Smoothed probability maps for the GM class

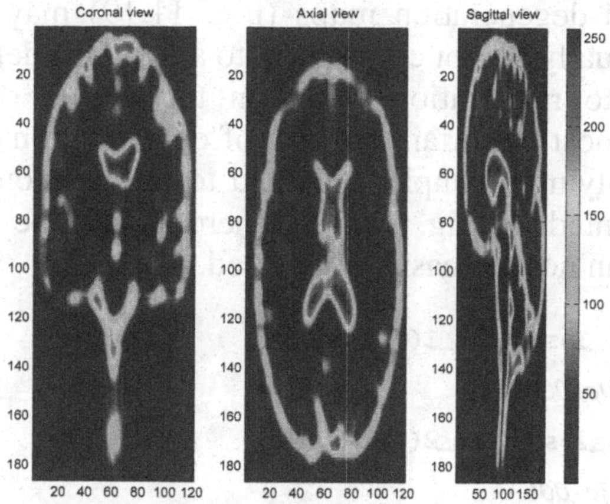

Fig. 11.18 Smoothed probability maps for the CSF class

Let us define the GM degeneration map as

$$D_{12} = \prod_{l} | C_1^l - C_2^l | \tag{11.30}$$

where C_t^l is the concentration map of tissue l at time t, for $l \in \{CSF, GM\}$, $t \in \{1,2\}$, and D_{12} shows the regions with more probable gray matter loss between images 1 and 2. The corresponding MATLAB code is

```
D12=abs(Ccsf_1-Ccsf_2).*abs(Cgm_1-Cgm_2);
imagesc(squeeze(D12(:,99,:)));
imagesc(squeeze(D12(84,:,:)));
```

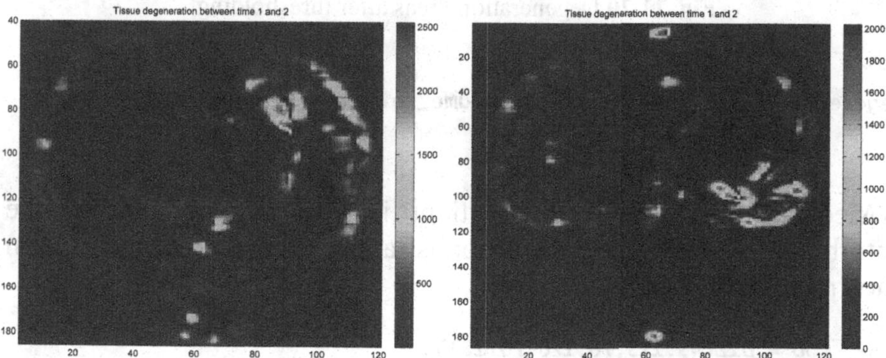

Fig. 11.19 Detected degeneration areas: *left* is an axial view, *right* is the coronal view. Highest values represent the highest degenerated regions

The obtained degeneration maps (Fig. 11.19) may contain some regions that actually do not correspond to a real GM degeneration (for instance, due to registration errors in the cortex or trunk area). However, the most important regions of degeneration (ROD) can be isolated by applying a simple threshold to the previous images. The result is presented in Fig. 11.20, where it can be observed that degeneration can now be easily quantified.

```
volume_roi_1=sum(Cgm_1(D12>500))
    7.4180e+006
volume_roi_2=sum(Cgm_2(D12>500))
    6.1643e+006
```

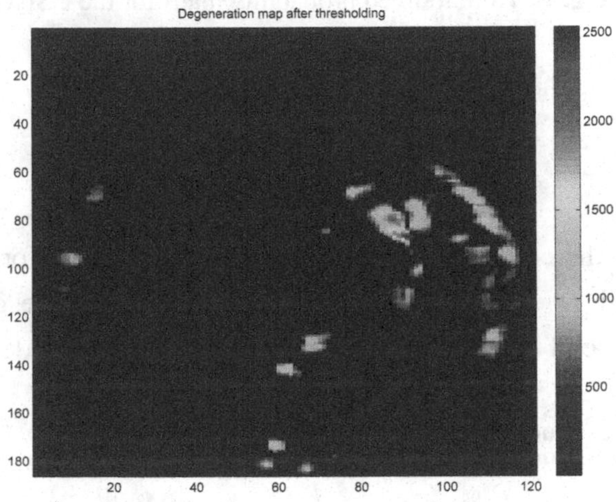

Fig. 11.20 Degeneration areas after thresholding

```
degeneration_roi=(volume_roi_1-volume_roi_2)/volume_roi_1*100
    16.9012
```

Medical diagnosis of this patient is GM degeneration in the entorhinal region. This corresponds approximately to the region shown in Fig. 11.21:

```
ROD = D12(45:115,70:120,70:121);
```

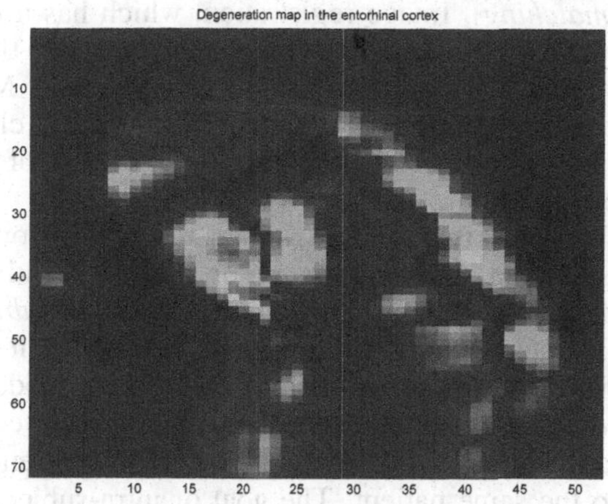

Fig. 11.21 Degeneration map in the entorhinal cortex

Once the candidates to define the GM degeneration regions have been detected, degeneration can be quantified more precisely within these regions:

```
volume_roi_1=sum(Cgm_1(ROD~=0))
        2.7792e+004
volume_roi_2=sum(Cgm_2(ROD~=0))

        2.3326e+004

degeneration_roi=(volume_roi_1-volume_roi_2)/volume_roi_1*100
     16.0672
```

This quantification gives an idea about the GM loss in percentage in these regions of interest. However, this longitudinal study is limited to two time instants only. It would be worth comparing the localization (`D12`) and quantification (`degeneration_roi`) of the degeneration at additional time instants, as well as statistically comparing it to a group of healthy subjects.

11.3 Going further

There is a large amount of different types of information available on medical image modalities (see Introduction in this chapter). All these information types can be efficiently combined by medical *image registration* (or *matching*). Image registration, which has not been handled in this chapter, consists of finding the transformation that brings two medical images into a voxel-to-voxel correspondence. Many variables participate in the registration paradigm, making the classification of registration techniques a difficult task. A wide overview of medical image registration is done in Maintz and Viergever (1998).

In image registration, two images are matched with one another: the target image, also called *reference image* or *scene* (f), and the image that will be transformed, also called *floating image* or *deformable model* (g). If the images to register belong to the same modality, it is said to be *monomodal* registration, and if they belong to different modalities, it is said to be *multimodal* registration. We can also distinguish between *intra-subject* and *inter-subject* registration. In the first case, both reference and floating images belong to the same patient. The goal of intra-subject registration is usually to compensate the variable positioning of the patient in the acquisition process as well as other possible geometrical artifacts in the images. This kind of registration is usually needed in surgical planning, lesion localization, or pathology evolution. In the second case, images are acquired from different patients. The goal of this kind of registration is to compensate the inter-subject anatomical variability in order to perform statistical studies or to profit from reference data,[13] for instance, for surgical planning.

The image modalities to register, as well as the envisaged application, determine the remaining characteristics of the registration process:

- the nature and domain of possible transformations τ,
- the features to be matched,
- the cost function to optimize.

The image registration problem can be formulated by the following minimization equation:

$$T^* = \underset{T \in \mathcal{T}}{\operatorname{argmin}} \operatorname{cost}(f, T \circ g) \tag{11.31}$$

Note that all transformations in τ have to follow some physical constraints in order to model a *realistic* deformation between two brains (we can deform a brain into an apple but it is not very likely to happen).

[13] Reference data is also called *Atlas* images, which represent the human anatomy.

11.3.1 Nature and domain of the transformation

In image registration, transformations are commonly either *rigid* (only translations and rotations are allowed), *affine* (parallel lines are mapped onto parallel lines), *projective* (lines are mapped onto lines), or *curved* (lines are mapped onto curves). Then, the transformation domain can be either *local* or *global*. A transformation is called global when a change in any one of the transformation parameters influences the transformation of the image as a whole. In a local transformation a change in any transformation parameter affects only a part of the image.

11.3.2 Features and cost function

The feature selection depends on the image modality and the definition of the cost function depends on the selected features.

The intensity image can be directly used as features for the matching process (this approach is called *voxel-based registration*) but other identifiable anatomical elements can be used such as point landmarks, lines, or surfaces, which are previously extracted in both reference and floating images (this is called *model-based registration*).

Once feature selection is done, the cost function can be determined, commonly using

$$\textit{Cost function} = -\textit{Similarity measure}$$

The similarity measure can be intensity-based (in voxel-based registration) or distance-based (in model-based registration). Some typical similarity measures based on voxel intensity (Bankman 2000) are absolute or squared intensity difference, normalized cross-correlation, measures based on the optical flow concept,[14] information theory measures such as mutual information. Some common distance-based measures are Procrustean metric, Euclidean distance, curvature, etc.

11.3.3 Optimization

An optimization algorithm (Fig. 11.22) is needed to either maximize a similarity measure or minimize an error measure, over the search space defined by the parameters of the transformation. Various optimization strategies, such as Powell's or simplex optimization, steepest gradient, or genetic algorithms, are available. In fact, the strategy selection depends on

[14] The optical flow concept has been presented in Chapter 9 within the context of motion estimation.

the selected similarity measure since it may not always be possible to compute the derivatives with respect to the transformation parameters.

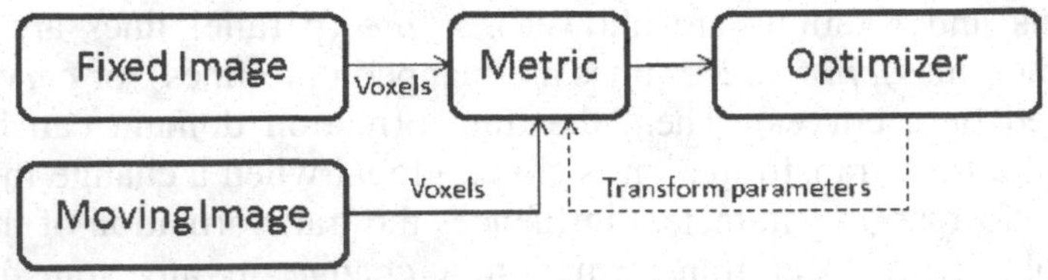

Fig. 11.22 Image registration scheme

11.4 Conclusions

In this chapter, we have seen how GMMs and HMRFs are used for classifying images into homogeneous regions. GMMs are used to model the image intensity response and HMRFs are used to increase the model performance by including the spatial interactions across image voxels.

The brain tissue segmentation problem has been addressed, specially the detection and quantification of GM loss. The goal of the proof-of-concept section was to perform a longitudinal study of two MR images. To this end, we have studied the image histogram and its fitting using a mixture of three Gaussian distributions. After that we have applied the HMRF model and used the resulting probability maps to localize and quantify GM loss.

Finally, let us highlight that studies on patterns of brain atrophy in medical images are usually transversal studies, i.e., they are performed statistically in comparison to a probabilistic pattern that represents the normal anatomy. Nevertheless, the statistical classification framework that has been presented in this chapter remains valid. It is important to note that registration, which has been briefly commented in Section 11.3, is needed in both longitudinal and transversal studies.

11.5 Acknowledgments

The authors would like to thank Prof. R. Meuli from the Radiology Department of the Lausanne University Hospital (CHUV) for providing the magnetic resonance images. This work is supported by the Center for Biomedical Imaging (CIBM) of the Geneva-Lausanne Universities, the EPFL, and the foundations Leenaards and Louis-Jeantet.

References

Ashburner, J. and K.J. Friston (2000). "Voxel-based morphometry – the methods." Neuroimage 11(6 Pt 1): 805–821.

Bach Cuadra, M., L. Cammoun, et al. (2005). "Comparison and validation of tissue modelization and statistical classification methods in T1-weighted MR brain images." IEEE Transactions on Medical Imaging 24(12): 1548–1565.

Bankman, I.N. (2000). Handbook of Medical Imaging. San Diego, Academic Press.

Besag, J. (1974). "Spatial interaction and the statistical analysis of lattice systems." Journal of Royal Statistical Society 36: 192–225.

Bilmes, J.A. (1998). A Gentle Tutorial of the EM Algorithm and its Applications T Pa-rameter Estimation for Gaussian Mixture and Hidden Markov models. Berkeley, ICSI.

Galton, C.J., K. Patterson, et al. (2001). "Differing patterns of temporal atrophy in Alzheimer's disease and semantic dementia." Neurology 57(2): 216–225.

Geman, S. and D. Geman (1984). "Stochastic relaxation, Gibbs distributions, and the Bayesian restoration of images." IEEE Transactions Pattern Analysis and Machine Intelligence 6: 721–741.

Gudbjartsson, H. and S. Patz (1995). "The Rician distribution of noisy MRI data." Magnetic Resonance in Medical Sciences 34(6): 910–914.

Haacke, E.M., R.W. Brown, et al. (1999). Magnetic Resonance Imaging: Physical Principles and Sequence Design, Wiley-Liss, New York.

Hajnal, J.V., D.L.G. Hill, et al. (2001). Medical Image Registration, CRC, London.

Li, S. Z. (2001). Markov Random Field Modeling in Image Analysis, Springer, Tokyo.

Maintz J.B.A., and M.A. Viergever. (1998). "A survey of medical image registration." Medical Image Analysis 2(1): 1–36.

Mazziotta, J.C., A.W. Toga, et al. (2000). Brain Mapping: The Disorders, San Diego, Academic Press.

Suri, J.S., D. Wilson, et al. (2005). Handbook of Biomedical Image Analysis: Segmentation Models (Volume 1 & 2), Springer, Heidelberg.

Van Leemput, K., F. Maes, et al. (2003). "A unifying framework for partial volume segmentation of brain MR images." IEEE Transactions on Medical Imaging 22(1): 105–119.

Zhang, Y., M. Brady, et al. (2001). "Segmentation of brain MR images through a Hidden Markov Random Field Model and the Expectation–Maximization algorithm." IEEE Transactions Medical Imaging 20(1): 45–57.

Index

N

O

P

Q

R

S

Printed in the United States of America.

Printed in the United States of America